Analytic Geometry

THE MACMILLAN COMPANY
NEW YORK • CHICAGO
DALLAS • ATLANTA • SAN FRANCISCO
LONDON • MANILA
IN CANADA
BRETT-MACMILLAN LTD.
GALT, ONTARIO

Analytic Geometry

FIFTH EDITION

CLYDE E. LOVE, Ph.D.
Professor Emeritus of Mathematics
in the University of Michigan

EARL D. RAINVILLE, Ph.D.
Professor of Mathematics
in the University of Michigan

New York
THE MACMILLAN COMPANY

Published simultaneously in Canada

Printed in the United States of America

Ninth Printing 1964

Preface

One major difference between the present edition and the fourth edition of this book is in the treatment of polar coordinates. Here polar coordinates are introduced in Chapter 2 and used in various later chapters. I have attempted to convey to the student the idea that polar coordinates are a tool, not an isolated topic, in analytic geometry.

The two chapters on calculus have been omitted. Simple differentiations naturally appear in problems involving tangents and are studied in Chapter 10. The basic concepts of curve tracing appear in Chapter 3, but a detailed treatment of the sketching of curves of degree greater than two is delayed until after the conics have been studied. The space devoted to solid geometry has been slightly enlarged.

New topics in the fifth edition are: the distance formula in polar coordinates, circles of Appolonius, radical axis, common chord, tangents to a conic from an external point, chord of contact, the shape of certain higher plane curves, parametric equations of lines, circles, conics, the method of least squares, parametric equations of lines in space, generation of surfaces of revolution.

Certain chapters may be omitted entirely, or in part, to make a shorter course. The chapters on tangents and normals to conics, families of curves, curve fitting are self contained. The chapters on parametric equations, trigonometric functions, exponentials and logarithms can also be omitted, if so stringent a cut is necessary. Attention is called to the fact that a shorter book, Love's *Elements of Analytic Geometry* also published by The Macmillan Company, was prepared for a short course in this subject.

I wish to thank Professor Fred Brafman of Wayne University for an independent check on material in the text and on most of the answers to exercises, Professor Ralph L. Shively, Oppenheim Professor of Mathematics at Manchester College, for an independent reading of the proofs, and Professor Donat K. Kazarinoff of the University of Michigan for various useful suggestions.

Earl D. Rainville

Contents

Plane Analytic Geometry

Chapter 6: The Circle

Chapter 7: Conic Sections. The Parabola

Chapter 8: The Central Conics

Chapter 9: The General Equation of Second Degree

Chapter 10: Tangents and Normals to Conics

Chapter 11: Algebraic Curves

Plane Analytic Geometry

CHAPTER 1 *Rectangular Coordinates*

1. Introduction. It is common practice to divide geometry into two kinds, synthetic and analytic. Synthetic geometry, usually first studied in high school, employs the straight edge and compass as its basic tools. Elementary analytic geometry, the subject of this course, uses algebra (equations, formulas, and their algebraic manipulation) as its main tool.

A fundamental goal of plane analytic geometry is the investigation of interesting and useful properties of configurations involving points, straight lines, and curves other than straight lines. Not only does the use of algebra contribute to the study of geometry, but geometric interpretation of algebraic equations and manipulations results in a fuller comprehension of many phases of algebra.

Analytic geometry can be developed from a system of axioms and definitions as is usually done with synthetic geometry. In this course, however, we do not attempt a complete separation of the analytic from the synthetic. We shall use freely a few theorems and concepts from earlier courses taken by the student.

2. Directed line segments. When a line segment is measured in a definite sense *from* one endpoint *to* the other, the segment is said to be *directed.* If the terminal points are A and B, we speak of the segment AB or the segment BA according as the sense is from A to B or from B to A.

A B

Figure 1

If one sense is chosen as positive, then the opposite sense is negative: thus

$$AB = -BA, \qquad \text{or} \qquad AB + BA = 0.$$

If C is any third point of the straight line through A and B,

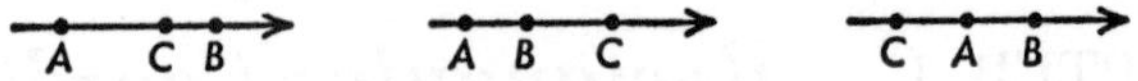

Figure 2

then for all possible positions of A, B, and C we have

$$AB + BC = AC, \tag{1}$$

$$AB + BC + CA = 0. \tag{2}$$

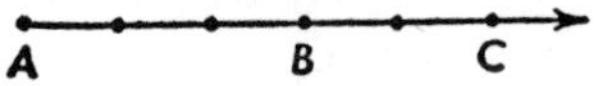

Figure 3

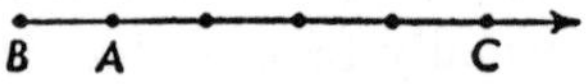

Figure 4

For example, in Fig. 3, with the positive direction to the right,

$$AB = 3, \qquad BC = 2, \qquad AC = 5, \qquad CA = -5;$$

in Fig. 4, $AB = -1$, $BC = 5$, $AC = 4$, $CA = -4$.

Two directed segments lying in the same line or in parallel lines are said to be *equal* if they have the same length and are measured in the same sense.

In ordinary affairs we think of distance as a directed or undirected quantity, according to circumstances. Say that we drive 5 miles, then have to return to the starting point for repairs. As regards gasoline consumption, we have traveled (undirected segments) 10 miles; the net advance toward our destination (directed segments) is zero.

3. Position of a point on a surface. If a point lies on a given surface, two magnitudes, or "coordinates," are necessary to determine its position, each coordinate being measured in a definite sense. Thus we may say that one town is 10 miles east and 8 miles south of another. Note that without the directions — east and south — the coordinates would be ambiguous and therefore useless.

4. Rectangular coordinates. Given a point P (Fig. 5) in a certain plane, let us draw in that plane two directed lines Ox, Oy, perpendicular to each other. The line Ox is called the *x-axis*, Oy the *y-axis*, and their point of intersection O is the *origin*. The position of P is known if its distances from the axes are given, each being measured in a definite sense, *from* the axis *to* the point. These directed segments are called the *rectangular coordinates*, or the *Cartesian* coordinates* of P: the distance from the y-axis (NP or its equal OM) is the *abscissa*, the distance MP from the x-axis is the *ordinate*.

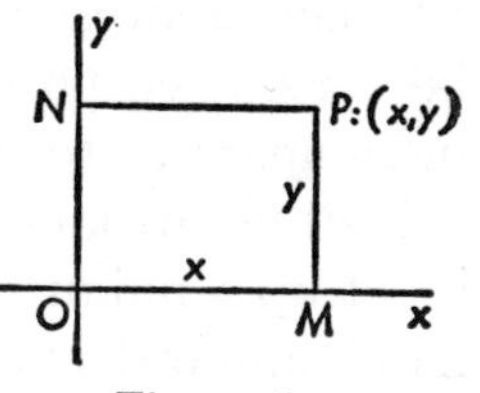

Figure 5

We shall ordinarily assume the axes as in Fig. 5, and shall consider abscissas *positive if measured to the right, negative if measured to the left;* ordinates *positive if measured upward, negative if measured downward.* Of course these conventions may be modified at any time, according to convenience — for example, by assuming the y-axis positive downward.

The coordinates of a point are always written in parentheses, with the abscissa first: thus in Fig. 6, with 3 spaces as the unit, the point P:(1, 3) has the abscissa 1 and the ordinate 3. Also shown are Q:(2, −1), R:(−2, −3), and S:($-\frac{10}{3}$, 0).

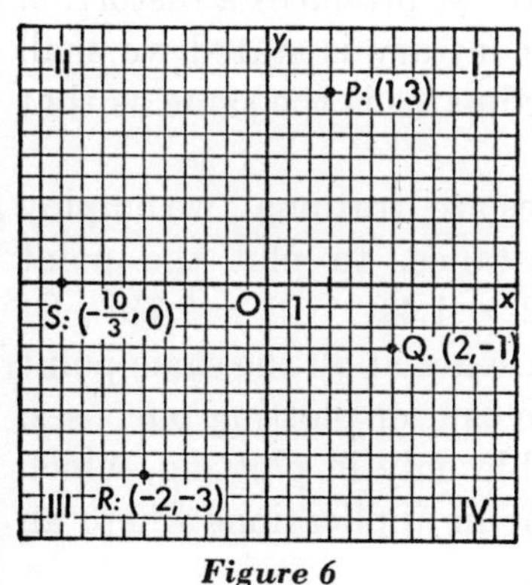

Figure 6

The axes divide the plane into four compartments, called *quadrants*, and numbered as in Fig. 6. Thus the abscissa is positive in the first and fourth quadrants, the ordinate positive in the first and second.

* This name was introduced as a tribute to the French mathematician René Descartes (1596–1650), who seems to have been the first to make an extensive study of geometry with the aid of a coordinate system.

Except where the contrary is indicated, it will be assumed that segments oblique to the axes are *undirected,* segments parallel to an axis are *directed.* Segments parallel to Ox will be considered positive to the right, negative to the left; segments parallel to Oy, positive upward, negative downward.

By the introduction of a Cartesian coordinate system there is set up a unique correspondence between points, on the one hand, and ordered pairs of real numbers, on the other: *to every ordered pair of real numbers there corresponds one and only one point in the plane, a d conversely.*

5. Units. Drawings involving rectangular coordinates are best made on square-ruled paper, called *coordinate paper* (see Fig. 6). The unit of measurement chosen need not, and usually should not, be the width of one space on the coordinate paper; the scale should be selected with regard to the nature of the drawing to be made — neither so large that some of the points fall beyond the limits of the paper, nor so small that the properties of the figure become obscured. The scale adopted should be clearly indicated.

Cases often arise in which it is convenient to adopt different scales on the two axes, which of course produces a distortion of the figure. Except where the contrary is stated we shall assume always that the unit for ordinates is the same as that for abscissas.

To plot a point whose coordinates are irrational, we employ decimal approximations. For instance, to plot the point $(\sqrt{2}, \sqrt{3})$, we might take $\sqrt{2} = 1.41$, $\sqrt{3} = 1.73$. Of course $(\sqrt{2}, \sqrt{3})$ and $(1.41, 1.73)$ are not at all the same point — we are merely doing the best we can for plotting purposes. Such approximations are permissible only in plotting, which is an inaccurate process at best, or in applications where an approximate result is satisfactory.

6. Distance between two points. The distance between two points P_1, P_2 can be expressed in terms of their coordinates by the theorem of Pythagoras. Let the coordinates of the two points be denoted by the letters x, y with subscripts: $P_1{:}(x_1, y_1)$, $P_2{:}(x_2, y_2)$. Now, in Fig. 7,

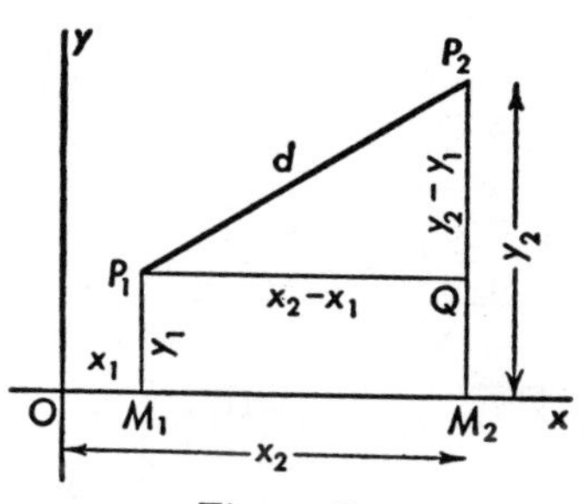

Figure 7

(1) $$d = \sqrt{\overline{P_1Q}^2 + \overline{QP_2}^2}.$$

But

$$P_1Q = M_1M_2 = OM_2 - OM_1,$$
$$QP_2 = M_2P_2 - M_2Q:$$

that is,

$$P_1Q = x_2 - x_1,$$
$$QP_2 = y_2 - y_1.$$

Substituting these expressions for P_1Q and QP_2 into equation (1), we obtain a formula for the distance between the points (x_1, y_1) and (x_2, y_2):

(2) $$d = \sqrt{(x_2 - x_1)^2 + (y_2 - y_1)^2}.$$

By drawing the figure in various positions, the student may convince himself that the formula holds no matter where the points P_1, P_2 may be situated.

Example (*a*): Find the distance between the points (3, 2) and (−5, 4).

By formula (2), we find

$$d = \sqrt{(-8)^2 + 2^2} = \sqrt{68} = 2\sqrt{17}.$$

Example (*b*): Show that the points $P_1{:}(5, 0)$, $P_2{:}(2, 1)$, $P_3{:}(4, 7)$ are the vertices of a right triangle.

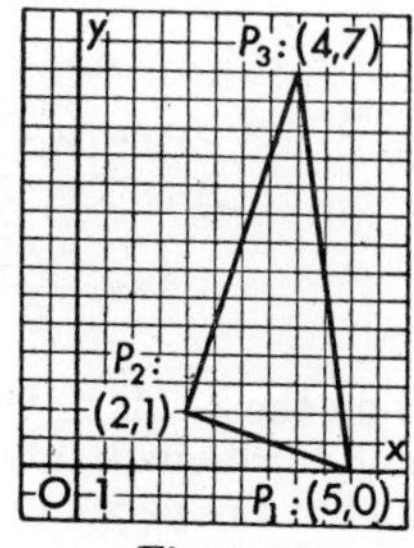

Figure 8

By (2),

$$P_1P_2 = \sqrt{9 + 1} = \sqrt{10},$$
$$P_2P_3 = \sqrt{4 + 36} = \sqrt{40},$$
$$P_1P_3 = \sqrt{1 + 49} = \sqrt{50},$$

so that

$$\overline{P_1P_2}^2 + \overline{P_2P_3}^2 = \overline{P_1P_3}^2.$$

Example (*c*): A moving point P remains always equidistant from P_1:$(-1, 0)$ and P_2:$(0, -2)$. Express this fact by an algebraic equation.

Let the coordinates of the moving point P be denoted by (x, y). Then, since by hypothesis

$$PP_1 = PP_2,$$

it follows that

$$\sqrt{(x + 1)^2 + y^2} = \sqrt{x^2 + (y + 2)^2}.$$

Square and simplify:

$$x^2 + 2x + 1 + y^2 = x^2 + y^2 + 4y + 4,$$

or

$$2x - 4y - 3 = 0.$$

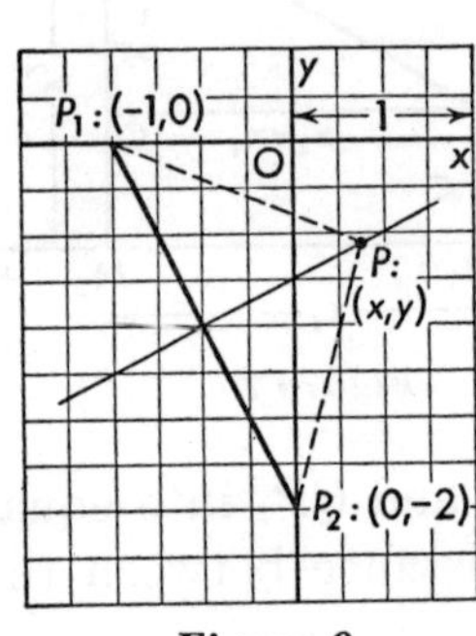

Figure 9

The locus of P is of course the perpendicular bisector of the segment P_1P_2.

Exercises

In Exs. 1–4, draw the figure on coordinate paper, choosing a suitable scale in each instance.

1. Triangle with vertices $(0, 3)$, $(-1, -2)$, $(4, 1)$.
2. Triangle with vertices $(3, 4)$, $(-2, 2)$, $(5, -2)$.
3. Quadrilateral with vertices $(-1, 2)$, $(-1, -1)$, $(3, -4)$, $(5, 4)$.
4. Quadrilateral with vertices $(2, 3)$, $(-2, -3)$, $(3, -5)$, $(4, 1)$.

In Exs. 5–8, find the distance between the given points.

5. $(2, 1)$, $(5, 5)$. *Ans.* 5.
6. $(-2, 4)$, $(3, 6)$. *Ans.* $\sqrt{29}$.
7. $(\frac{1}{2}, 1)$, $(-\frac{3}{2}, -3)$. *Ans.* $2\sqrt{5}$.
8. $(\frac{2}{3}, -\frac{1}{3})$, $(\frac{5}{6}, -\frac{1}{2})$. *Ans.* $\frac{\sqrt{2}}{6}$.

9. What can be said of the coordinates of all points on the x-axis? On the y-axis? On the line through O bisecting the first and third quadrants? The second and fourth quadrants? On the line parallel to the y-axis two units to the right of it? Two units to the left?

10. Where does a point lie if its abscissa is zero? If its ordinate is zero? If abscissa and ordinate are equal? Are numerically equal but of opposite sign?

11. Show that the points $(-1, -2)$, $(5, 4)$, $(-3, 0)$ are the vertices of a right triangle, and find its area. *Ans.* 12.

12. Show that the points $(4, 0)$, $(2, 1)$, $(-1, -5)$ are the vertices of a right triangle and find its area. *Ans.* $\frac{15}{2}$.

13. Show that the points $(-2, 4)$, $(3, -1)$, $(-1, -3)$ are the vertices of an isosceles triangle, and find its area. *Ans.* 15.

14. Show that the points $(1, -3)$, $(3, 2)$, $(-2, 4)$ are the vertices of an isosceles triangle, and find its area. *Ans.* $\frac{29}{2}$.

15. Show that the points $(1, 4)$, $(7, 0)$, $(5, -3)$, $(-1, 1)$ are the vertices of a rectangle, and find its area. *Ans.* 26.

16. Show that the points $(-1, -3)$, $(-2, 0)$, $(1, 6)$, $(2, 3)$ are the vertices of a parallelogram. Is the parallelogram a rectangle?

17. Explain why it is impossible to show graphically that a given point lies on a given circle, although it may be feasible to show graphically that a given point does not lie on a given circle. See Exs. 18–20.

18. Draw the circle with center at $(1, 2)$ and passing through $(8, 3)$. Does this circle pass through $(-4, -3)$? Through $(5, 8)$? Through $(0, 9)$?

19. Draw the circle with center at $(-3, 1)$ and passing through $(5, 4)$. Does this circle pass through $(3, 7)$? Through $(-6, 9)$?

20. Draw the circle with center at $(-5, -2)$ and tangent to the y-axis. Does this circle pass through $(-2, 2)$? Through $(-4, 3)$?

21. At what points does the circle of Ex. 20 cut the x-axis?

22. Find the radius of a circle with center at $(2, 3)$, if a chord of length 8 is bisected at $(-1, 4)$. *Ans.* $\sqrt{26}$.

23. Find the radius of a circle with center at $(-2, 1)$, if a chord of length 10 is bisected at $(-3, 0)$. *Ans.* $3\sqrt{3}$.

24. The center of a circle is at $(4, 2)$ and its radius is 5. Find the length of the chord which is bisected at $(2, -1)$. *Ans.* $4\sqrt{3}$.

25. The center of a circle is at $(-3, -2)$ and its radius is 7. Find the length of the chord which is bisected at $(3, 1)$. *Ans.* 4.

26. At what points does the circle of Ex. 25 cut the y-axis?

In Exs. 27–30, do the given points lie in a straight line?

27. $(1, -1)$, $(-1, -5)$, $(2, 1)$.

28. $(-3, -2)$, $(23, 15)$, $(-24, -16)$.

29. $(-2, 2)$, $(5, -8)$, $(-7, 9)$.

30. $(9, -14)$, $(5, -8)$, $(-9, 13)$.

In Exs. 31–34, express the given statement by an algebraic equation. What is the locus of the point (x, y) in each exercise? Draw the figure.

31. The point (x, y) is equidistant from $(0, 0)$ and $(4, -2)$.

Ans. $2x - y = 5$.

32. The point (x, y) is equidistant from $(4, -1)$ and $(-2, 3)$.
Ans. $3x - 2y = 1$.

33. The point (x, y) is at a distance 5 from $(0, -3)$.
Ans. $x^2 + y^2 + 6y - 16 = 0$.

34. The point (x, y) is at a distance 4 from $(5, -2)$.
Ans. $x^2 + y^2 - 10x + 4y + 13 = 0$.

In Exs. 35–38, locate the point which satisfies the given conditions.

35. Equidistant from $(3, 8)$, $(5, 2)$, and $(-3, -4)$. *Ans.* $(-2, 3)$.

36. Equidistant from $(3, 1)$ and $(-2, 2)$; also equidistant from $(1, 2)$ and $(3, 0)$. *Ans.* $(0, -1)$.

37. Equidistant from $(-3, 0)$ and $(1, 4)$, and at a distance 5 from $(-1, 7)$.
Ans. $(-1, 2)$, $(-6, 7)$.

38. At a distance 4 from $(7, 4)$ and at a distance $\sqrt{26}$ from $(2, -1)$.

39. Show that the quadrilateral with vertices $(0, 4)$, $(7, -7)$, $(2, -2)$, $(1, -9)$ consists of two equal triangles base to base. Find its area.
Ans. $A = 20$.

40. Show that the quadrilateral with vertices $(4, 12)$, $(2, 26)$, $(-12, 24)$, $(-16, 2)$ is kite-shaped, and find its area. *Ans.* $A = 300$.

7. Midpoint of a line segment. Let $P: (x, y)$ be the midpoint of the segment joining the points $P_1: (x_1, y_1)$, $P_2: (x_2, y_2)$. In Fig. 10,

$$OM = OM_1 + \tfrac{1}{2}(OM_2 - OM_1),$$

or

$$x = x_1 + \tfrac{1}{2}(x_2 - x_1) = \tfrac{1}{2}(x_1 + x_2).$$

A similar formula for y is easily obtained. Hence:

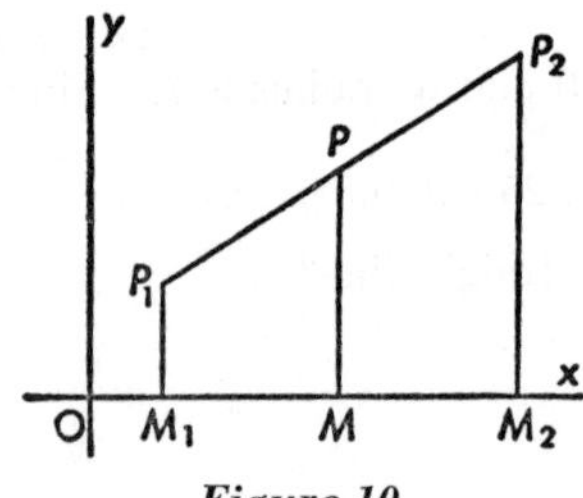

Figure 10

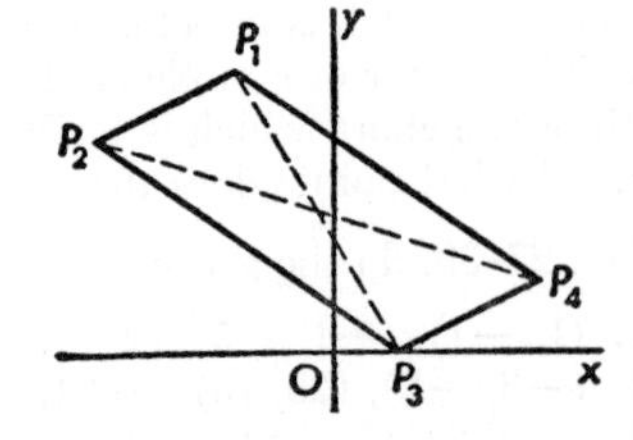

Figure 11

The coordinates of the point (x, y) midway between (x_1, y_1), (x_2, y_2) are

$$(1) \qquad x = \tfrac{1}{2}(x_1 + x_2), \qquad y = \tfrac{1}{2}(y_1 + y_2).$$

Example (*a*): Show that the quadrilateral with vertices $P_1:(-\frac{3}{2}, 4), P_2:(-\frac{7}{2}, 3), P_3:(1, 0), P_4:(3, 1)$ is a parallelogram.

Let us use the theorem that a quadrilateral whose diagonals bisect each other is a parallelogram. For the midpoint of P_1P_3, we have (Fig. 11)

$$x = \tfrac{1}{2}(-\tfrac{3}{2} + 1) = -\tfrac{1}{4}, \qquad y = \tfrac{1}{2}(4 + 0) = 2;$$

for the midpoint of P_2P_4,

$$x = \tfrac{1}{2}(-\tfrac{7}{2} + 3) = -\tfrac{1}{4}, \qquad y = \tfrac{1}{2}(3 + 1) = 2.$$

Since the two midpoints coincide, our problem is solved.

Example (*b*): The directed line segment from (1, 3) to (4, 8) is extended its own length. Find the terminal point.

Let the terminal point be (x, y). Then (4, 8) is the midpoint of the segment from (1, 3) to (x, y). Therefore,

$$\frac{x+1}{2} = 4, \qquad \frac{y+3}{2} = 8,$$

from which it follows that the terminal point is (7, 13).

8. Division of a line segment. Let $P:(x, y)$ be a point in the straight line through $P_1:(x_1, y_1)$, $P_2:(x_2, y_2)$. If the segments P_1P, PP_2 are such that

$$\frac{P_1P}{PP_2} = \frac{r_1}{r_2}, \tag{1}$$

the point P is said to divide the segment P_1P_2 in the ratio $r_1:r_2$ — internally if P lies in the segment P_1P_2, externally if P lies in the segment P_1P_2 produced (in either direction).

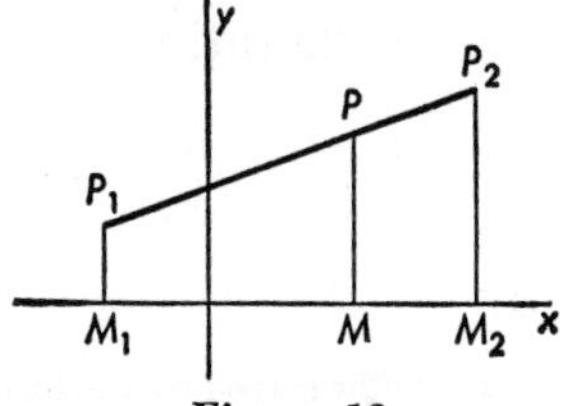

Figure 12

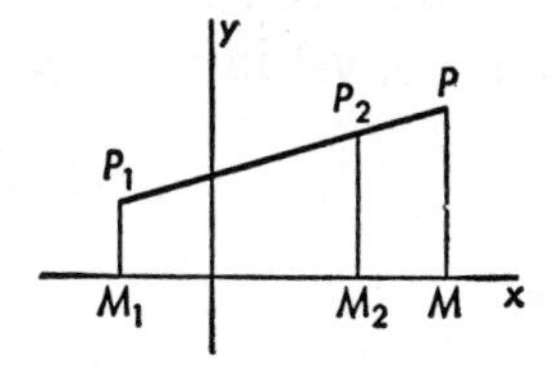

Figure 13

If the division of the segment P_1P_2 is internal, as in Fig. 12, then

$$\frac{P_1P}{PP_2} = \frac{M_1M}{MM_2} = \frac{x - x_1}{x_2 - x}.$$

Hence, with the aid of (1), we find that

$$\frac{x - x_1}{x_2 - x} = \frac{r_1}{r_2}.$$

Clear of fractions and solve for x:

$$r_2x - r_2x_1 = r_1x_2 - r_1x,$$
$$(r_1 + r_2)x = r_2x_1 + r_1x_2,$$
$$(2) \qquad x = \frac{r_2x_1 + r_1x_2}{r_1 + r_2}.$$

The formula for y is derived similarly. In fact, since our notation is symmetric in x and y, the formula for y need not be actually worked out at all — it can be obtained at once from (2) by merely changing each x to the corresponding y.

For the case of external division (Fig. 13), it will be convenient to consider the line P_1P_2 as *directed*. Then, since P_1P and PP_2 are measured in opposite directions, their ratio must be negative — i.e., of the numbers r_1, r_2, one must be taken positive, the other negative. With this understanding, the above argument applies without change. (The reader should verify this.)

Hence, in summary:

If the point $P:(x, y)$ divides the segment joining $P_1:(x_1, y_1)$ and $P_2:(x_2, y_2)$ internally or externally in the ratio

$$\frac{P_1P}{PP_2} = \frac{r_1}{r_2},$$

then*

* If preferred, the examples and exercises following may be solved by the formulas of Ex. 19 below, instead of by (3).

$$x = \frac{r_2x_1 + r_1x_2}{r_1 + r_2}, \qquad y = \frac{r_2y_1 + r_1y_2}{r_1 + r_2}. \tag{3}$$

When P is the midpoint of P_1P_2, so that $r_1 = r_2$, formulas (3) evidently reduce to those of § 7.

Example (*a*): The segment joining P_1:(1, 3), P_2:(5, −2) is trisected. Find the point of trisection nearer to P_1.

In Fig. 14, with

$$P_1P = \frac{1}{2}PP_2, \qquad \frac{P_1P}{PP_2} = \frac{1}{2},$$

we may take $r_1 = 1$, $r_2 = 2$:

$$x = \frac{2 \cdot 1 + 1 \cdot 5}{3} = \frac{7}{3}, \qquad y = \frac{2 \cdot 3 + 1(-2)}{3} = \frac{4}{3}.$$

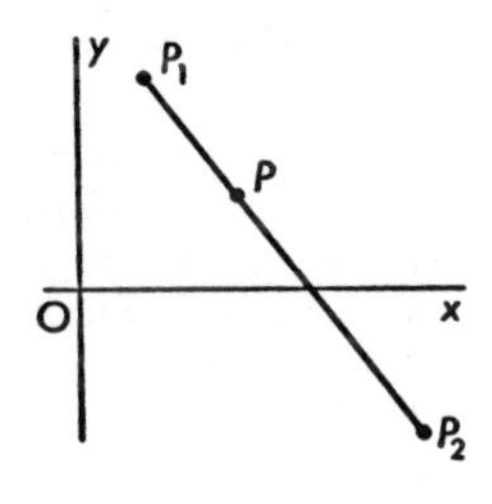

Figure 14

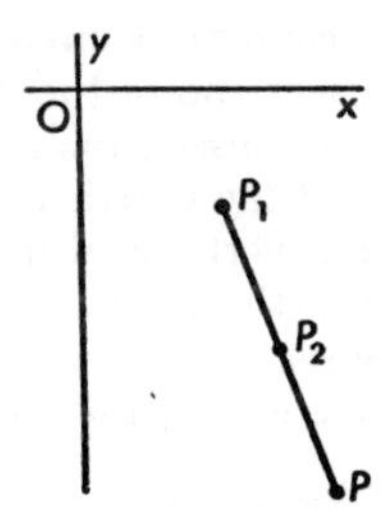

Figure 15

Example (*b*): The segment from P_1:(5, −4) to P_2:(7, −9) is extended beyond P_2 so that its length is doubled. Find the terminal point P (Fig. 15).

Here P_1P is twice the length of PP_2, but the two segments are in opposite directions: i.e.,

$$P_1P = -2PP_2, \qquad \frac{P_1P}{PP_2} = -2 = \frac{-2}{1}.$$

Take $r_1 = -2$, $r_2 = 1$:

$$x = \frac{1 \cdot 5 - 2 \cdot 7}{-1} = 9, \qquad y = \frac{1(-4) - 2(-9)}{-1} = -14.$$

See also the method used in Example (b), § 7.

Exercises

In Exs. 1–4, find the point midway between the given points.

1. $(5, 6), (3, -2)$.
2. $(4, 8), (-4, -3)$.
3. $(3, 0), (-\frac{1}{2}, 4)$.
4. $(\frac{3}{2}, -6), (-\frac{1}{3}, \frac{1}{2})$.

5. Show in two ways that the quadrilateral with vertices $(0, -1)$, $(1, 2)$, $(-4, 7)$, $(-5, 4)$ is a parallelogram.

6. Show in two ways that the quadrilateral with vertices $(11, 1)$, $(0, 0)$, $(-1, -3)$, $(10, -2)$ is a parallelogram.

7. Show in a new way that the points $(-1, -2)$, $(5, 4)$, $(-3, 0)$ are the vertices of a right triangle. (Ex. 11, p. 7.)

8. Show in a new way that the points $(4, 0)$, $(2, 1)$, $(-1, -5)$ are the vertices of a right triangle. (Ex. 12, p. 7.)

9. The segment joining $(5, 11)$ and $(-3, -1)$ is to be divided into four equal parts. Find the points of division.

10. Trisect the segment joining $(-2, 3)$ and $(7, 1)$. *Ans.* $(1, \frac{7}{3})$, $(4, \frac{5}{3})$.

11. The center of a circle is at $(2, -5)$; one point on the circle is $(-4, 2)$. Find the other end of the diameter through $(-4, 2)$.

12. Three consecutive vertices of a parallelogram are $(1, -3)$, $(-3, -1)$, $(3, 5)$. Find the fourth vertex. *Ans.* $(7, 3)$.

13. Three vertices of a parallelogram are $(1, -3)$, $(-3, -1)$, $(3, 5)$. Find the fourth vertex. *Ans.* $(-5, -9)$, $(-1, 7)$, or $(7, 3)$.

14. The segment joining $(-4, 7)$, $(5, -2)$ is divided into two segments, one of which is five times as long as the other. Find the point of division.
Ans. $(-\frac{5}{2}, \frac{11}{2})$; $(\frac{7}{2}, -\frac{1}{2})$.

15. The segment joining $(2, -4)$, $(9, 3)$ is divided into two segments, one of which is three-fourths as long as the other. Find the point of division. *Ans.* $(5, -1)$; $(6, 0)$.

16. The segment from $(-1, 4)$ to $(2, -2)$ is extended three times its own length. Find the terminal point. *Ans.* $(11, -20)$.

17. The segment joining $(-2, -3)$, $(6, 1)$ is extended each way a distance equal to one-fourth its own length. Find the terminal points.
Ans. $(8, 2)$; $(-4, -4)$.

18. The segment joining $(4, 0)$, $(3, -2)$ is extended each way a distance equal to three times its own length. Find the terminal points.

19. In Fig. 12 or Fig. 13, if $P_1P = k \cdot P_1P_2$, show that

$$x = x_1 + k(x_2 - x_1), \qquad y = y_1 + k(y_2 - y_1).$$

Solve the following exercises by using the formulas of Ex. 19.

20. Example (*a*), § 8.
21. Example (*b*), § 8.
22. Ex. 14.
23. Ex. 15.
24. Ex. 16.
25. Ex. 17.

9. Inclination; slope. The *angle of inclination*, also called simply the *inclination*, of a straight line is the smallest positive angle from the positive x-axis to the line — the angles α in Figs. 16–17. By special definition, the inclination of a line parallel to Ox is zero.

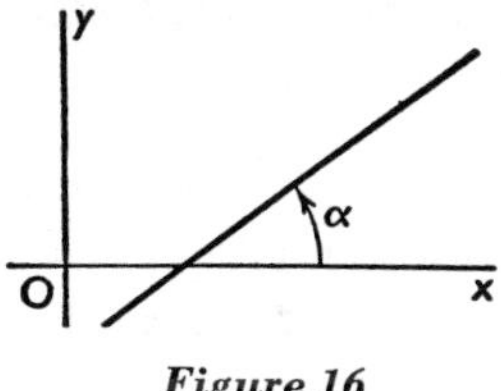

Figure 16

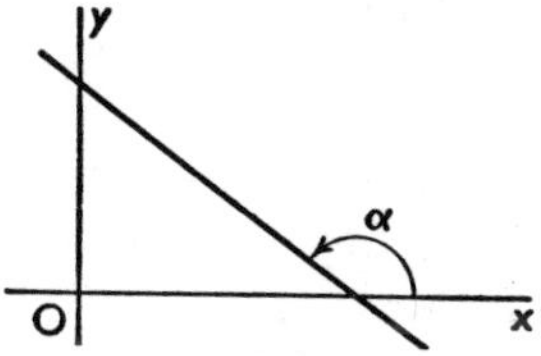

Figure 17

The *slope* of a line is the *tangent of the angle of inclination.* Slope is usually denoted by m:

$$m = \tan \alpha. \tag{1}$$

If the axes are in the conventional position, a line sloping *upward to the right* has *positive slope,* since the tangent of a positive acute angle is positive; a line sloping *downward to the right* has *negative slope.* The slope of a line parallel to the x-axis is zero.

It should be noted that the idea of slope is meaningless in the case of a line parallel to the y-axis (including the y-axis itself), since $\tan \alpha$ "approaches infinity" (i.e., exceeds all bounds) as α approaches 90°. Therefore, *in all discussions involving slopes, lines parallel to the y-axis are excluded.*

Figure 18

From Fig. 18 we obtain

$$m = \tan \alpha = \frac{QP_2}{P_1Q}.$$

The slope of the line joining the points $P_1:(x_1, y_1)$ *and* $P_2:(x_2, y_2)$ *is*

$$m = \frac{y_2 - y_1}{x_2 - x_1} \qquad (x_2 \neq x_1) \tag{2}$$

In ordinary language, the slope of a line means the ratio of "rise" to "run" — i.e., the ratio of the vertical distance to the horizontal distance covered in traversing any segment of the line. Thus a road with 10% slope, or "grade," rises 10 ft. for every 100 ft. horizontally.

When the scales in the two directions are not the same, the defining formula (1) is *replaced by* (2).

10. Parallel and perpendicular lines. If two lines are parallel, they have the same slope; and conversely.

Given two perpendicular lines L_1, L_2, with slopes

$$m_1 = \tan \alpha_1, \qquad m_2 = \tan \alpha_2,$$

let L_1 denote the one with positive slope, so that

$$\alpha_2 = 90° + \alpha_1.$$

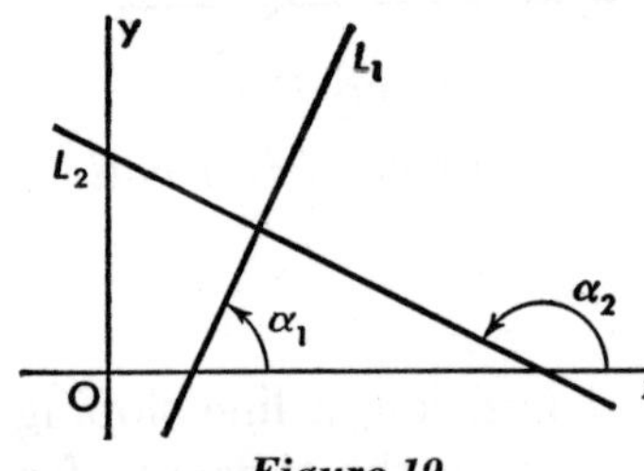

Figure 19

By trigonometry,

$$\tan \alpha_2 = \tan (90° + \alpha_1) = - \cot \alpha_1 = - \frac{1}{\tan \alpha_1},$$

whence

$$(1) \qquad m_2 = - \frac{1}{m_1}.$$

On the other hand, if (1) is true, then

$$(2) \qquad \tan \alpha_2 = - \cot \alpha_1.$$

Since α_1 and α_2 are each positive and less than 180°, it follows from (2) that they differ by 90°. Hence, if (1) holds, the lines L_1 and L_2 are perpendicular.

Of course, any line parallel to the x-axis is perpendicular to any line parallel to the y-axis.

THEOREM: *For lines not parallel to the axes, two lines are perpendicular if and only if their slopes are negative reciprocals.*

Example (*a*): Verify that the points $P_1\colon(-1, 3)$, $P_2\colon(0, 5)$, $P_3\colon(3, 1)$ are the vertices of a right triangle (Fig. 20).

From the figure we see that if there is a right angle it must be at P_1. The slopes of P_1P_2, P_1P_3 are respectively

$$m_1 = \frac{5-3}{0+1} = 2, \qquad m_2 = \frac{1-3}{3+1} = -\frac{1}{2},$$

whence it follows that P_1P_2 and P_1P_3 are perpendicular.

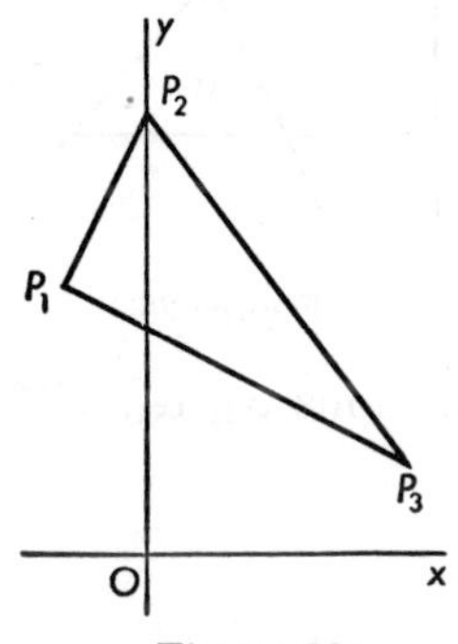

Figure 20

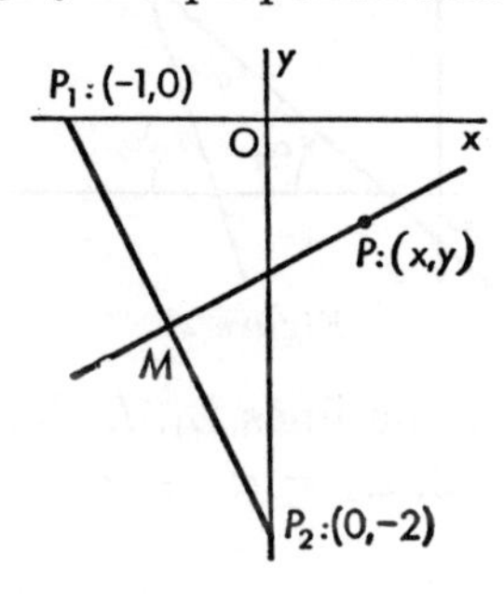

Figure 21

Example (*b*): A moving point $P\colon(x, y)$ remains always equidistant from $P_1\colon(-1, 0)$ and $P_2\colon(0, -2)$. Express this fact by an algebraic equation. [Example (*c*), § 6.]

We know that P must lie on the perpendicular bisector of the line segment joining P_1 and P_2. Let M (Fig. 21) be the midpoint of P_1P_2. Then M is the point $(-\frac{1}{2}, -1)$.

The slope of MP is

$$m_1 = \frac{y+1}{x+\frac{1}{2}}.$$

The slope of P_1P_2 is

$$m_2 = \frac{-2-0}{0+1} = -2.$$

By the theorem,

$$\frac{y+1}{x+\frac{1}{2}} = \frac{1}{2}, \qquad \text{or} \qquad 2x - 4y - 3 = 0.$$

11. Angle between two lines. By the angle *from* a line L_1 *to* a line L_2 we shall understand the positive angle through which L_1 must be rotated to come to coincidence with L_2 (the angles φ in Figs. 22–23).

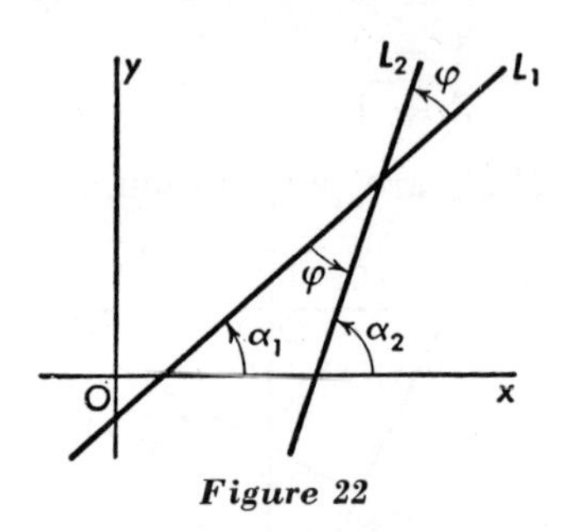

Figure 22

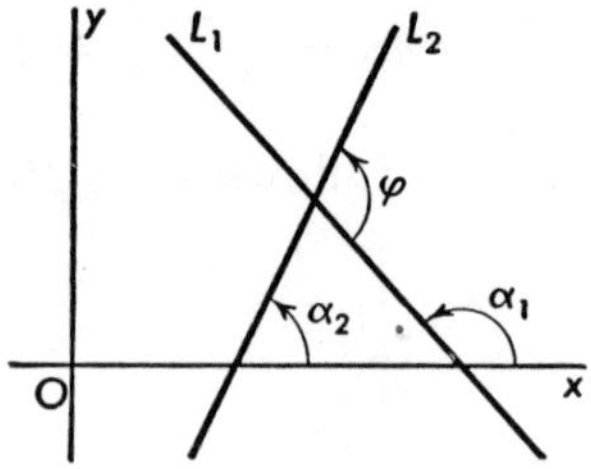

Figure 23

Let the lines L_1, L_2 have the inclinations α_1, α_2. Then in Fig. 22, $\alpha_2 = \alpha_1 + \varphi$, so that

$$\varphi = \alpha_2 - \alpha_1. \tag{1}$$

In Fig. 23, $\alpha_1 = \alpha_2 + (180° - \varphi)$, from which

$$\varphi = 180° + (\alpha_2 - \alpha_1). \tag{2}$$

From either (1) or (2) it follows that

$$\tan \varphi = \tan (\alpha_2 - \alpha_1) = \frac{\tan \alpha_2 - \tan \alpha_1}{1 + \tan \alpha_1 \tan \alpha_2}.$$

But the slopes of the lines are

$$\tan \alpha_1 = m_1, \qquad \tan \alpha_2 = m_2,$$

so that:

The angle from a line of slope m_1 to a line of slope m_2 is given by the formula

$$\tan \varphi = \frac{m_2 - m_1}{1 + m_1 m_2}. \tag{3}$$

This result will be more easily remembered if we realize that it is not, properly speaking, a new formula at all, but merely a restatement of the formula for the tangent of the difference of two angles.

Formula (3) fails if one line, say L_2, is parallel to Oy, but in that case $\varphi = 90° - \alpha_1$, so that $\tan \varphi = \cot \alpha_1 = \dfrac{1}{m_1}$.

Example: Find the interior angles of the triangle with vertices (1, 1), (4, 3), (5, 2). (We shall be content to find the tangents, without troubling to look up the angles themselves in a table of "natural tangents.")

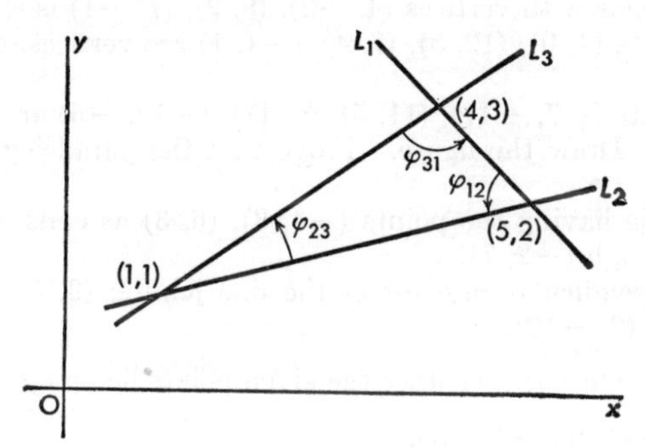

Figure 24

Let the lines forming the triangle be denoted by L_1, L_2, L_3 as shown in Fig. 24. We readily find the slopes of these lines to be

$$m_1 = -1, \qquad m_2 = \tfrac{1}{4}, \qquad m_3 = \tfrac{2}{3}.$$

Let φ_{12} (read φ one two) denote the angle from L_1 to L_2, φ_{23} from L_2 to L_3, φ_{31} from L_3 to L_1. By (3),

$$\tan \varphi_{12} = \frac{\frac{1}{4} + 1}{1 - \frac{1}{4}} = \frac{5}{3}.$$

Similarly,

$$\tan \varphi_{23} = \tfrac{5}{14}, \qquad \tan \varphi_{31} = -5.$$

Check:

$$\tan (\varphi_{12} + \varphi_{23}) = \frac{\frac{5}{3} + \frac{5}{14}}{1 - \frac{5}{3} \cdot \frac{5}{14}} = 5 = -\tan \varphi_{31},$$

$$\varphi_{12} + \varphi_{23} = 180° - \varphi_{31}, \qquad \varphi_{12} + \varphi_{23} + \varphi_{31} = 180°.$$

Exercises

In Exs. 1–6, find the slope of the line joining the given points.

1. (6, 2), (3, −4). **2.** (2, −1), (−3, 4).
3. (4, 0), (1, −2). **4.** (−2, −1), (5, 3).
5. $(\frac{1}{2}, -\frac{1}{2})$, $(0, \frac{1}{3})$. **6.** $(\frac{3}{2}, \frac{1}{3})$, $(\frac{5}{6}, \frac{1}{2})$.

Verify the statements in Exs. 7–14 by methods based on §§ 9–10.*

7. The points (6, −1), (3, 0), (5, 6) are the vertices of a right triangle.
8. The points $(-1, \frac{1}{2})$, $(0, -\frac{5}{2})$, $(5, \frac{5}{2})$ are the vertices of a right triangle.
9. The triangle with vertices (−3, 3), (−1, −1), (3, −3) is isosceles.
10. The triangle with vertices (4, −2), (8, 2), (7, −1) is isosceles.
11. The points (4, 0), (12, 3), (7, 4), (−1, 1) are vertices of a parallelogram.
12. The points (−7, −11), (11, 5), (6, 11), (−12, −5) are vertices of a parallelogram. Draw the figure. Prove that the parallelogram is not a rectangle.
13. The circle having the points (−4, 3), (6, 3) as ends of a diameter also passes through (−2, 7).
14. The perpendicular bisector of the line joining (9, 5) and (−7, 3) passes through (3, −12).

In Exs. 15–18, determine whether the given points lie on a straight line.

15. (−10, −4), (0, −1), (30, 8).
16. (−22, −12), (−1, 2), (25, 19).
17. (9, 10), (4, 3), (−3, −7).
18. (−4, 9), (−2, 6), (8, −9).
19. Show that the quadrilateral with vertices (0, 1), (4, 2), (3, 6), (−5, 4) has two right angles. Find the area. *Ans.* $A = \frac{51}{2}$.
20. Show that the quadrilateral with vertices (10, 10), (−14, −2), (−10, −10), (4, −24) can be divided into two right triangles. Find the area. *Ans.* $A = 400$.

In Exs. 21–24, express the given statement by an algebraic equation.

21. The point (x, y) is equidistant from (0, 0) and (4, −2). (Ex. 31, p. 7.)
22. The point (x, y) is equidistant from (4, −1) and (−2, 3). (Ex. 32, p. 8.)
23. The point (x, y) lies on a circle which has the segment from (2, −4) to (5, 6) as a diameter. *Ans.* $x^2 + y^2 - 7x - 2y - 14 = 0$.
24. The point (x, y) lies on a circle which has the segment from (0, −1) to $(\frac{1}{3}, \frac{4}{3})$ as a diameter. *Ans.* $3x^2 + 3y^2 - x - y - 4 = 0$.

* The student will frequently be asked to solve a previous exercise in a new way. This is so that, by weighing the merits of the various methods, he may learn how to pick the best one in a given case.

In Exs. 25–30, if the line L_1 passes through the first pair of points, L_2 through the second pair, find the angle from L_1 to L_2.

25. (2, 0), (3, 5); (5, 2), (4, 6). *Ans.* $\tan \varphi = \frac{9}{19}$.
26. (4, 3), (6, −2); (9, 5), (6, −2). *Ans.* 135°.
27. (1, 9), (2, 6); (3, 3), (−1, 5). *Ans.* 45°.
28. (−5, −3), (2, 6); (6, 4), (8, 2). *Ans.* $\tan \varphi = 8$.
29. (4, 0), (0, 7); (4, −5), (1, 10). *Ans.* $\tan \varphi = -\frac{1}{3}$.
30. (−2, 0), (8, 6); (0, 6), (7, 3). *Ans.* $\tan \varphi = -\frac{18}{13}$.

In Exs. 31–34, find the interior angles; check the answers.

31. The triangle of Ex. 7.
32. The triangle of Ex. 8.
33. The triangle of Ex. 9.
34. The triangle of Ex. 10.

12. Area of a triangle. Consider a triangle with vertices $P_1:(x_1, y_1)$, $P_2:(x_2, y_2)$, $P_3:(x_3, y_3)$. In Fig. 25, if from the area of the trapezoid $M_3M_2P_2P_3$ we subtract the areas of the trapezoids $M_3M_1P_1P_3$ and $M_1M_2P_2P_1$, there remains the area of the triangle. Now, from the figure,

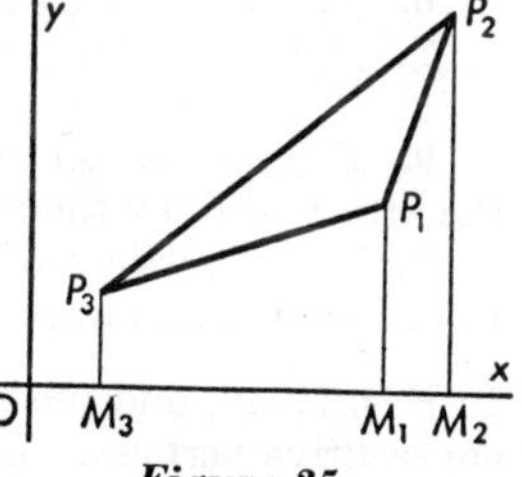

Figure 25

$$M_3M_2P_2P_3 = \tfrac{1}{2}(x_2 - x_3)(y_2 + y_3),$$
$$M_3M_1P_1P_3 = \tfrac{1}{2}(x_1 - x_3)(y_1 + y_3),$$
$$M_1M_2P_2P_1 = \tfrac{1}{2}(x_2 - x_1)(y_2 + y_1).$$

Thus

$$A = \tfrac{1}{2}[(x_2 - x_3)(y_2 + y_3) - (x_1 - x_3)(y_1 + y_3) - (x_2 - x_1)(y_2 + y_1)],$$

or after some rearrangement,

$$(1) \qquad A = \tfrac{1}{2}[x_1(y_2 - y_3) + x_2(y_3 - y_1) + x_3(y_1 - y_2)].$$

Since the right member of (1) is the expansion, by minors of the first column, of the determinant below, we have proved that:

The area of the triangle with vertices (x_1, y_1), (x_2, y_2), (x_3, y_3) *is*

$$(2) \qquad A = \frac{1}{2}\begin{vmatrix} x_1 & y_1 & 1 \\ x_2 & y_2 & 1 \\ x_3 & y_3 & 1 \end{vmatrix}.$$

One qualifying remark is necessary. It can be shown that the formula gives a positive or negative answer according as motion around the triangle in the order P_1, P_2, P_3 is counterclockwise or clockwise. To avoid confusion, it is best to arrange the vertices in the counterclockwise order, so that the formula will always yield a positive result.

Exercises

In Exs. 1–8, find the area of the triangle with vertices as given.

1. (2, −5), (6, 2), (4, 1). *Ans.* 5.
2. (2, 3), (1, 1), (−2, −2). *Ans.* $\frac{3}{2}$.
3. (−3, −2), (7, 4), (−8, −1).
4. (1, −2), (−1, 6), (−4, −1).
5. (3, 7), (5, −2), (6, 1).
6. (9, 2), (4, 6), (−4, 0).
7. (2, 0), (11, 6), (−4, −4). Interpret the result.
8. (1, 1), (−5, 9), (4, −3). Interpret the result.

9. Three vertices of a parallelogram are (3, 2), (−1, 7), (4, −3). Find the area and draw the appropriate parallelograms.

10. Three vertices of a parallelogram are (0, −1), (2, 5), (−5, −3). Find the area and draw the appropriate parallelograms.

In Exs. 11–14, find the area of the quadrilateral having the given points as consecutive vertices. Check by dividing into triangles in two ways.

11. (5, 2), (4, 3), (2, 4), (−8, −1).
12. (3, −5), (9, 5), (0, 5), (−2, 1).
13. (3, −2), (−1, −3), (7, −2), (5, 3).
14. (8, −2), (5, 6), (4, 1), (−7, 4).

In Exs. 15–18, find the distance of the point from the line.

15. Point (12, 7), line through (5, 3), (2, −3). *Ans.* $2\sqrt{5}$.
16. Point (5, −3), line through (1, −1), (−3, −4). *Ans.* 4.
17. Point (27, 20), line through (0, 0), (40, 30). Plot the points. *Ans.* $\frac{1}{5}$.
18. Point (2, 5), line through (−5, −5), (7, 12). Plot the points.

19. Prove that if the coordinates of the vertices of a parallelogram are whole numbers, the parallelogram contains a whole number of units of area. Is this theorem true for any polygon?

20. Prove that if the coordinates of the vertices of a polygon are rational numbers, the area of the polygon is also a rational number.

21. Prove that if the coordinates of the vertices of a triangle are even numbers, the area of the triangle is an even number. Is this theorem true for any polygon? Is the converse true?

22. Prove that it is impossible to construct an equilateral triangle such that the coordinates of the vertices are all rational numbers. (Find the area by § 12 and by the formula $A = \frac{1}{2}bh$.)

23. Prove the rule of signs (§ 12) for the triangle with vertices P_1, P_2, O. (After obtaining A, note the slopes of OP_1, OP_2.)

13. Theorems of elementary geometry.

We have already used several theorems from elementary synthetic geometry. Therefore, we cannot logically apply our analytic tools to the proof of those theorems.*

Interesting and instructive exercises are obtained by disregarding the logical order of development and proving theorems of synthetic geometry by application of the analytic tools studied in §§ 6–12.

Example (*a*): Prove that the diagonals of a parallelogram bisect each other (Fig. 26, p. 22).

Take one vertex as origin, with the x-axis passing through another vertex. We may take P_1 as $(x_1, 0)$, P_2 as (x_2, y_2), whence the coordinates of P_3 *must be* $(x_1 + x_2, y_2)$. Thus the midpoint of P_1P_2 is $\left(\frac{x_1 + x_2}{2}, \frac{y_2}{2}\right)$; likewise the midpoint of OP_3 is $\left(\frac{x_1 + x_2}{2}, \frac{y_2}{2}\right)$.

The following general remarks should be carefully noted.

(I) When (as here) the proposition to be proved embodies an intrinsic geometric property of the figure, independent of any auxiliary device such as a coordinate system, we may without loss of generality *place the axes in any convenient position with reference to the figure.*

(II) Since general theorems cannot be deduced from special cases, *the figure itself must not be made special in any way:* for instance, in the above example, by assigning numerical coordinates to the vertices, or by choosing as the given parallelogram a rhombus or a rectangle.

* As mentioned earlier, it is possible to develop analytic geometry in a manner independent of synthetic geometry, if the sacrifice of the work done in high school geometry is deemed worthwhile.

(III) While in elementary geometry the proof is obtained by studying the properties of the figure, in analysis the proof arises from algebraic work involving the coordinates: the figure is of no use at all beyond helping us to bear in mind the nature of the problem. Thus, *our notation must be such that the coordinates themselves express the data* of the problem. In the above example, when coordinates have been assigned to three of the vertices, those of the fourth *are determined*, and must be correctly expressed in terms of those already assigned.

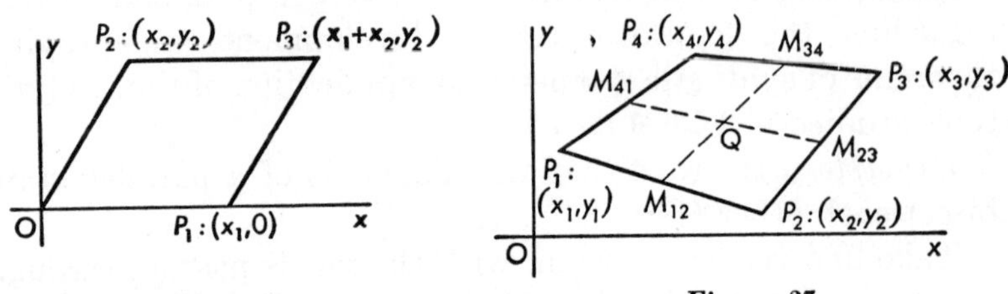

Figure 26 *Figure 27*

Example (*b*): Prove that the lines joining the midpoints of opposite sides of a quadrilateral bisect each other (Fig. 27).

Let M_{12} be the midpoint of P_1P_2, etc. We have to show that the midpoint of $M_{12}M_{34}$ coincides with the midpoint of $M_{41}M_{23}$. The abscissas of M_{12} and M_{34} are $\dfrac{x_1 + x_2}{2}$ and $\dfrac{x_3 + x_4}{2}$, so that the abscissa of the point Q midway between M_{12} and M_{34} is

$$
\begin{aligned}
x &= \frac{1}{2}\left(\frac{x_1 + x_2}{2} + \frac{x_3 + x_4}{2}\right) \\
&= \tfrac{1}{4}(x_1 + x_2 + x_3 + x_4). \qquad (1)
\end{aligned}
$$

Now from this one result the theorem can be deduced. For, first, the right member of (1) is unchanged under any permutation of the subscripts 1, 2, 3, 4; hence the midpoints of $M_{12}M_{34}$ and $M_{23}M_{41}$ must have the same abscissa. Second, since our notation is symmetric in x and y, any result that is true for abscissas must be true for ordinates.

Exercises

In Exs. 1–21, prove the theorem.

1. The midpoint of the hypotenuse of a right triangle is equidistant from the three vertices.

2. The distance between the midpoints of the non-parallel sides of a trapezoid is half the sum of the parallel sides.

3. An isosceles triangle has two equal medians.

4. A triangle having two equal medians is isosceles.

5. The diagonals of a rhombus are perpendicular.

6. If the sum of the squares of two sides of a triangle equals the square of the third side, the triangle is right-angled.

7. A quadrilateral whose diagonals bisect each other is a parallelogram.

8. A rectangle whose diagonals are perpendicular is a square.

9. A quadrilateral whose diagonals bisect each other at right angles is a rhombus.

10. The diagonals of an isosceles trapezoid are equal.

11. A trapezoid whose diagonals are equal is isosceles.

12. The medians of a triangle intersect in a trisection point of each.

13. If a convex quadrilateral has two opposite sides equal and parallel, it is a parallelogram.

14. The lines joining the midpoints of the sides of a triangle divide it into four equal triangles.

15. The line segments joining the midpoints of adjacent sides of a quadrilateral form a parallelogram.

16. The sum of the squares of the sides of a parallelogram equals the sum of the squares of the diagonals.

17. In any triangle, the sum of the squares of any two sides equals twice the square of the median to the third side plus half the square of the third side.

18. In any triangle, the sum of the squares of the three medians equals three-fourths the sum of the squares of the sides.

19. The sum of the squares of the sides of a quadrilateral equals the sum of the squares of the diagonals plus four times the square of the line segment joining the midpoints of the diagonals.

20. If one of the parallel sides of a trapezoid is twice the other, the diagonals intersect in a point of trisection.

21. The angles at either base of an isosceles trapezoid are equal.

22. In Fig. 27, prove that the segment joining the midpoints of the diagonals P_1P_3, P_2P_4 is bisected at Q.

CHAPTER 2 *Polar Coordinates*

14. Distance and bearing. Instead of locating a point by its distance from two perpendicular lines (§ 4), we frequently, in ordinary usage, locate it by its distance and "bearing" from some fixed point: one town is 5 miles southeast of another; one boundary marker is 90 ft. N. 10° E. of another; etc. This method also has its counterpart in analytic geometry.

15. Polar coordinates. Let us choose a fixed line Ox in the coordinate plane, and a point O on this line. The position of any point P (Fig. 28) in the plane is determined if we know the length of the line OP together with the angle that this line makes with Ox, both the distance and the angle being measured in a definite sense. The segment OP and the angle xOP are the *polar coordinates* of P: they are called the *radius vector* and the *polar angle* respectively, and are denoted by r, θ. The fixed line Ox is the *initial line*, or *polar axis*, and the point O is the *pole*, or *origin*.

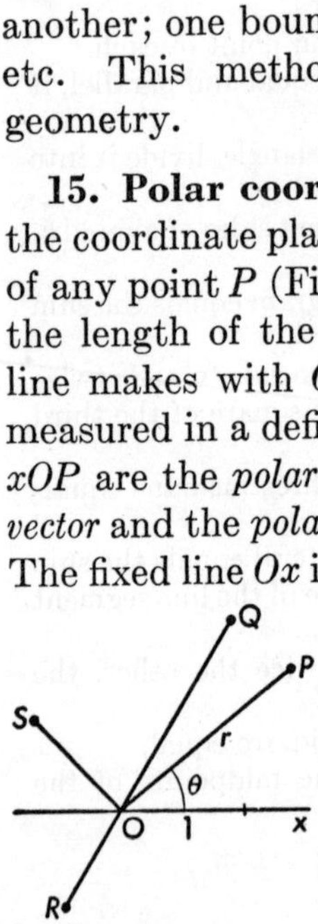

Figure 28

The polar coordinates of a point are written as $P:(r, \theta)$, or simply (r, θ). The polar angle is *positive* when measured *counterclockwise*, *negative clockwise;* the radius vector is *positive* if laid off *on the terminal side of* θ, *negative* in the opposite direction, i.e., *on the terminal side produced through* O. Figure 28 shows $Q:(2, 60°)$, $R:(-1, 60°)$, $S:(-1, -45°)$.

To plot a point whose polar coordinates are given, we begin by drawing the line on which the radius vector lies — i.e., the line making an angle θ with Ox — and then lay off on that line, in the proper sense, the distance r.

Point plotting using polar coordinates is most easily done on "polar coordinate paper," which is paper ruled in concentric circles and straight lines.

Although to every pair of polar coordinates corresponds a single point, the same point may be represented by various pairs of coordinates. For instance, in Fig. 28, the coordinates (2, 60°), (−2, 240°), (−2, −120°) all represent the point Q.

16. Distance between two points. We already know a formula for the distance between two points in terms of the rectangular coordinates of those points. Let us now obtain the corresponding formula in terms of the polar coordinates of the two points. In Fig. 29, the required distance D, between the points (r_1, θ_1) and (r_2, θ_2), can be found by using the law of cosines from trigonometry:

$$D^2 = r_1^2 + r_2^2 - 2r_1r_2 \cos (\theta_2 - \theta_1). \tag{1}$$

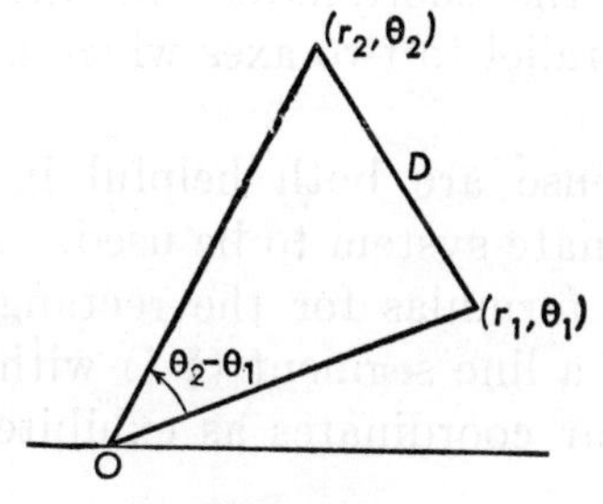

Figure 29

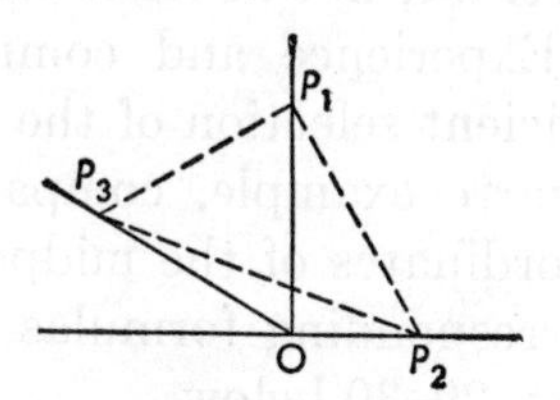

Figure 30

Equation (1) is equivalent to the earlier formula

$$D^2 = (x_1 - x_2)^2 + (y_1 - y_2)^2. \tag{2}$$

Example: Show that the points (3, 90°), ($\sqrt{3}$, 0°), (3, 150°) are the vertices of a right triangle. (Fig. 30.)

Designate the points by P_1, P_2, P_3, in the order given. By the distance formula, we obtain

$$\overline{P_1P_2}^2 = 9 + 3 - 6\sqrt{3}\cos 90° = 12 - 0 = 12,$$
$$\overline{P_1P_3}^2 = 9 + 9 - 18\cos 60° = 18 - 9 = 9,$$
$$\overline{P_2P_3}^2 = 3 + 9 - 6\sqrt{3}\cos 150° = 12 - 6\sqrt{3}\left(-\frac{\sqrt{3}}{2}\right) = 21.$$

Thus the triangle is a right triangle with its right angle at P_1.

17. Choice of coordinate system. A coordinate system is a tool; it is not inherently present in a specific geometric problem. With both polar and rectangular coordinates available, we use whichever appears simpler for the particular problem under consideration. At times it will prove beneficial to change from one coordinate system to the other. The requisite formulas will be obtained (§ 43) when we first need them.

Polar and rectangular coordinates are the only ones employed in the plane geometry portion of this book. Many other systems have been devised and used. In one of those systems (oblique coordinates) the coordinates are directed distances measured on lines parallel to two axes which intersect, but not at right angles.

Experience and common sense are both helpful in the efficient selection of the coordinate system to be used. As a bizarre example, compare the formulas for the rectangular coordinates of the midpoint of a line segment (§ 7) with the corresponding formulas in polar coordinates as exhibited in Exs. 29–30 below.

Exercises

In Exs. 1–4, plot the points whose polar coordinates are given.

1. $(3, 45°)$, $(2, -30°)$, $(-2, 135°)$.
2. $(1, 90°)$, $(-3, 180°)$, $(\frac{1}{2}, 120°)$.
3. $(4, 270°)$, $(-2, 60°)$, $(3, 0°)$.
4. $(2, 225°)$, $(1, -60°)$, $(0, 20°)$.

In Exs. 5–12, find the distance between the given points.

5. $(4, 0°)$, $(2\sqrt{3}, 30°)$. *Ans.* 2.
6. $(3, 30°)$, $(8, 90°)$. *Ans.* 7.
7. $(3, 15°)$, $(5, 135°)$. *Ans.* 7.
8. $(7, -20°)$, $(8, 100°)$. *Ans.* 13.
9. $(7, 30°)$, $(3\sqrt{2}, 75°)$. *Ans.* 5.
10. $(7, 15°)$, $(5\sqrt{2}, 150°)$. *Ans.* 13.
11. $(5, 0°)$, $(3\sqrt{2}, 45°)$. *Ans.* $\sqrt{13}$.
12. $(-5, 0°)$, $(3\sqrt{2}, 45°)$. *Ans.* $\sqrt{73}$.
13. Does the circle with center at $(3, 10°)$ and radius 7 pass through the point $(5, 130°)$? *Ans.* Yes.
14. Does the circle with center at $(8, 20°)$ and radius 7 pass through the point $(5, 80°)$? *Ans.* Yes.

In Exs. 15–18, show that the given points lie on a straight line.

15. $(6, 90°)$, $(3, 30°)$, $(2\sqrt{3}, 0°)$.
16. $(1, 30°)$, $(2, 90°)$, $(-2, 150°)$.
17. $(-6, 15°)$, $(3, 135°)$, $(2\sqrt{3}, 105°)$.
18. $(6, 0°)$, $(12, 60°)$, $(4\sqrt{3}, -30°)$.

In Exs. 19–21, show that the given points are the vertices of a right triangle, and find its area.

19. $(2, 45°)$, $(\sqrt{2}, 90°)$, $(-2, 135°)$. *Ans.* 2.
20. $(2, 60°)$, $(2\sqrt{3}, 90°)$, $(1, 120°)$. *Ans.* $\sqrt{3}$.
21. $(4, 45°)$, $(\sqrt{2}, 90°)$, $(-12, 135°)$. *Ans.* 20.

In Exs. 22–28, express the statement by an equation.

22. The point (r, θ) lies on a circle of radius 5 with center at $(5, 0°)$.
Ans. $r = 10 \cos \theta$.
23. The point (r, θ) lies on a circle of radius 3 with center at $(-3, 0°)$.
Ans. $r = -6 \cos \theta$.
24. The point (r, θ) is at a distance 4 from $(3, 30°)$.
Ans. $r^2 - 6r \cos (\theta - 30°) - 7 = 0$.
25. A right triangle has its right angle at (r, θ) and its other vertices fixed at $(4, 30°)$ and $(3, 120°)$.
Ans. $r = 4 \cos (\theta - 30°) + 3 \cos (\theta - 120°)$.
26. A right triangle has its right angle at (r, θ) and its other vertices fixed at $(5, 30°)$ and $(2, 90°)$.
Ans. $r^2 - 2r \sin \theta - 5r \cos (\theta - 30°) + 5 = 0$.
27. The point (r, θ) is equidistant from $(2, 90°)$ and $(-2, 150°)$.
Ans. $\theta = 30°$.
28. The point (r, θ) is equidistant from $(2, 30°)$ and $(3, 60°)$.
Ans. $2r[3 \cos (\theta - 60°) - 2 \cos (\theta - 30°)] = 5$.

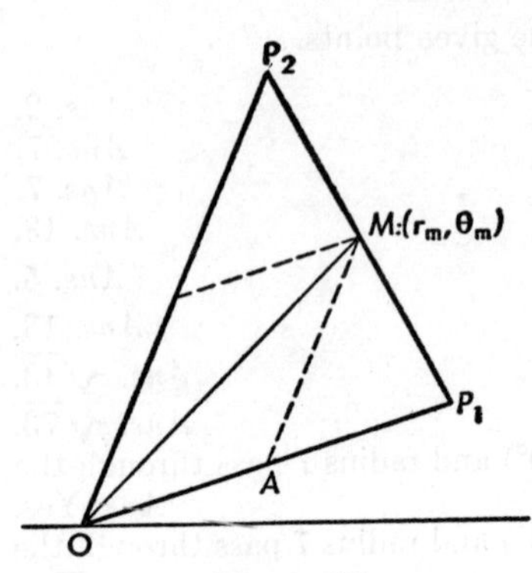

Figure 31

In Fig. 31, the point $M:(r_m, \theta_m)$ is the midpoint of the line segment joining $P_1: (r_1, \theta_1)$ to $P_2: (r_2, \theta_2)$. The point A is the midpoint of OP_1. By elementary geometry we know that

$$\angle OAM = 180° - (\theta_2 - \theta_1),$$

$AM = \frac{1}{2}OP_2$, etc.

29. Use the law of sines to show that the polar coordinate θ_m for the point M in Fig. 31 is given by

$$\tan \theta_m = \frac{r_1 \sin \theta_1 + r_2 \sin \theta_2}{r_1 \cos \theta_1 + r_2 \cos \theta_2}.$$

30. Use the law of cosines to show that the polar coordinate r_m for the point M in Fig. 31 is given by

$$r_m{}^2 = \tfrac{1}{4}r_1{}^2 + \tfrac{1}{4}r_2{}^2 + \tfrac{1}{2}r_1r_2 \cos (\theta_2 - \theta_1).$$

CHAPTER 3 *Curves. Functions*

18. Constants; variables. In analytic geometry we have to deal with two kinds of quantities — "constants" and "variables."

A *constant* is a quantity whose value remains unchanged throughout any given problem. Examples are the coordinates of a fixed point, the radius of a given circle, the slope of a given line, etc. Coordinates of fixed points are usually denoted by attaching subscripts to the letters used for variable coordinates, as (x_1, y_1), (x_2, y_2), (r_1, θ_1), (r_2, θ_2); other letters, such as a, b, m, etc., are also used to denote constants.

A *variable* is a quantity that may take different values (usually an infinite number of them) in the same problem. The variables most frequently occurring in analytic geometry are the coordinates (x, y) or (r, θ) of a point moving along a definite path.

For example, the fact that a moving point (x, y) remains always at the constant distance a from the origin is expressed (§ 6) by the equation

$$x^2 + y^2 = a^2. \tag{1}$$

This equation remains true if the point (x, y) moves along the circle of radius a with center at the origin.

The same example produces an even simpler equation in polar coordinates. The fact that a moving point (r, θ) remains always at the constant distance a from the origin is expressed by the equation $r = a$.

In elementary analytic geometry, all quantities occurring — both constants and variables — are restricted to *real values.* Thus in (1), putting $x = \frac{1}{2}a$, we find $y = \pm\frac{1}{2}\sqrt{3}a$, which says that the points $(\frac{1}{2}a, \pm\frac{1}{2}\sqrt{3}a)$ are on the circle; putting $x = 2a$, we find $y = \pm\sqrt{3}ai$, which merely verifies the fact that there is on the circle no point with abscissa $2a$.

19. The locus of an equation. Let two variables x and y be connected by an equation — for instance,

$$y = \tfrac{1}{2}x + 2, \qquad y^3 = 2x, \text{ etc.}$$

If any value be assigned to either variable, one or more values of the other are determined, in general, by the equation; thus there exist infinitely many pairs of values of x and y that satisfy the equation. Each pair of numbers may be represented geometrically by a point. The points so determined are not scattered at random throughout the plane, but form in the aggregate a definite *curve.*

The locus of an equation is a curve containing those points, and only those points, whose coordinates satisfy the equation.

The curve corresponding to a given equation is said to *represent the equation geometrically,* while the equation *represents the curve analytically.*

It should be remarked that in exceptional cases an equation may represent only a single point, or it may have no locus whatever. For example, the equation $x^2 + y^2 = 0$ is satisfied only by the coordinates (0, 0); the equation $x^2 + y^2 = -1$ is satisfied by no real values of x and y.

Example (*a*): Trace the curve $y = \dfrac{x}{2} + 2.$

We assign values to x and compute the values of y:

x	0	1	2	-1	-2	-3	-4	-5
y	2	$\frac{5}{2}$	3	$\frac{3}{2}$	1	$\frac{1}{2}$	0	$-\frac{1}{2}$

We now plot these points, choosing two spaces on the coordinate paper as the unit, and draw a smooth curve through them. The "curve" in this case appears to be a straight line: it will be proved in § 36, and assumed meanwhile, that *the locus of every equation of the first degree is a straight line.*

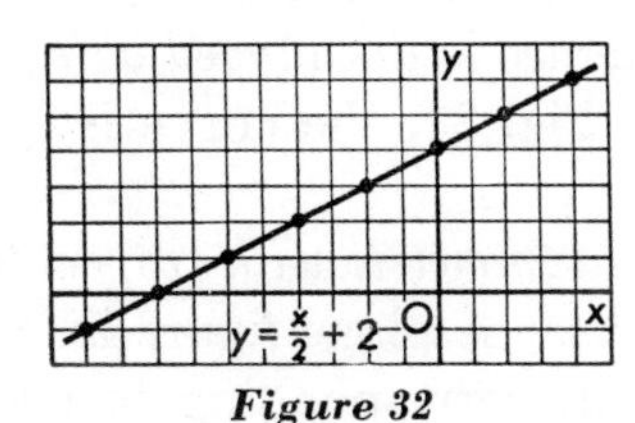

Figure 32

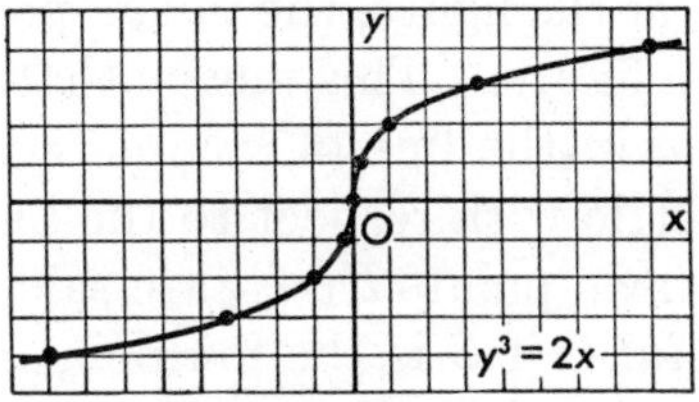

Figure 33

Example (*b*): Plot the curve $y^3 = 2x$.

Here, to avoid the extraction of cube roots, let us write the equation in the form

$$x = \tfrac{1}{2}y^3$$

and assign values to y. Since x becomes very large as y increases, we will assign to y moderately small values:

y	0	$\frac{1}{2}$	1	$\frac{3}{2}$	2	$-\frac{1}{2}$	-1	$-\frac{3}{2}$	-2
x	0	$\frac{1}{16}$	$\frac{1}{2}$	$\frac{27}{16}$	4	$-\frac{1}{16}$	$-\frac{1}{2}$	$-\frac{27}{16}$	-4

20. Intercepts on the axes. To find the points where the curve crosses Ox we must evidently *put* $y = 0$ *and solve for* x; to find the intersections with Oy we *put* $x = 0$ *and solve for* y. Of course in some cases the solution cannot be easily carried out on account of algebraic difficulties.

The directed distances from the origin cut off by the curve on the x-axis and y-axis — i.e., the abscissas and ordinates, respectively, of the intersections with the axes — are called the x- and *y-intercepts*. Thus in Example (*a*), § 19, the x-intercept is -4, the y-intercept 2; in Example (*b*), each intercept is zero.

21. Symmetry. Two points P_1, P_2 are said to be *symmetric* with respect to a line, if that line is the perpendicular bisector of the segment P_1P_2; the line is then called a *line of symmetry*. Each of the points P_1, P_2 is the *image*, or *reflection*, of the other in the line L. A curve or other plane figure is *symmetric with respect to a line* if, corresponding to every point P_1 of the figure, the image-point P_2 in that line also belongs to the figure. This means that the figure is *unchanged by reflection* in the line of symmetry. In Fig. 34, the curve is symmetric with respect to the line L.

Two points P_1, P_2 are said to be *symmetric with respect to a point* C if C is the midpoint of P_1P_2. A plane figure is symmetric with respect to a point C if, corresponding to every point P_1 of the figure, there is a point P_2, also belonging to the figure, such that C is the midpoint of P_1P_2. The point C is called the *center of symmetry*, or simply the *center* (Fig. 35).

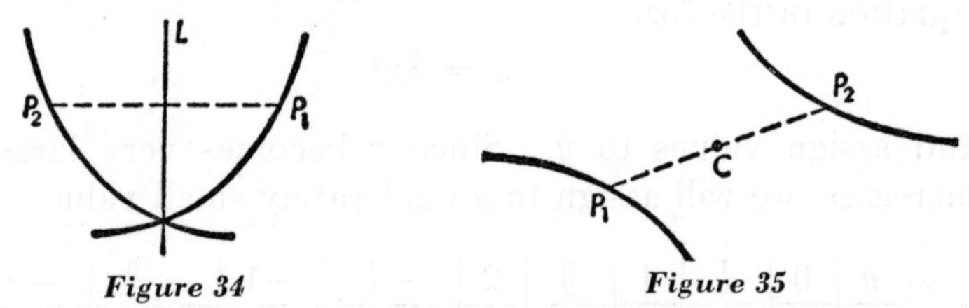

Figure 34 *Figure 35*

THEOREM I: *A curve is symmetric with respect to the x-axis if its equation is unchanged* when y is replaced by* $-y$; *and conversely. A curve is symmetric with respect to the y-axis if its equation is unchanged when x is replaced by* $-x$; *and conversely.*

THEOREM II: *A curve is symmetric with respect to the origin if its equation is unchanged when x is replaced by* $-x$ *and y by* $-y$ *simultaneously; and conversely.*

Proof of I: By hypothesis, if any pair of coordinates (x, y) satisfy the equation, the coordinates $(x, -y)$ of the image-point with respect to the x-axis also satisfy the equation. Proof of the converse and of II is left to the student.

* More precisely, if the new form of the equation is *equivalent* to the original form —i.e., if each form is satisfied by all the values of the variables that satisfy the other.

Example (*a*): The curve (Fig. 119, p. 161)

$$y^2 = (x - 2)(x^2 - 1)$$

is symmetric with respect to Ox, because y may be changed to $-y$ without changing the equation.

Example (*b*): The curve (Fig. 33, p. 31)

$$y^3 = 2x$$

is symmetric with respect to the origin, because the signs of x and y may be changed simultaneously without changing the equation. With this information in hand, it becomes unnecessary to assign negative values to x: after the curve has been plotted in the first quadrant, the other half is obtained by reflection in the origin.

Example (*c*): Trace the curve $r = 2a \cos \theta$.

θ	0	$\frac{1}{6}\pi$	$\frac{1}{4}\pi$	$\frac{1}{3}\pi$	$\frac{1}{2}\pi$
r	$2a$	$\sqrt{3}\,a$	$\sqrt{2}\,a$	a	0

Plotting these points, we get the upper half of the curve. Since for values of θ in the *second* quadrant $\cos \theta$, and hence r, is negative, the curve falls in the *fourth* quadrant; the values of $\cos \theta$ are numerically the same as in the first quadrant, but in reverse order, so that the lower half of the curve is symmetric to the upper half. In the third quadrant $\cos \theta$ takes the same values as in the first, but negative: the upper half is repeated. Similarly, for θ in the fourth quadrant, the lower half is repeated. Since

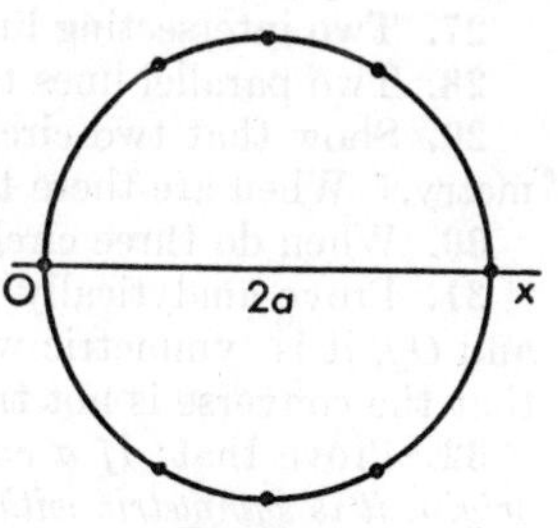

Figure 36

$$\cos (\theta \pm 2n\pi) = \cos \theta, \qquad n = 1, 2, \cdots,$$

values of $\theta > 2\pi$ or $\theta < 0$ will merely repeat the same curve.

Symmetry tests for equations in polar coordinates will be taken up in § 102.

Exercises

In Exs. 1–8, plot three points (the third as a check), and draw the line.

1. $x + 2y = 7.$
2. $3x - y = 8.$
3. $3x = 10.$
4. $2y + 5 = 0.$
5. $5x - 3y = 13.$
6. $4x + 3y = -11.$
7. $2x + 3y = 0.$
8. $5x - y = 0.$

In Exs. 9–18, find the intercepts on the axes, test for symmetry, plot an adequate number of points, and draw the curve on a suitable scale.

9. $y^2 = x + 4.$
10. $x^2 = y - 2.$
11. $y = x^2 - x.$
12. $y = 6 - x - x^2.$
13. $xy = 4.$
14. $y = x^3 - 9x.$
15. $4y^2 - 2y + x = 0.$
16. $x = (y^2 - 4)^2.$
17. $y = x^3 + 3x^2 + 2x.$
18. $y = x^3 - 3x^2 - 4x.$

In Exs. 19–22, plot the curve with due regard to the symmetry properties of the sine and cosine.

19. $r = 2a \sin \theta.$
20. $r = a \cos^2 \theta.$
21. $r = a \sin^2 \theta.$
22. $r = a(1 + \sin \theta).$

Without formal proofs, state how many lines of symmetry are possessed by each of the figures in Exs. 23–28.

23. (*a*) A circle; (*b*) a circular arc.
24. (*a*) A straight line; (*b*) a straight line segment. *Ans.* (*b*) Two.
25. A triangle. Discuss special cases fully.
26. A quadrilateral. Discuss all special cases.
27. Two intersecting lines taken together.
28. Two parallel lines taken together.
29. Show that two circles taken together always have one line of symmetry. When are there two such lines? When more than two?
30. When do three circles have one or more lines of symmetry?
31. Prove analytically that if a curve is symmetric with respect to Ox and Oy, it is symmetric with respect to the origin. Show by an example that the converse is not true.
32. Prove that: *If a curve is symmetric with respect to one axis and the origin, it is symmetric with respect to the other axis also.*
33. By finding the midpoint and slope of the line segment joining the points (h, k) and (k, h), prove that: *In any equation, the effect of interchanging x and y is to reflect the curve in the line $y = x$.*

22. Functions. If two variables y and x are so related that, *when the value of x is given, the value of y is determined,* then y is said to be a *function of x.* The variable x, to which values are assigned at pleasure, is the *independent variable*, or *argument*.

Numerous examples of functional relationship are already familiar to the reader.

Example (a): The area of a circle is determined by the radius — i.e., the area is a function of the radius:

$$A = \pi r^2.$$

Example (b): The distance s traveled by a car in time t at 40 m.p.h. is

$$s = 40t. \tag{1}$$

In general, if y is a function of x, then x is also a function of y, and either may be chosen as independent variable. Thus in (b), to find where he will be at a given time, a driver uses formula (1); but to find when he will reach a given point, he computes

$$t = \frac{s}{40}.$$

To express the fact that y is a function of x, without specifying the precise form of the function, we may write

$$y = f(x)$$

(read "y equals f of x"). Other letters may be used in the functional symbol, as $F(x)$, $P(x)$, etc.

23. Graph of a function. Any function may be represented graphically by merely plotting the argument as abscissa and the function as ordinate: i.e., given

$$y = f(x), \tag{1}$$

we interpret the pairs of values of x and y as rectangular coordinates of points in a plane, and draw the curve through those points. This curve is called the *graph* of the function. Thus in Example (a), § 22, the graph is the parabola (see § 60) of Fig. 37; in (b) it is the straight line in Fig. 38, p. 36.

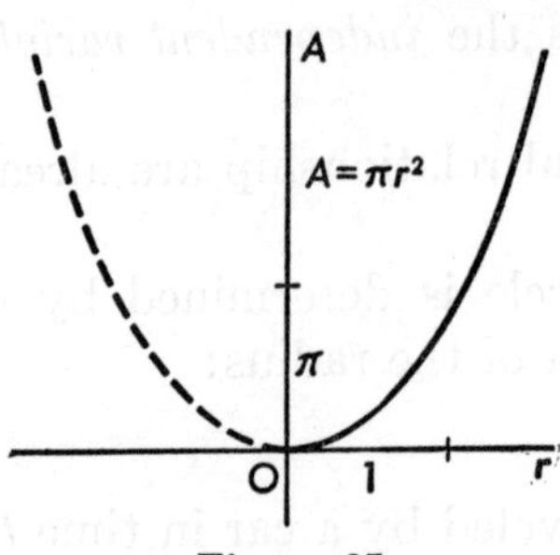

Figure 37

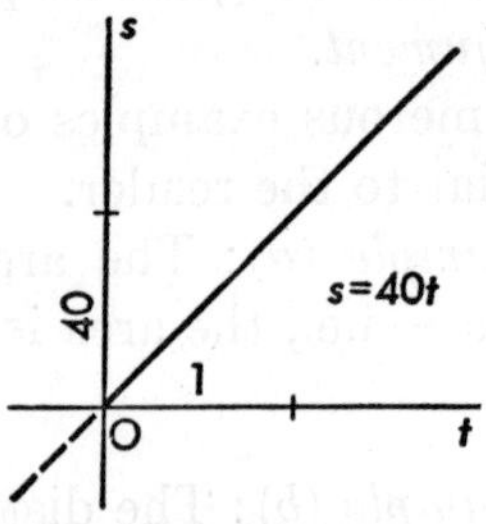

Figure 38

The graph of the function $f(x)$ is identical with the locus of equation (1) as defined in § 19. But in many applications, by the very nature of the problem, the independent variable is limited to a definite *range*, or *interval:* for instance, in Figs. 37–38, the dotted portions have no meaning, since r and t are necessarily positive. We may disregard those portions of the curve lying outside the range where the function is defined.

In some applications the function is represented by different formulas in different intervals. This means that the graph will consist of segments which may be connected (Ex. 16 below) or disconnected (Ex. 17). The graph may even consist of a set of isolated points (Ex. 18).

Exercises

In Exs. 1–16, express the function by a formula and draw the graph, indicating the portion of the graph that has a meaning.

1. The distance covered at 10 yd. per sec., as a function of time.

2. The value of a farm at $60 per acre, with buildings worth $4000, as a function of the number of acres.

3. The value of a consignment of wheat at $1.75 per bu., as a function of the number of bushels.

4. The function of Ex. 3, if $150 must be deducted for fixed charges.

5. The surface area of a sphere as a function of the radius.

6. The surface area of a cube as a function of the length of one edge.

7. The diagonal of a square as a function of the perimeter.

8. The total area (including both bases) of a circular cylinder of radius unity, as a function of the altitude.

9. The volume of a circular cone of radius unity, as a function of the altitude.

10. The total surface area (including both ends) of a cylindrical can of altitude unity, as a function of the radius.

11. The total surface area of a box of dimensions a, $a - 2$, $a + 2$.

12. The volume of the box in Ex. 11.

13. The altitude of a right triangle of area unity, as a function of the base.

14. The width of a rectangle of area unity, as a function of the length.

15. The lateral surface area of a circular cone of radius unity, as a function of the height.

16. The difference in length between a line segment of length 4 and one of length c.

17. Letter postage in the U.S. is 3¢ per oz. or fraction thereof. Graph the cost as a function of the weight.

18. A certain telegram costs 70¢ for the first fifteen words or less, plus 5¢ for each additional word. Graph the cost as a function of the number of words.

19. A driver, starting with 10 gal. of gasoline, buys 5 gal. at the end of each 100 mi. If the car travels 20 mi. per gal., graph the quantity of gasoline in the tank, as a function of distance.

20. Solve Ex. 19 if the car travels 15 mi. per gal. Determine analytically and graphically how far the car can go. *Ans.* 375 mi.

24. Consequences of the definition of locus. From the definition of the locus of an equation (§ 19) we may deduce certain rules which are so important that in spite of their obviousness it seems worth while to state them explicitly.

By the definition, a point lies on a curve *if and only if its coordinates satisfy the equation of the curve.*

RULE I: *To find out whether a point lies on a given curve, substitute its coordinates for x and y in the equation of the curve, and note whether the equation holds.*

Example (*a*): The point (2, 12) lies on the curve

$$y = 3x^2$$

because $12 = 3 \cdot 4$; the point $(-1, -3)$ does not lie on the curve because $-3 \neq 3 \cdot 1$.

RULE II: *To express analytically the condition that a point shall lie on a curve, write the equation of the curve with the coordinates of the point substituted for x and y.*

Example (*b*): The point (x_1, y_1) lies on the curve $y^2 = 4x$ if and only if $y_1{}^2 = 4x_1$.

Example (*c*): Determine a and b so that the curve

$$y = ax^2 + bx \tag{1}$$

shall pass through the points P_1: (2, 0), P_2: (3, 6).

Substituting in (1) the coordinates (2, 0) and (3, 6) in turn, we get

$$\begin{aligned} 0 &= 4a + 2b, \qquad \text{or} \qquad 2a + b = 0, \\ 6 &= 9a + 3b, \qquad \text{or} \qquad 3a + b = 2. \end{aligned}$$

These simultaneous equations give

$$a = 2, \qquad b = -4,$$

whence the equation of the curve is

$$y = 2x^2 - 4x.$$

Rule III: *To find the ordinate of a point on a curve when the abscissa is given, substitute the given abscissa for x in the equation of the curve and solve for y.* Similarly we may find the abscissa when the ordinate is given.

25. Number of points required to determine a curve. In Example (*c*), § 24, we were given an equation containing two undetermined constants a, b, and we undertook to make this curve pass through two given points. Substituting the coordinates of the points in the equation of the curve, we obtained two equations for a and b, exactly the right number to determine two unknowns. It is easy to see that if there had been any number n of constants and an equal number n of given points, the result would have been similar.

Theorem: *The number of points required to determine a curve is equal to the number of independent constants in the equation of the curve.*

26. Factorable equations. When the product of two or more factors is equated to zero, the equation thus formed is satisfied whenever any one of the factors is zero. For instance, the equation

(1) $$3(x + 2y)(x^2 - y^2) = 0$$

is satisfied if and only if

(2) $$x + 2y = 0,$$
(3) $$x + y = 0,$$

or

(4) $$x - y = 0.$$

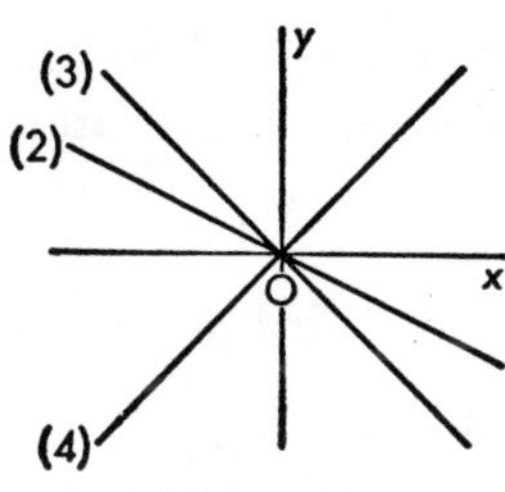

Figure 39

Hence, the locus of an equation whose right member is zero and whose left member can be factored consists of all points whose coordinates when substituted in the equation cause any one of the factors to vanish. That is, the curve represented by such an equation breaks up into several distinct "branches," the loci of the equations formed by equating the several factors separately to zero. Thus (1) represents the three lines (2), (3), (4).

27. Classification of curves. An algebraic plane curve is one whose equation in rectangular coordinates is a polynomial in x and y, equated to zero: for example,

$$x^2 - xy + x + 3y - 5 = 0,$$
$$x^3 + 2x^2y - y^3 + 5y = 0.$$

Any non-algebraic function is called *transcendental.* A *transcendental curve* is one whose equation in rectangular coordinates involves transcendental functions: for example,

$$y = \sin x, \qquad y = \log_{10} x, \qquad y = 2^x.$$

For the present we confine our attention to algebraic curves. Transcendental curves will be studied in Chaps. 13, 14.

The classification of curves is based exclusively on their equations in rectangular coordinates. To classify a curve whose equation is given in polar coordinates, we first transform the equation to one in rectangular coordinates by the formulas of § 43.

28. Degree of an algebraic curve. An algebraic curve whose equation is of the nth degree is called a *curve of the nth degree.* For example, the curve

$$3x^2y + 2x^2 - y = 6$$

is of third degree; the curve

$$\frac{y}{x} = 1 \tag{1}$$

is of first degree; the curve

$$y^2 = \frac{x+3}{x^2+2x}$$

is of fourth degree.

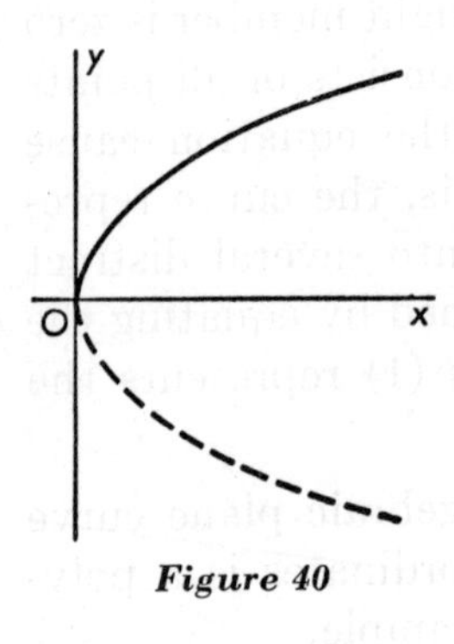

Figure 40

Since an irrational equation may not be fully equivalent to the one obtained by rationalizing, its locus may be only a part of the curve corresponding to the rational form. For instance, the equation

$$y = 2\sqrt{x}$$

represents only the upper half of the curve (Fig. 40)

$$y^2 = 4x.$$

Of course, when a rational equation is cleared of fractions, certain pairs of values of x and y may be introduced. For instance: since division by zero is rigidly excluded, the coordinates (0, 0) do not satisfy equation (1). Hence, strictly, the locus of (1) is the line $y = x$ with one point missing. But such distinctions would cause endless bother and serve no purpose, so we tacitly agree to include the missing point in the locus unless the contrary is indicated.

29. Points of intersection of two curves. In rectangular coordinates, the points of intersection of two curves are points whose coordinates *satisfy both equations*, and there are no other points having this property. Hence *the points of intersection of two curves are found by solving the equations of the curves as simultaneous equations.* In polar coordinates the location of all points of intersection of two curves can involve slight additional complications. See § 103 for details.

It may happen that all the values of x and y, found by solving two simultaneous equations, are imaginary; or the equations may be "incompatible," not satisfied by any pairs of values either real or imaginary (Ex. 31 below). Of course in either case the result means that the curves have no intersection.

A check should always be made by substituting the values of x and y in both equations and noting whether the equations hold.

Example: Find the points of intersection of the straight line

$$2x + y = 10 \tag{1}$$

and the circle

$$x^2 + y^2 = 25. \tag{2}$$

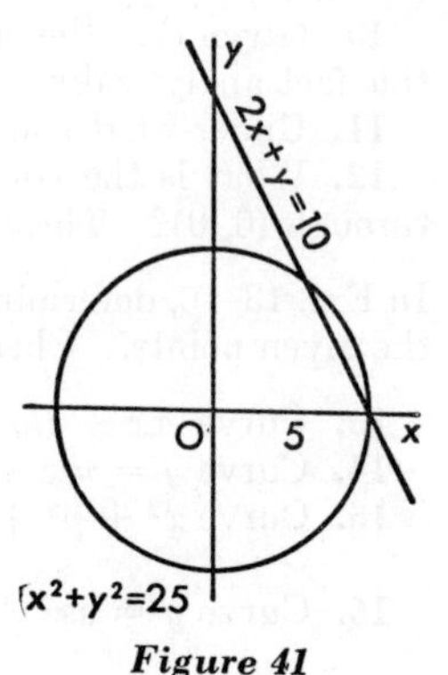

Figure 41

Substituting the value of y from (1) in (2), we get

$$x^2 + (10 - 2x)^2 = 25,$$

or

$$5x^2 - 40x + 75 = 0,$$
$$x^2 - 8x + 15 = 0,$$

whence

$$x = 3 \text{ or } 5.$$

By (1),

$$y = 4 \text{ or } 0,$$

and the points are (3, 4), (5, 0).

We know that when two equations in x and y are solved as simultaneous, the number of solutions (i.e., the number of pairs of values of x and y satisfying both equations) is not greater than the product of the degrees of the equations.

THEOREM: *The number of points of intersection of two curves is not greater than the product of the degrees of their equations.*

Exercises

In Exs. 1–6, determine whether the points lie on the curve.

1. Curve $x + 3y = 7$; points $(1, 2)$, $(2, -3)$, $(10, -1)$, $(0, 2)$.
2. Curve $3x - 2y + 1 = 0$; points $(1, 1)$, $(1, 2)$, $(3, 5)$, $(\frac{1}{2}, \frac{5}{4})$.
3. Curve $y^2 = 4ax$; points $(2a, a)$, $(-a, -2a)$, $(4a, -4a)$, $(\frac{1}{4}a, a)$.
4. Curve $x^2 + y^2 = 5$; points $(-2, -1)$, $(\sqrt{2}, \sqrt{3})$, $(1.41, 1.73)$.
5. Curve $y = x^3 + 2x^2 - x - 3$; points $(1, -1)$, $(-2, -1)$, $(-4, -30)$.
6. Curve $x^2 - xy + 4y = 3$; points $(3, -6)$, $(-2, -1)$, $(2, -\frac{1}{2})$.
7. Determine k so that the straight line $x + 3y = k$ shall pass (*a*) through $(1, 2)$; (*b*) through $(0, 1)$; (*c*) through $(2, -4)$.
8. For what values of m does the line $y = mx - 3$ pass through (*a*) $(1, 2)$; (*b*) $(2, 0)$; (*c*) $(5, 4)$; (*d*) $(0, 1)$; (*e*) $(0, 4)$?
9. For what values of c does the curve $cy = x^3$ pass through (*a*) $(1, 2)$; (*b*) $(2, 4)$; (*c*) $(0, 3)$; (*d*) $(3, 0)$; (*e*) $(0, 0)$?
10. Given that the point (x_1, y_1) lies on the curve $x^2 - 4y^2 = 25$, express this fact analytically.
11. Under what condition does (h, k) lie on the line $y = 7x - 3$?
12. What is the condition that the curve $y = ax^2 + bx + c$ shall pass through $(0, 0)$? Through $(1, 2)$? Through $(-2, 3)$?

In Exs. 13–16, determine the constants so that the curve shall pass through the given points. Check the answers.

13. Curve $Ax + By = 5$; points $(3, 1)$, $(-1, -2)$. *Ans.* $3x - 4y = 5$.
14. Curve $y = mx + b$; points $(1, 2)$, $(5, -6)$.
15. Curve $x^2 + y^2 + ax + by = 0$; points $(4, 2)$, $(6, 8)$.
Ans. $x^2 + y^2 + 2x - 14y = 0$.
16. Curve $y = ax^2 + bx + c$; points $(1, 6)$, $(-2, -6)$, $(0, 4)$.
Ans. $y = 4 + 3x - x^2$.
17. On the curve $y^2 = x^3$, find the points (*a*) whose abscissa is 1; (*b*) whose abscissa is -3; (*c*) whose ordinate is -8. *Ans.* (*b*) None.
18. On the curve $y^2 = 4ax$, find the points (*a*) whose abscissa is a; (*b*) whose ordinate is $-a$; (*c*) whose ordinate is zero.
19. On the curve $y^2 = 4y - x + 2$, find the points (*a*) whose abscissa is -3; (*b*) whose abscissa is 7; (*c*) whose ordinate is 6.

20. On the curve $y^2 = x^3 - 7x + 10$, find the points (*a*) whose abscissa is 3; (*b*) whose ordinate is 2.

In Exs. 21–26, draw the curves.

21. $y^2 + 4xy = 0$.
22. $4x^3y = xy^3$.
23. $x^2 - 2xy + y^2 = 9$.
24. $4x^2 + 4xy + y^2 = 0$.
25. $y^2 = x^4$.
26. $x^2y = y^2$.

27. Show that the equation $ax^2 + bxy + cy^2 = 0$ represents two distinct lines through the origin if $b^2 - 4ac > 0$; two coincident lines through the origin if $b^2 - 4ac = 0$; the point (0, 0) if $b^2 - 4ac < 0$. (Solve for y in terms of x; investigate specially the cases $a = b = 0$, $b = c = 0$, $a = c = 0$.)

28. What is the locus of $ax^2 + bx + c = 0$? Of $ay^2 + by + c = 0$? (Ex. 27.)

In Exs. 29–40, find the points of intersection of the given curves; check your answers.

29. $x + y = 7$, $3x - y = 5$.
30. $2x - 3y = 13$, $5x + y = 7$.
31. $x + 2y + 1 = 0$, $4y = 5 - 2x$. Draw the lines.
32. $2x - 3y = 0$, $6y - 4x = 3$. Draw the lines.
33. $x^2 + y^2 = 50$, $3y = x + 20$.
34. $x^2 + y^2 = 25$, $3x + y = 5$.
35. $x^2 + y^2 = 5$, $2x - y = 5$. *Ans.* (2, −1) twice.
36. $x^2 + y^2 = 10$, $x + 3y = 10$. *Ans.* (1, 3) twice.
37. $y = x^2$, $y^2 = 3y - 2x$. *Ans.* (0, 0), (−2, 4), (1, 1) twice.
38. $x^2 + y^2 = 5$, $y^2 = 2x + 5$.
39. $y = x^3 - 3x^2 + x$, $2x + y = 1$. Make an accurate detail of the curves in the interval $0 \leqq x \leqq 2$. *Ans.* (1, −1) three times.
40. $y = x^4 - 4x^3 + 6x^2 - 3$, $y = 4x - 4$. Make a large scale detail of the curves in the interval $0 \leqq x \leqq 2$. *Ans.* (1, 0) four times.

In Exs. 41–44, determine the degree of the equation.

41. $x^{\frac{1}{2}} + y^{\frac{1}{2}} = a^{\frac{1}{2}}$. *Ans.* 2.
42. $x^{\frac{1}{3}} + y^{\frac{1}{3}} = a^{\frac{1}{3}}$. *Ans.* 3.
43. $x^{\frac{2}{3}} + y^{\frac{2}{3}} = a^{\frac{2}{3}}$. *Ans.* 6.
44. $x^{\frac{3}{2}} + y^{\frac{3}{2}} = a^{\frac{3}{2}}$. *Ans.* 6.

CHAPTER 4 *The Equation of a Locus*

30. Path of a moving point. Very often a curve is defined as the *path*, or *locus*, *of a point which moves according to a given law.* In such a case, the statement of the law of motion usually suggests an equality between certain distances or other quantities involving the coordinates of the moving point. We always *denote the coordinates of the moving point by* (x, y) or (r, θ), and try to obtain (by the formulas of analytic geometry) expressions for the distances or other quantities involved in the statement of the law of motion, after which it is usually a simple matter to write out the equation of the locus.

Example (*a*): A point $P:(x, y)$ moves so as to remain always equidistant from the points $P_1:(3, 2)$ and $P_2:(-1, 5)$. Find the equation of its locus. (Fig. 42.)

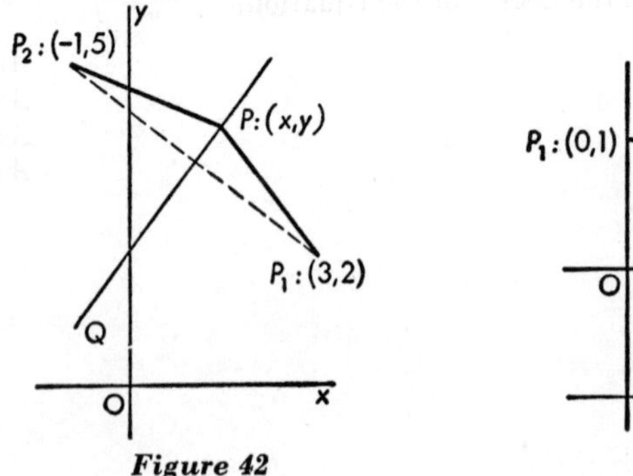

Figure 42

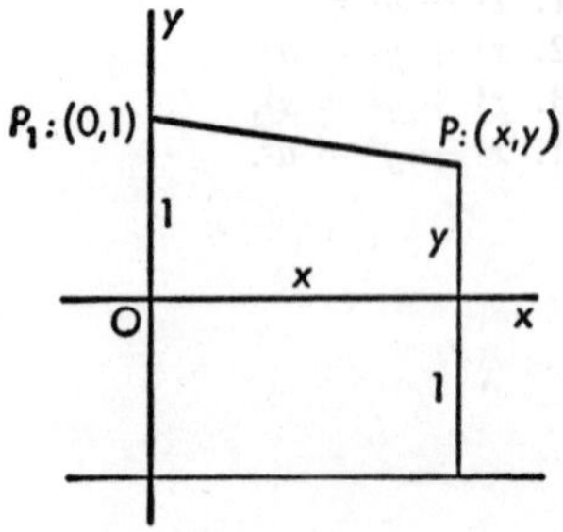

Figure 43

Since $$PP_1 = PP_2,$$

the equation of the locus is

$$\sqrt{(x-3)^2 + (y-2)^2} = \sqrt{(x+1)^2 + (y-5)^2}.$$

Squaring both members, we get

$$x^2 - 6x + 9 + y^2 - 4y + 4 = x^2 + 2x + 1 + y^2 - 10y + 25,$$

or

$$8x - 6y + 13 = 0,$$

which shows that the locus is a straight line. It is the perpendicular bisector PQ of the line joining the fixed points.

Example (*b*): A point moves so as to remain equidistant from the line $y = -1$ and the point P_1:(0, 1). Find the equation of its locus. (Fig. 43.)

The distance of P:(x, y) from the line $y = -1$ is $y + 1$, as is evident from Fig. 43. Hence

$$y + 1 = \sqrt{x^2 + (y-1)^2},$$
$$y^2 + 2y + 1 = x^2 + y^2 - 2y + 1,$$
$$x^2 = 4y.$$

To prove that a curve and its equation as found in the present chapter correspond to each other in the manner required by the definition of "locus of an equation" (§ 19), it is clearly necessary to show (*a*) that the coordinates of every point of the locus satisfy the equation; (*b*) that every point whose coordinates satisfy the equation lies on the locus. The method outlined above establishes only the first of these statements. However, in most cases the second part of the proof can be carried out by merely reversing the steps already taken, and therefore will usually be omitted.

To find the equation of a locus, using polar coordinates, we assume a point P:(r, θ) in a general position on the curve, and from the statement of the problem or by means of some characteristic property of the figure try to obtain an equation involving r, θ, and the constants of the problem: this must be the equation of the given locus.

Polar coordinates are strongly indicated in that numerous class of problems where the distance of the moving point from a fixed point varies according to some simple law. The fixed point should usually be taken as pole and a line of symmetry (if such exists) as polar axis.

Example (*c*): Determine the path of a point which is equidistant from a given point and a given line.

Let us use polar coordinates, taking the given point as pole, the given line as perpendicular to Ox at $Q:(2a, 0)$, and the variable point as $P:(r, \theta)$.

With the notation in Fig. 44, the statement of our problem yields

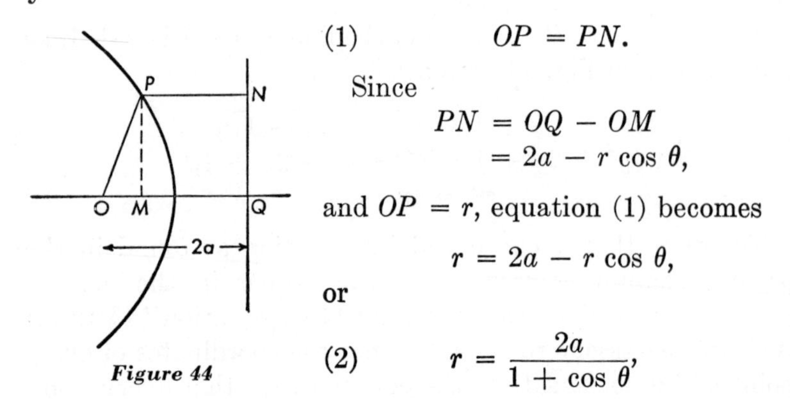

Figure 44

$$(1) \qquad OP = PN.$$

Since

$$PN = OQ - OM$$
$$= 2a - r\cos\theta,$$

and $OP = r$, equation (1) becomes

$$r = 2a - r\cos\theta,$$

or

$$(2) \qquad r = \frac{2a}{1 + \cos\theta},$$

which is the equation of the desired locus of P.

Exercises

In Exs. 1–8, a point $P:(x, y)$ moves according to the given law. Find the equation of its locus, and draw the curve.

1. Equidistant from $(2, -1)$, $(3, 2)$. *Ans.* $x + 3y = 4$.
2. Equidistant from $(0, 4)$, $(-3, -5)$. *Ans.* $x + 3y = -3$.
3. Its distance from $(4, 0)$ is always twice its distance from $(1, 0)$. *Ans.* $x^2 + y^2 = 4$.

4. Its distance from $(0, -4)$ is two-thirds of its distance from $(0, -9)$.
Ans. $x^2 + y^2 = 36$.

5. At distance 5 from $(-3, 0)$. *Ans.* $x^2 + y^2 + 6x - 16 = 0$.

6. At distance 3 from $(1, -4)$. *Ans.* $x^2 + y^2 - 2x + 8y + 8 = 0$.

7. The sum of the squares of its distances from $(2a, 0)$, $(-2a, 0)$ is $10a^2$.

8. The sum of the squares of its distances from $(0, a)$, $(0, -a)$ is $6a^2$

In Exs. 9–15, find the polar equation of the locus, and draw the curve.

9. A line through O making an angle α with Ox.

10. A circle of radius a with center at $(a, 0)$. *Ans.* $r = 2a \cos \theta$.

11. A circle of radius a with center at $(a, \frac{1}{2}\pi)$.

12. A line through $(a, 0)$ perpendicular to Ox.

13. A line through $(a, \frac{1}{2}\pi)$ parallel to Ox. *Ans.* $r = a \csc \theta$.

14. A line through (a, π) perpendicular to Ox. *Ans.* $r = -a \sec \theta$.

15. A circle of radius a with center at (r_1, θ_1).
Ans. $r^2 - 2r_1 r \cos (\theta - \theta_1) + r_1^2 - a^2 = 0$.

16. Find the locus of a point (r, θ) whose distance from O is twice its distance from $(c, 0)$. *Ans.* $3r^2 - 8rc \cos \theta + 4c^2 = 0$.

17. Find the locus of a point (r, θ) whose distance from O is one-half its distance from $(c, \frac{1}{2}\pi)$. *Ans.* $3r^2 + 2rc \sin \theta - c^2 = 0$.

18. Find the locus of a point equidistant from O and the line through $(2a, \frac{1}{2}\pi)$ parallel to Ox. *Ans.* $r = \dfrac{2a}{1 + \sin \theta}$.

19. Solve Example (c) above, if the moving point is half as far from the given point as from the given line. *Ans.* $r(2 + \cos \theta) = 2a$.

20. Solve Example (c) above, if the moving point is twice as far from the given point as from the given line. *Ans.* $r(1 + 2 \cos \theta) = 4a$.

In Exs. 21–32, a point $P:(x, y)$ moves according to the given law. Find the equation of its locus.

21. Equidistant from the line $y = 3$ and the point $(0, -3)$.
Ans. $x^2 + 12y = 0$.

22. Equidistant from the y-axis and the point $(4, 0)$.

23. Equidistant from the line $x + 3 = 0$ and the point $(2, 1)$.
Ans. $y^2 - 10x - 2y - 4 = 0$.

24. Equidistant from the line $y + 4 = 0$ and the point $(-3, -2)$.
Ans. $x^2 + 6x - 4y - 3 = 0$.

25. Its distance from the point $(9, 0)$ is three times its distance from the line $x = 1$. *Ans.* $8x^2 - y^2 = 72$.

26. Its distance from the point $(0, 1)$ is one-half of its distance from the line $y = 4$. *Ans.* $4x^2 + 3y^2 = 12$.

27. Its distance from $(0, 5)$ is two-thirds of its distance from the x-axis.
Ans. $9x^2 + 5y^2 - 90y + 225 = 0$.

28. Its distance from $(3, 2)$ is twice its distance from the y-axis.
Ans. $3x^2 - y^2 + 6x + 4y - 13 = 0$.

29. The sum of its distances from (4, 0) and (−4, 0) is 10.
Ans. $9x^2 + 25y^2 = 225$.

30. The sum of its distances from (0, −1) and (0, 1) is $2\sqrt{2}$.
Ans. $2x^2 + y^2 = 2$.

31. The difference of its distances from (0, 5) and (0, −5) is 2.
Ans. $24y^2 - x^2 = 24$.

32. The difference of its distances from (2, 0) and (−2, 0) is 2.
Ans. $3x^2 - y^2 = 3$.

31. Loci defined geometrically. Sometimes we have a curve actually drawn, and are required to find its equation from the known geometric properties of the figure, as in (*a*) below; or a geometric construction may be given by means of which the points of the curve are determined, as in (*b*) and (*c*).

In such cases we *assume a point of the curve in a general position, and denote its coordinates by* (x, y) or (r, θ). Then the problem is merely to express some characteristic property of the curve by means of an equation involving x and y, or r and θ, and the constants of the problem. (By "characteristic property" is meant, of course, one that holds for all the points of the curve, and for no other points.) *No matter what property is used,* the result must be the equation of the curve. (See Exs. 21–24 below.)

Example (*a*): Find the equation of the straight line whose intercepts on the axes are $OA = 3$ and $OB = 2$.

Assume a point $P: (x, y)$ in a general position on the line. It is clear that the triangles MAP and OAB are similar if and only if P is on the line. Hence if the fact that these triangles are similar be expressed by an equation involving x and y, that equation must represent the line. Now

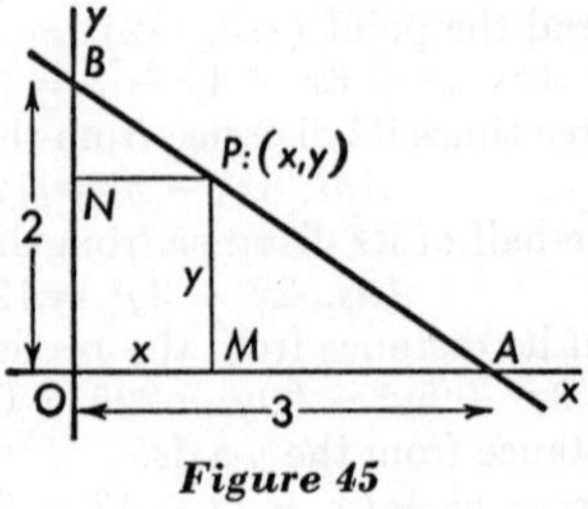

$$\frac{MP}{OB} = \frac{MA}{OA};$$

that is,

$$\frac{y}{2} = \frac{3 - x}{3}, \quad \text{or} \quad 2x + 3y = 6.$$

Figure 45

Example (*b*): Find the locus of the centers of circles passing through the point P_1: (0, 1) and tangent to the line $y + 1 = 0$.

Let us draw one* of the circles in question (Fig. 46), and let its center P be the point (x, y). We note that P is equidistant from the line $y + 1 = 0$ and the point (0, 1), so that the locus is the same as in Example (*b*), § 30:

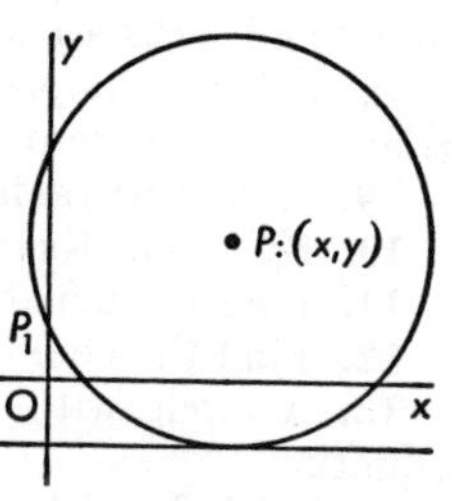

Figure 46

$$y + 1 = \sqrt{x^2 + (y - 1)^2},$$
$$x^2 = 4y.$$

Example (*c*): In Fig. 47, a random line is drawn through O intersecting the circle at P; P is projected to M; a length $OQ = OM$ is laid off on OP. Find the locus of Q.

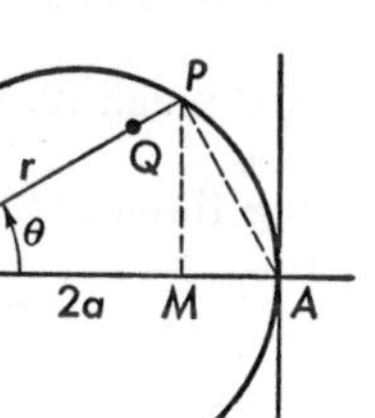

Figure 47

By trigonometry,

$$OP = 2a \cos \theta,$$
$$OM = OP \cos \theta = 2a \cos^2 \theta;$$

since

$$OM = OQ = r,$$
$$r = 2a \cos^2 \theta. \quad \text{(Ex. 20, p. 34.)}$$

Exercises

In Exs. 1–33, use rectangular coordinates.

1. Find the equation of the perpendicular bisector of the line segment joining (2, 4), (−1, 6). *Ans.* $6x - 4y = -17$.

2. Find the equation of the perpendicular bisector of the line segment joining (3, 4), (0, −5). *Ans.* $x + 3y = 0$.

3. Find the locus of the centers of circles passing through the points (−5, 5), (1, 7). *Ans.* $3x + y = 0$.

4. Find the locus of the centers of circles passing through the points (−5, −1), (4, 0). *Ans.* $9x + y = -5$.

5. The base of an isosceles triangle is the line from (4, −3) to (−4, 5). Find the locus of the third vertex in two ways.

* To draw more than one would merely obscure the figure, without helping us to find the equation of the locus.

6. The base of an isosceles triangle joins the points $(0, 7)$, $(-2, 3)$. Find the locus of the third vertex in two ways.

7. Solve Example (*a*), § 31, by making use of the fact that the triangles MAP and NPB are similar.

8. Solve Example (*a*), § 31, by means of the fact that the area of the triangle APB is zero. (§ 12.)

9. In two ways, find the equation of the line through $(4, 1)$, $(-6, 3)$.

10. By two methods, find the equation of the line through $(-5, 2)$, $(4, 6)$.

11. Find the equation of a circle of radius 5 with center at $(-1, 4)$.

12. Find the equation of a circle of radius 3 with center at $(0, -3)$.

13. A circle with center at $(-1, 4)$ passes through $(3, 3)$. Find its equation.

14. A circle with center at $(-2, -3)$ passes through $(4, 1)$. Find its equation.

15. A circle is drawn having the points $(2, 2)$, $(-4, 4)$ as ends of a diameter. Find its equation by three methods.

16. A circle is drawn having the points $(1, 4)$, $(3, 2)$ as ends of a diameter. Find its equation by three methods.

17. Find the locus of the centers of circles passing through $(2, -3)$ and tangent to the line $x + 4 = 0$.

18. Find the locus of the centers of circles passing through $(2, 3)$ and tangent to the line $y + 2 = 0$.

19. One of the equal sides of an isosceles triangle joins the points $(2, 4)$, $(1, -3)$. Find the locus of the third vertex.

Ans. $x^2 + y^2 - 4x - 8y - 30 = 0$; $x^2 + y^2 - 2x + 6y - 40 = 0$.

20. One of the equal sides of an isosceles triangle joins the points $(2, -1)$, $(3, 4)$. Find the locus of the third vertex.

21. Two vertices of a right triangle are at $(2, 1)$, $(3, -4)$, with the right angle at $(2, 1)$. Find the locus of the third vertex by using § 10.

22. Solve Ex. 21 by using § 6.

23. Solve Ex. 21 by using § 7.

24. Solve Ex. 21 by using § 12.

25. Two vertices of a triangle are $(2, 1)$, $(4, 3)$; the area is 2. Find the locus of the third vertex. *Ans.* $x - y = 3$; $x - y + 1 = 0$.

26. Two vertices of a triangle are $(3, 5)$, $(1, 4)$; the area is 1. Find the locus of the third vertex. *Ans.* $x - 2y + 5 = 0$; $x - 2y + 9 = 0$.

27. The base of a triangle joins $(4, 3)$, $(0, 5)$; the altitude is 2. Find the locus of the third vertex. (§ 12.) *Ans.* $x + 2y = 10 \pm 2\sqrt{5}$.

28. The base of a triangle joins $(2, 5)$, $(-1, 3)$; the altitude is 1. Find the locus of the third vertex. (§ 12.) *Ans.* $2x - 3y + 11 = \pm\sqrt{13}$.

29. The hypotenuse of a right triangle joins the points $(5, -2)$, $(3, 6)$. Find the locus of the third vertex in various ways.

30. The hypotenuse of a right triangle joins the points $(2c, 0)$, $(0, 2c)$. Find the locus of the third vertex in various ways.

31. Two circles of radius 4 are drawn with centers at (2, 1), (−1, −3). Find the equations of their common tangents. (See Exs. 25–26.)

Ans. $4x - 3y = 25$; $4x - 3y = -15$.

32. Two circles of radius 5 are drawn with centers at (6, 0), (0, −8). Find the equations of their common tangents.

Ans. $4x - 3y = 24 \pm 25$; $3x + 4y + 7 = 0$.

33. A line segment of length $2a$ moves with its ends in the coordinate axes. Find the locus of its midpoint. *Ans.* $x^2 + y^2 = a^2$.

Exs. 34–41 refer to Fig. 47 which immediately precedes this set of exercises.

34. A distance $OR = MA$ is laid off on OP. Find the locus of R.

35. A distance $OS = OM - MA$ is laid off on OP. Find the locus of S.

36. The line OP is produced to a point R such that $PR = 2a$. Find the locus of R.

37. The line OP is produced to R such that $PR = AP$. Find the locus of R.

38. A point R is marked on OP such that $OR = OP - OM$. Find the locus of R.

39. A distance $AT = AP$ is laid off on the vertical line L; a distance $OR = OT$ is then laid off on OP produced. Find the locus of R.

40. A point R is marked on OP such that $OR = MP$. Find the locus of R.

41. Let OP produced intersect L at a point T: find the locus of the midpoint of PT. *Ans.* $r = a(\sec\theta + \cos\theta)$.

42. A tangent drawn to a circle of radius c with center at O intersects the axes at points A and B. Use polar coordinates to find the locus of the midpoint of AB. *Ans.* $r \sin 2\theta = c$.

43. The center of a circle of fixed radius a moves along Ox; tangents to the circle are drawn through O. Find the locus of the points of tangency.

Ans. $r = a \cot\theta$.

CHAPTER 5 *The Straight Line*

32. Line parallel to a coordinate axis. If a straight line is parallel to the y-axis, its equation is

$$x = k,$$

where k is the directed distance of the line from the axis; and conversely. For all points on that line, and no other points, have the abscissa k.

Similarly, a line parallel to and at a directed distance k from the x-axis has the equation

$$y = k.$$

Figure 48 shows the lines $x = 2$, $y = -3$.

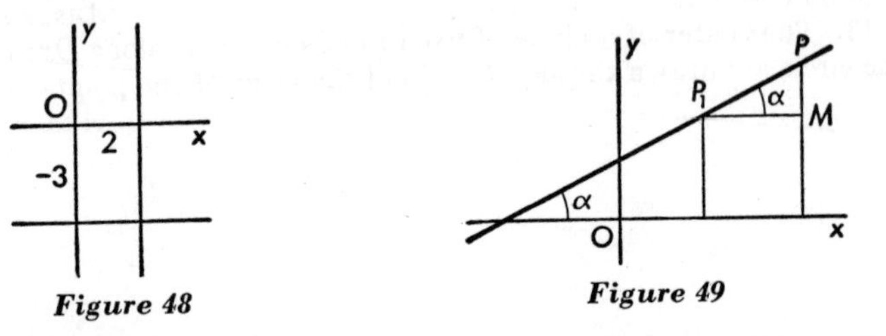

Figure 48

Figure 49

33. Point-slope form. Let us find the equation of a straight line passing through a given point $P_1:(x_1, y_1)$ in a given direction — i.e., having a given slope m.

Assuming a point $P:(x, y)$ in a general position on the line, we note that

$$\tan \alpha = \frac{MP}{P_1M} = m, \qquad \text{or} \qquad \frac{y - y_1}{x - x_1} = m.$$

Hence: *The equation of the straight line of slope m through the point* (x_1, y_1) *is*

$$y - y_1 = m(x - x_1). \tag{1}$$

This is the *point-slope form* of the equation of the straight line. It is one of several so-called "standard forms" that will be developed in this chapter.

The point-slope form fails in case the line is parallel to the y-axis, since for such a line $m = \tan 90°$, which is nonexistent. But by § 32, the equation of a line through (x_1, y_1) parallel to the y-axis is simply

$$x = x_1.$$

Example: Find the equation of the line through $(3, -6)$ perpendicular to the line joining $(4, 1)$ and $(2, 5)$.

By § 9, the slope of the line joining $(4, 1)$ and $(2, 5)$ is -2, whence the slope of the required line is $\frac{1}{2}$, and by (1) its equation is

$$y + 6 = \tfrac{1}{2}(x - 3),$$

or

$$x - 2y = 15.$$

When the slope and one point of a line are given, the line can be drawn, if desired, *without writing the equation.* For example, to draw the line of slope $\frac{3}{2}$ through the point $(4, 2)$: starting at $(4, 2)$, we measure off 2 units to the right and then 3 upward (or 4 to the right and 6 upward, or 2 to the left and 3 downward, etc.); through the point thus reached and $(4, 2)$ we draw the line. (Fig. 50.)

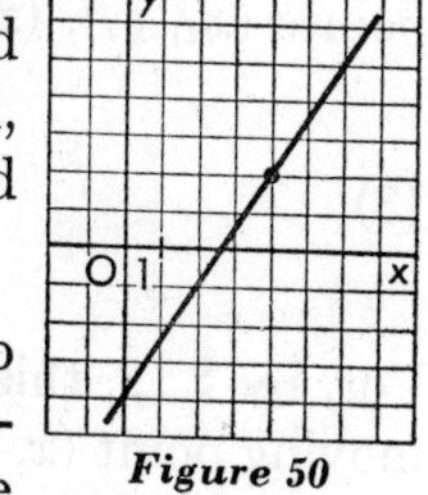

Figure 50

34. Line through two points. If two points have the same abscissa, the line joining them is parallel to the y-axis. Thus the equation of the line through the points (k, y_1) and (k, y_2) is

$$x = k.$$

If two points have different abscissas, the slope of the line joining them can be written down at once by § 9. Then the equation of the line joining the two points follows by means of the point-slope form obtained in the previous section. Thus, the equation of the line through the points (x_1, y_1) and (x_2, y_2), for which $x_2 \neq x_1$, is found by computing

$$m = \frac{y_2 - y_1}{x_2 - x_1} \tag{1}$$

and then using the point-slope form

$$y - y_1 = m(x - x_1). \tag{2}$$

Example: Find the equation of the line through the points $(3, -5)$, $(-6, -2)$.

In this case

$$m = \frac{-2 + 5}{-6 - 3} = -\frac{1}{3},$$

whence the equation of the line is

$$y + 5 = -\tfrac{1}{3}(x - 3),$$

or

$$x + 3y + 12 = 0.$$

Check: $3 - 15 + 12 = 0$, $-6 - 6 + 12 = 0$.

Alternate method: The equation of the line through any two points (x_1, y_1), (x_2, y_2) may be written in the form

$$\begin{vmatrix} x & y & 1 \\ x_1 & y_1 & 1 \\ x_2 & y_2 & 1 \end{vmatrix} = 0. \tag{3}$$

For, by § 12, this equation merely expresses the fact that the moving point (x, y) forms with the fixed points (x_1, y_1), (x_2, y_2) a triangle of area zero. See also Ex. 29 below.

It is to be noted that formula (3) holds in all cases — even when the line is parallel to the y-axis.

Exercises

In Exs. 1–6, draw the line.

1. $y = 0.$

2. $x = 0.$

3. $x = 5.$

4. $2y = 3.$

5. $4y + 1 = 0.$

6. $3x - 4 = 0.$

In Exs. 7–10, draw the line; then write its equation.

7. Of slope $\frac{1}{4}$ through (3, 1).

8. Of slope $\frac{3}{2}$ through (−2, 4).

9. Of slope $-\frac{7}{3}$ through (0, −5).

10. Of slope $-\frac{1}{2}$ through (−4, 0).

In Exs. 11–19, write the equation of the line.

11. Through (4, 1) and (*a*) parallel, (*b*) perpendicular to the line through (7, 3), (5, −1). *Ans.* (*a*) $2x - y = 7$, (*b*) $x + 2y = 6$.

12. Through (2, −3) and (*a*) parallel, (*b*) perpendicular to the line through (0, −4), (3, −2). *Ans.* (*a*) $2x - 3y = 13$, (*b*) $3x + 2y = 0$.

13. Through (4, 0) and (*a*) parallel, (*b*) perpendicular to the line through (−1, −5), (2, −4).

14. Through (1, −5) and (*a*) parallel, (*b*) perpendicular to the line through (−3, −2), (−5, 3).

15. Through (6, 1), (4, 4).

16. Through (−4, −1), (5, −2).

17. Through (−2, 3), (3, −2).

18. Through (3, 5), (3, −7).

19. Through (−1, 5) and (*a*) parallel, (*b*) perpendicular to the line through (1, 3), (1, −4).

20. Show that *three points* (x_1, y_1), (x_2, y_2), (x_3, y_3) *lie in a straight line if and only if*

$$\begin{vmatrix} x_1 & y_1 & 1 \\ x_2 & y_2 & 1 \\ x_3 & y_3 & 1 \end{vmatrix} = 0.$$

21. In the triangle with vertices (2, 0), (3, 2), (4, −3), find the equations of the altitudes, and their point of intersection. *Ans.* $(-\frac{6}{7}, -\frac{4}{7})$.

22. In the triangle with vertices (0, 6), (−2, −2), (4, 2), find the equations of the altitudes, and their point of intersection. *Ans.* $(\frac{12}{5}, \frac{12}{5})$.

23. In the triangle of Ex. 21, find the equations of the medians, and their point of intersection. *Ans.* $(3, -\frac{1}{3})$.

24. In the triangle of Ex. 22, find the equations of the medians, and their point of intersection. *Ans.* $(\frac{2}{3}, 2)$.

25. In the triangle of Ex. 21, find the equations of the perpendicular bisectors of the sides, and their point of intersection. *Ans.* $(\frac{69}{14}, -\frac{3}{14})$.

26. In the triangle of Ex. 22, find the equations of the perpendicular bisectors of the sides, and their point of intersection. *Ans.* $(-\frac{1}{5}, \frac{9}{5})$.

27. Show that the three points found in Exs. 21, 23, 25 lie in a straight line.

28. Show that the three points found in Exs. 22, 24, 26 lie in a straight line.

29. Without expanding the determinant, verify formula (3), § 34, by showing (*a*) that the equation is of first degree; (*b*) that it is satisfied by the coordinates of the given points.

35. Slope-intercept form. Given a line of slope m whose y-intercept is b, let us assume a point $P:(x, y)$ on the line. Then

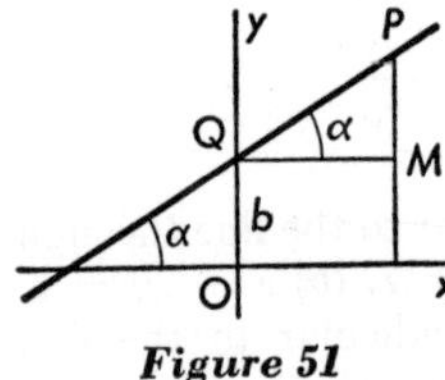

Figure 51

$$m = \frac{MP}{QM} = \frac{y - b}{x},$$

whence: *The equation of the straight line with slope m and y-intercept b is*

$$y = mx + b. \tag{1}$$

This is the *slope-intercept* form. It is evidently a special case of the point-slope form.

Note that (1) contains two constants, which checks (§ 25) with the fact that a straight line is determined by two points.

36. General equation of first degree. If an equation is of the first degree in x and y, it can contain at most a term in x, a term in y, and a constant term: i.e., it can be written in the form

$$Ax + By + C = 0. \tag{1}$$

If $B = 0$, the equation evidently represents a line parallel to the y-axis. If $B \neq 0$, the equation can be solved for y: i.e., it is reducible to the form

$$y = mx + b. \tag{2}$$

Now for any value of m and any value of b there exists a line with slope m and y-intercept b; thus it follows from § 35 that equation (2) represents a line. Hence we have proved that:

The locus of every equation of first degree is a straight line.

Next we show that:

Every straight line may be represented by an equation of the first degree.

For, if the line is parallel to the y-axis, its equation is

$$x = k,$$

which is of the first degree. If it intersects the y-axis, it must have a slope and a y-intercept (either or both may of course be zero), and by § 35 its equation may be written in the form

$$y = mx + b,$$

which is also of the first degree.

It might seem at first thought that equation (1) contains three constants. But if we divide through by any one of them, say A, there remain only two constants, $\frac{B}{A}$ and $\frac{C}{A}$. Thus the essential constants are not the coefficients, but the ratios of any two of them to the third. For example, the equations

$$y = 2x - 3, \qquad 4x - 2y = 6, \qquad 6y - 12x + 18 = 0$$

all represent the same line.

37. Reduction to the slope-intercept form. The foregoing theory establishes the

RULE: *To reduce the equation of any line (not parallel to the y-axis) to the slope-intercept form, solve the equation for y. When this has been done, the coefficient of x is the slope and the constant term is the y-intercept.*

Example (*a*): To reduce the equation

$$3x + 4y - 6 = 0$$

to the slope-intercept form, write

$$4y = -3x + 6,$$

$$(1) \qquad y = -\tfrac{3}{4}x + \tfrac{3}{2},$$

whence the slope is $-\frac{3}{4}$ and the y-intercept $\frac{3}{2}$.

Example (*b*): Find the angle from the line

$$(2) \qquad 5x - 2y - 8 = 0$$

to the line

$$(3) \qquad 2x + 3y - 6 = 0.$$

Using the formula of § 11, we can find the desired angle as soon as we know the slopes of the lines. Not being interested in the y-intercepts, we need to apply the above rule only in part. In (2), mentally transposing the term in x to the right member and dividing by the coefficient of y, we find

$$m_1 = \tfrac{5}{2}.$$

Similarly, the slope of the line (3) is

$$m_2 = -\tfrac{2}{3}.$$

Substitute in formula (3), § 11:

$$\tan \varphi = \frac{-\frac{2}{3} - \frac{5}{2}}{1 + \frac{5}{2}(-\frac{2}{3})} = \frac{19}{4}.$$

The student should prove the following theorem with the aid of the slope-intercept form.

THEOREM: *The lines*

$$A_1x + B_1y + C_1 = 0,$$
$$A_2x + B_2y + C_2 = 0,$$

are parallel if and only if

$$\begin{vmatrix} A_1 & B_1 \\ A_2 & B_2 \end{vmatrix} = 0.$$

38. Parallel and perpendicular lines. By reduction to the slope-intercept form, it is easily seen that the lines

$$Ax + By + C = 0,$$
$$Ax + By + K = 0$$

are parallel, while (Theorem, § 10) the lines

$$Ax + By + C = 0,$$
$$Bx - Ay + K = 0$$

are perpendicular. Hence, if a line is to be parallel to a given line, the coefficients of x and y in the required equation may be taken *the same as those in the given equation;* if a line is to be perpendicular to a given line, the coefficients of x and y in the required equation may be found by *interchanging the coefficients of x and y and changing the sign of one of them.* In each case, of course, the constant term must be determined by an additional condition.

Example: Write the equation of a line through the point $(3, -1)$ perpendicular to the line $3x + 2y = 6$.

The left member of the required equation will be $2x - 3y$; if the new equation is to be satisfied by the coordinates $(3, -1)$, the right member must be what the left member becomes when we put 3 for x and (-1) for y. Hence the required equation is

$$2x - 3y = 9.$$

39. Three concurrent lines. Consider three lines

$$A_1x + B_1y + C_1 = 0,\quad A_2x + B_2y + C_2 = 0,\quad A_3x + B_3y + C_3 = 0, \tag{1}$$

where it is assumed that no two of the lines are parallel. We wish to determine under what condition the three lines meet in a point; that is, are concurrent.

Since no two of the lines (1) are parallel, the second order determinants

$$\begin{vmatrix} A_1 & B_1 \\ A_2 & B_2 \end{vmatrix}, \qquad \begin{vmatrix} A_1 & B_1 \\ A_3 & B_3 \end{vmatrix}, \qquad \begin{vmatrix} A_2 & B_2 \\ A_3 & B_3 \end{vmatrix}$$

are each different from zero (Theorem, § 37). If we look upon equations (1) as a system of three linear equations in the two unknowns x and y, a theorem of elementary algebra states that the system (1) will have a solution if and only if the determinant of the coefficients vanishes:

$$\begin{vmatrix} A_1 & B_1 & C_1 \\ A_2 & B_2 & C_2 \\ A_3 & B_3 & C_3 \end{vmatrix} = 0. \tag{2}$$

THEOREM: *If no two of the three lines (1) are parallel, then the three lines are concurrent if and only if equation (2) is satisfied.*

40. Concurrence of the medians. We can now prove that the medians of any triangle are concurrent, without resorting, as was done in Ex. 12, p. 23, to previous knowledge that the medians intersect in a trisection point.

Take any triangle ABC, place the x-axis along the side AB and use the midpoint of AB as origin. Using $2a$ to denote the length AB, the coordinates of the vertices A, B, C may be written $(-a, 0)$, $(a, 0)$, (b, c) as shown in Fig. 52. The coordinates of the midpoints D and E are determined at once and then the equations of the three

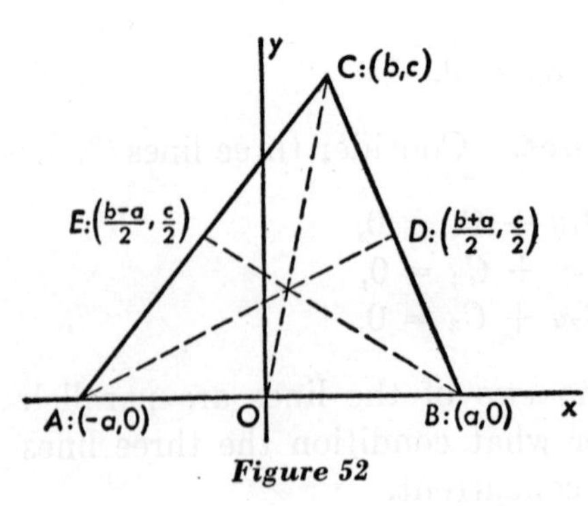

Figure 52

medians can be found by methods of § 34. These equations are

$$\begin{aligned} AD&: cx - (b + 3a)y + ac = 0, \\ BE&: cx - (b - 3a)y - ac = 0, \\ OC&: cx - \quad by \quad = 0. \end{aligned}$$

Form the determinant

$$\Delta = \begin{vmatrix} c & -(b + 3a) & ac \\ c & -(b - 3a) & -ac \\ c & -b & 0 \end{vmatrix}.$$

It is very easy to show that $\Delta = 0$, which by § 39 proves that the medians are concurrent.

Exercises

In Exs. 1–6, reduce the equation to the slope-intercept form.

1. $2x - 7y + 5 = 0$.
2. $x + 3y - 4 = 0$.
3. $3x + 6y - 2 = 0$.
4. $4x - 2y + 3 = 0$.
5. $x + y = 7$.
6. $5x + 3y = 0$.

In Exs. 7–12, find the angle from the first line to the second.

7. $2x + 3y = 0$, $7x - y = 4$. *Ans.* $\tan \varphi = -\frac{23}{11}$.
8. $x + 4y = 5$, $3x - 4y = 8$. *Ans.* $\tan \varphi = \frac{16}{13}$.
9. $x - 4y = 7$, $3x + 5y = 4$.
10. $3x + 4y = 1$, $7x + y = -5$.
11. $2x + 3 = 0$, $x - 3y = 7$.
12. $3x = 4$, $3x + 2y = -8$.

In Exs. 13–20, write, at sight, the equation of the line through the given point (*a*) parallel, (*b*) perpendicular to the given line. (§ 38.)

13. Point (3, 5), line $4x - y = 5$.
Ans. (*a*) $4x - y = 7$; (*b*) $x + 4y = 23$.
14. Point (2, 4), line $2x + 3y = 6$.
Ans. (*a*) $2x + 3y = 16$; (*b*) $3x - 2y = -2$.
15. Point (1, 2), line $2x + y = 7$.
16. Point (−2, 4), line $3x + 5y = 8$.
17. Point (−1, −4), line $4x - 2y = 3$.
18. Point (0, −2), line $x + 3y = 7$.
19. Point (0, −1), line $2y + 3 = 0$.
20. Point (2, 4), line $4x + 3 = 0$.
21. Show that the lines $3x - y = 7$, $2x + 6y = 5$, $9x = 3y - 4$, and $x + 3y = 8$ form a rectangle.

22. Show that the lines $x = 2y + 4$, $3x + 2y = 5$, $3x - 6y = 7$, and $3x + 2y = -1$ form a parallelogram.

23. Find the locus of the centers of circles touching the line $3x + 7y = 4$ at $(-1, 1)$.

24. Find the locus of the centers of circles touching the line $5x + y = 6$ at $(2, -4)$.

25. Show that a circle can be drawn which touches the lines $x - y + 1 = 0$ and $7x + y = 13$ at $(4, 5)$ and $(2, -1)$ respectively.

26. Can a circle be drawn touching the lines $x + y = 6$ and $7x + 20y = 73$ at $(7, -1)$ and $(-1, 4)$ respectively?

27. A point P moves so that $\overline{AP}^2 + \overline{BP}^2 = 2\overline{CP}^2$, where A, B, C are fixed points. Determine the locus of P. Choose the axes and notation so that the fixed points A, B, C become respectively $(-a, 0)$, $(a, 0)$, (b, c). Investigate specially the cases $b = 0$, $c = 0$, $b = c = 0$, $b^2 + c^2 = a^2$.

28. Given a quadrilateral $ABCD$, a point P moves so that

$$\overline{AP}^2 + \overline{CP}^2 = \overline{BP}^2 + \overline{DP}^2.$$

Show that the locus of P is in general a straight line. Investigate special cases: (*a*) $ABCD$ a parallelogram; (*b*) $ABCD$ a rectangle.

29. Prove that the perpendicular bisectors of the sides of a triangle are concurrent. Choose axes as in Fig. 52.

30. Prove that the altitudes of a triangle are concurrent. Choose axes as in Fig. 52.

31. In Ex. 29, find the point of intersection of the perpendicular bisectors. *Ans.* $\left(0, \dfrac{b^2 + c^2 - a^2}{2c}\right)$.

32. In Ex. 30, find the point of intersection of the altitudes. *Ans.* $\left(b, \dfrac{a^2 - b^2}{c}\right)$.

33. In § 40, find the point of intersection of the medians. *Ans.* $(\frac{1}{3}b, \frac{1}{3}c)$.

34. Prove that, in any triangle, the point of intersection of the perpendicular bisectors of the sides, the point of intersection of the altitudes, and the point of intersection of the medians lie in a straight line. Use the results of Exs. 31–33 above and Ex. 20, p. 55. Try to obtain the result without expanding the determinant.

35. Show that the determinant in (2) § 39 is always zero if the three lines are parallel, and never zero if two are parallel but not coincident, with the third intersecting them.

36. Show, preferably without expanding, that the determinant Δ, § 40, is zero.

37. A point P moves so that $\overline{AP}^2 - \overline{BP}^2 = c^2$, where A, B are fixed points. Let the distance between A and B be called $2a$ and choose the axes so that A, B become $(-a, 0)$, $(a, 0)$. Find the locus of P. Investigate also the special cases $c = 0$, $c = 2a$.

41. Intercept form. Let the intercepts of a straight line on the axes be $OA = a$ and $OB = b$. Choosing a point $P:(x, y)$ in a general position on the line, we see that the triangles MAP and OAB are similar. Whence

$$\frac{MP}{OB} = \frac{MA}{OA},$$

or

$$\frac{y}{b} = \frac{a - x}{a}.$$

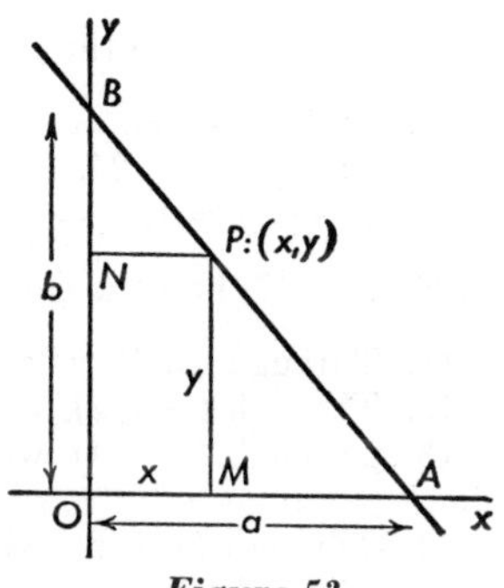

Figure 53

With a little rearrangement we find:

The equation of the straight line with x-intercept a and y-intercept b is

$$(1) \qquad \boldsymbol{\frac{x}{a} + \frac{y}{b} = 1.}$$

This equation, called the *intercept form*, fails in case the line passes through the origin or is parallel to either axis.

As an instance of the kind of problem where the intercept form is indicated, consider an

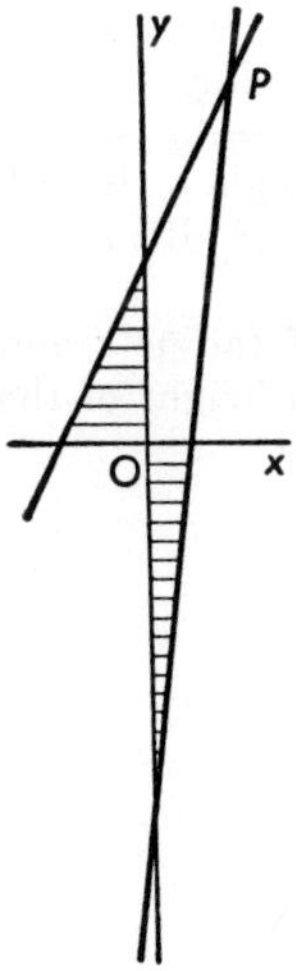

Example: A line forms with the axes a triangle of area 1 and passes through the point $P:(1, 4)$. Find the dimensions of the triangle.

Use of the intercept form is suggested by the fact that we can then express the first condition at once. It is apparent that the intercepts of the line must be of opposite sign: therefore

$$(2) \qquad -\tfrac{1}{2}ab = 1.$$

Substituting the coordinates (1, 4) in (1), we get

$$(3) \qquad \frac{1}{a} + \frac{4}{b} = 1.$$

Figure 54

Equations (2) and (3) give

$$a = -1,\ b = 2, \qquad \text{or} \qquad a = \tfrac{1}{2},\ b = -4.$$

Exercises

In Exs. 1–4, draw the line with the given numbers as x- and y-intercepts, respectively. Write the equation of the line.

1. $2, -3$.
2. $-5, 4$.
3. $\frac{1}{3}, \frac{2}{5}$.
4. $-\frac{2}{3}, -\frac{1}{4}$.

In Exs. 5–10, reduce the equation to the intercept form. Draw the line.

5. $3x + 5y = 30$.
6. $5x - 2y = -10$.
7. $3x - 4y = -24$.
8. $x + 2y = 9$.
9. $2x + 3y = 1$.
10. $7x - y = -1$.

In Exs. 11–17, find the equation of the line. Check the answer.

11. Through (3, 4), with equal intercepts.
12. Through (−5, 2), with equal intercepts.
13. Through (−7, 4), with intercepts numerically equal but of opposite sign.
14. Through (−8, 6), with x-intercept twice the y-intercept.
15. Through $(\frac{12}{5}, 1)$, forming with the axes a triangle of area 5.
Ans. $5x - 2y = 10$; $5x + 8y = 20$; $5x + 18y = 30$; $5x - 72y = -60$.
16. Through (−2, 4), forming with the axes a triangle of area 9. Why are there only two solutions?
17. Through (1, 3), forming with the axes a triangle of area 6. (Three solutions.)
18. A line passes through (2, 2), and the segment of the line intercepted between the axes is of length $\sqrt{5}$. Find the equation of the line.
Ans. $2x - y = 2$; $x - 2y = -2$.
19. A rectangle is inscribed, base to base, in a right triangle of base b and height h. What relation must hold between the base and altitude of the rectangle?
20. A circular cylinder is inscribed in a circular cone of radius r and height h. What relation must hold between the radius and height of the cylinder?

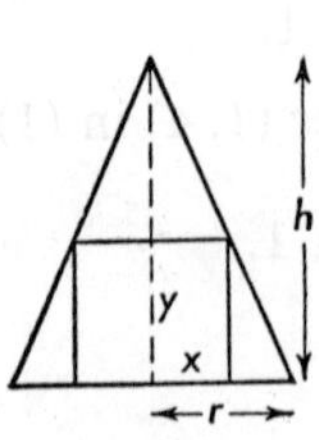

Figure 55

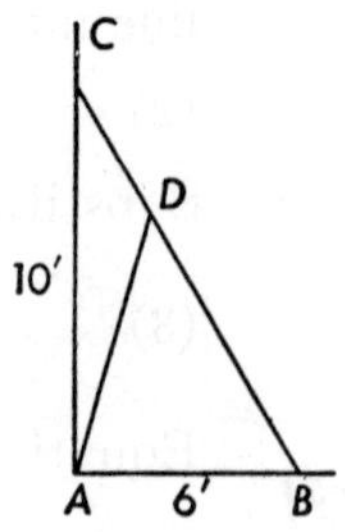

Figure 56

21. A beam BC leans against a wall AC and is stayed by a strut AD. If D is 2 ft. out from the wall, find the length of the strut. (Fig. 56.)

Ans. 6 ft. 11.5 in.

22. In Ex. 21, if a strut 6 ft. long is used, at what height above the ground will it reach the beam? *Ans.* 5 ft. 3.5 in.

23. Derive the intercept form from the fact that, in Fig. 53, the area of the triangle APB is zero.

24. Derive the intercept form from the fact that in Fig. 53, the area of the rectangle plus the area of the small triangles equals the area of the large triangle.

25. Derive the intercept form from the fact that, in Fig. 53, $AP + PB = AB$.

26. Derive the intercept form with P lying in the segment AB produced in either direction; also when a or b, or both, are negative.

42. Polar equation of the straight line. First, let L be any line not through the origin (pole). Let the perpendicular to L through the origin intersect L at a point N, whose polar coordinates are (p, β) with $0 \leqq \beta < 180°$. The number

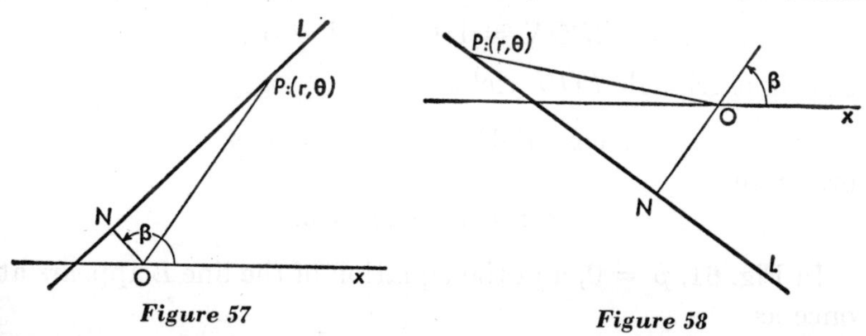

Figure 57 *Figure 58*

p may be positive or negative. There are then four representative positions for the line L as shown in Figs. 57–60. If $P:(r, \theta)$ is a variable point on L, it will be shown that, for all positions of L, the coordinates (r, θ) must satisfy

$$r \cos (\beta - \theta) = p. \qquad (1)$$

Equation (1) is called the *polar equation of the straight line.*

Next let L be a line through the origin. With $p = 0$ and β defined as the angle from the initial line Ox to the perpendicular to L through O, as in Fig. 61, equation (1) still holds for any point (r, θ) on L.

We shall obtain equation (1) for L in the positions indicated in Figs. 57, 58, 61, and leave the balance of the proof as an exercise.

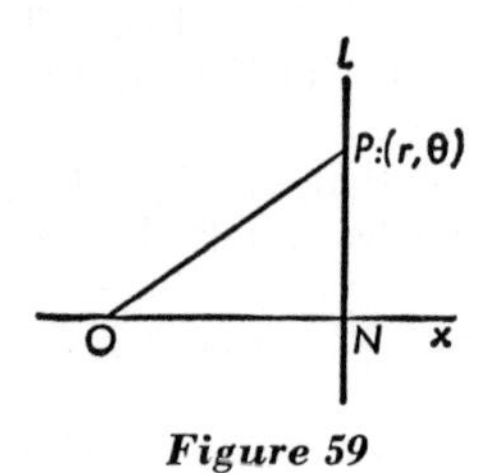

Figure 59

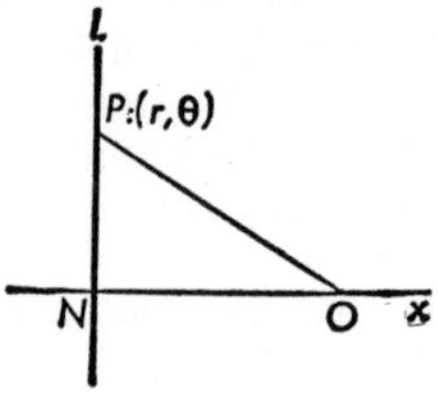

Figure 60

In Fig. 57, $\angle PON = \beta - \theta$, so that from the right triangle PON, we obtain

$$r \cos (\beta - \theta) = p.$$

In Fig. 58, p is negative, so $ON = -p$. Since

$$\angle PON = 180° - (\theta - \beta),$$

the right triangle PON yields

$$r \cos (180° + \beta - \theta) = -p,$$

or, again,

$$r \cos (\beta - \theta) = p.$$

In Fig. 61, $p = 0$, and the equation of the line L appears at once as

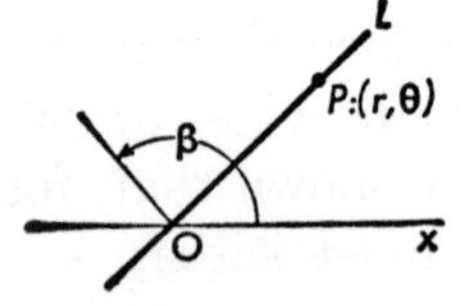

Figure 61

$$\theta = \beta - 90°. \tag{2}$$

From (2) we obtain $\beta - \theta = 90°$, so that

$$r \cos (\beta - \theta) = 0,$$

which is equation (1) for $p = 0$.

Thus, together with Ex. 33 below, we have obtained the polar equation (1) for any line.

43. Change of coordinate system. It is convenient in many types of problems to be able to change from rectangular to polar coordinates, or vice versa.

With the axes placed as in Fig. 62, let a point P have the rectangular coordinates x, y and the polar coordinates r, θ. Then, from the figure,

$$(1)\qquad \begin{cases} x = r\cos\theta, \\ y = r\sin\theta, \\ x^2 + y^2 = r^2, \end{cases}$$

$$(2)\qquad \begin{cases} r = \sqrt{x^2+y^2}, \\ \cos\theta = \dfrac{x}{\sqrt{x^2+y^2}}, \\ \sin\theta = \dfrac{y}{\sqrt{x^2+y^2}}. \end{cases}$$

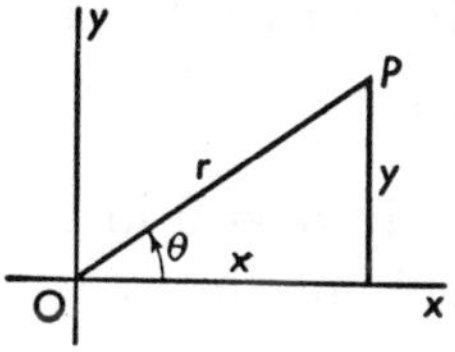

Figure 62

44. Normal form. The formulas of § 43 prove particularly useful in the study of algebraic curves in Chapter 11. At present we employ those formulas to transform to rectangular coordinates the polar equation of the straight line,

$$(1)\qquad r\cos(\beta - \theta) = p.$$

From trigonometry we borrow the formula

$$(2)\qquad \cos(A - B) = \cos A\cos B + \sin A\sin B.$$

With the aid of (2), equation (1) may be written

$$r\cos\beta\cos\theta + r\sin\beta\sin\theta = p,$$

or, in view of the formulas of § 43,

$$(3)\qquad x\cos\beta + y\sin\beta = p.$$

Equation (3) is called the *normal* form* of the equation of a straight line. Because of the restriction $0 \leqq \beta < 180°$ imposed in § 42, we know that either $\sin\beta$ is positive, or $\sin\beta = 0$ and $\cos\beta = 1$. Hence, in the normal form (3), either the coefficient of y is positive or the equation is $x = p$.

* The word "normal" is used in more advanced mathematics as meaning "perpendicular." It is in this sense rather than in the sense of "natural" or "usual" that the word is used here.

45. Reduction to the normal form. Given the equation of any straight line

(1) $$Ax + By + C = 0,$$

we wish to develop a method for putting that equation into its normal form

(2) $$x \cos \beta + y \sin \beta = p.$$

If $B = 0$ in equation (1), the normal form is obtained by solving for x:

$$x = -\frac{C}{A}.$$

If $B \neq 0$ in equation (1), then since (1) and (2) are to represent the same line, the coefficients in one equation must be proportional to those in the other:

$$\cos \beta = kA, \qquad \sin \beta = kB, \qquad p = -kC.$$

From the first two of these equations we find

$$\cos^2 \beta + \sin^2 \beta = k^2(A^2 + B^2),$$

whence

$$k^2(A^2 + B^2) = 1, \qquad \text{and} \qquad k = \pm \frac{1}{\sqrt{A^2 + B^2}}.$$

Since kB must be positive, the sign of k must be chosen the same as the sign of B.

RULE: *To reduce to the normal form*

(*a*) *the equation*

$$Ax + By + C = 0 \qquad (B \neq 0),$$

divide through by $\pm \sqrt{A^2 + B^2}$, *using the sign of* B *in front of the square root, then transfer the constant term to the other side of the equation;*

(*b*) *the equation*

$$Ax + C = 0,$$

solve for x.

Example (*a*): Put the equation

$$(3) \qquad 3x - 4y + 30 = 0$$

into its normal form.

Here the coefficient of y is negative. Therefore we divide each member of (3) by

$$-\sqrt{3^2 + 4^2} = -5,$$

obtaining the desired normal form,

$$(4) \qquad -\tfrac{3}{5}x + \tfrac{4}{5}y = 6.$$

Example (*b*): Find the locus of points at the distance $\sqrt{5}$ from the line L

$$x - 2y - 2 = 0.$$

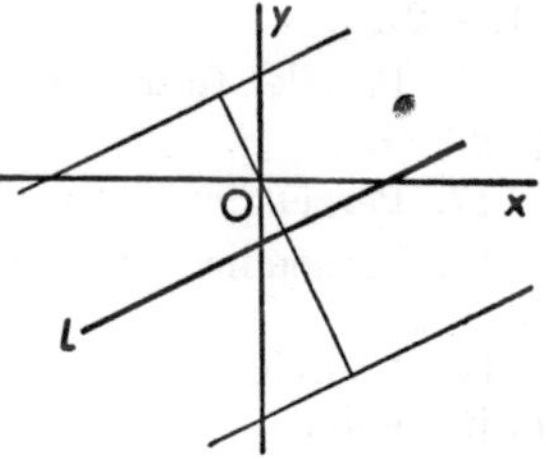

Figure 63

By the rule, the given equation is, in the normal form,

$$-\frac{x}{\sqrt{5}} + \frac{2y}{\sqrt{5}} = -\frac{2}{\sqrt{5}},$$

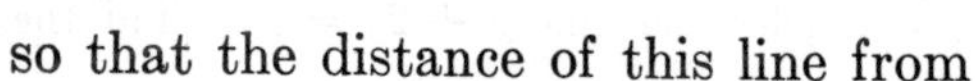
so that the distance of this line from the origin is $p = -\dfrac{2}{\sqrt{5}}$. The required locus consists of the two lines parallel to, and at a distance $\sqrt{5}$ from, the given line. Hence the required lines are

$$-\frac{x}{\sqrt{5}} + \frac{2y}{\sqrt{5}} = -\frac{2}{\sqrt{5}} \pm \sqrt{5}, \qquad \text{or} \qquad x - 2y = 2 \pm 5.$$

Exercises

In Exs. 1–10, reduce the equation to the normal form, and find the distance of the line from the origin.

1. $4x + 3y + 20 = 0.$ **2.** $x - 3y = 20.$

3. $5x - 12y = 26.$ **4.** $5x + 12y + 39 = 0.$

5. $x + 4y = 0.$ **6.** $2x = y.$

7. $5x + 4 = 0.$ **8.** $4y - 7 = 0.$

9. $y = mx + b.$ **10.** $y - y_1 = m(x - x_1).$

In Exs. 11–18, find the equation of the line.

11. Parallel to $3x + 4y = 20$ and passing (*a*) at a distance 3 from the origin; (*b*) 2 units farther from the origin; (*c*) at a distance 5 from the given line. *Ans.* (*b*) $3x + 4y = \pm 30$; (*c*) $3x + 4y = 20 \pm 25$.

12. Parallel to $x + y = 4$ and passing (*a*) at a distance 3 from the origin; (*b*) $\sqrt{2}$ units farther from the origin; (*c*) at a distance $3\sqrt{2}$ from the given line. *Ans.* (*b*) $x + y = \pm 6$; (*c*) $x + y = 4 \pm 6$.

13. Parallel to the line $x - 3y = 2$ and passing (*a*) twice as far from the origin; (*b*) at a distance 4 from the origin; (*c*) at a distance $2\sqrt{10}$ from the given line.

14. Parallel to the line $3x - 2y = 26$ and passing (*a*) half as far from the origin; (*b*) at a distance 3 from the origin; (*c*) at a distance $3\sqrt{13}$ from the given line.

15. Parallel to $4x - 3y = 7$ and passing at a distance 4 from the point $(1, -2)$. *Ans.* $4x - 3y = 10 \pm 20$.

16. Parallel to $x - y = 9$ and passing at a distance $5\sqrt{2}$ from the point $(4, 1)$. *Ans.* $x - y = 3 \pm 10$.

17. Perpendicular to $y = x$ and passing at a distance $3\sqrt{2}$ from $(4, 1)$.

18. Perpendicular to $y = 7x + 1$ and passing at a distance $\sqrt{2}$ from $(4, -2)$.

19. A circle of radius 4 touches the line $6x + 8y = 1$. Find the locus of its center.

20. A circle of radius $2\sqrt{5}$ touches the line $x + 2y = 3$. Find the locus of its center.

21. One side of a square is the line from $(1, 3)$ to $(4, -1)$. Find the other vertices. *Ans.* $(5, 6), (8, 2)$; $(-3, 0), (0, -4)$.

22. The base of a triangle is the line from $(4, 1)$ to $(3, -2)$; the area is 7. Find the locus of the third vertex, and check by § 12.
Ans. $3x - y = -3$; $3x - y = 25$.

23. The base of a triangle joins the points $(-2, 4), (-1, 3)$; the area is $\frac{9}{2}$. Find the locus of the third vertex, and check by § 12.

24. In Ex. 23, if the triangle is isosceles, find the third vertex.
Ans. $(-6, -1)$; $(3, 8)$.

In Exs. 25–28, find the distance between the given lines.

25. $x + 3y = 5$, $x + 3y = 25$. *Ans.* $2\sqrt{10}$.

26. $x + 3y = -5$, $x + 3y = 25$. *Ans.* $3\sqrt{10}$.

27. $4x - 3y = 12$, $4x - 3y = -8$.

28. $4x + 3y = 7$, $4x + 3y = 8$.

29. Two sides of a square lie along the lines $2y = 20 - 3x$, $3x + 2y = -6$. Find the area of the square. *Ans.* 52.

30. Two sides of a square lie along the lines $x = 4y - 11$, $x = 4y + 40$. Find the area of the square. *Ans.* 153.

31. Find the area of the rectangle bounded by the lines $3x - y - 1 = 0$, $x + 3y = 7$, $y = 3x - 3$, $x + 3y + 8 = 0$.

32. A circle touches the lines $y = 3x - 7$, $y = 3x + 13$. Find its area, and the locus of its center.

33. Obtain equation (1), § 42, for the line L in the positions indicated in Figs. 59, 60.

46. Directed distance from line to point. To find the directed distance from the line

$$Ax + By + C = 0 \tag{1}$$

(the line L in Fig. 64) to any point $P:(x_1, y_1)$ not on that line, let us write the equation of the line through P parallel to (1): by § 38, that equation is

$$Ax + By = Ax_1 + By_1. \tag{2}$$

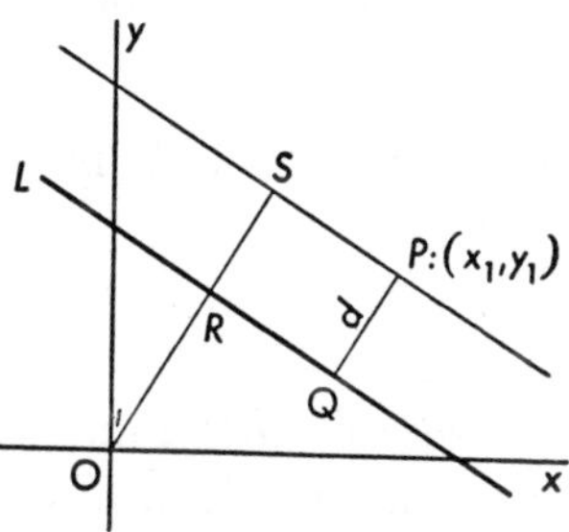

Figure 64

By reduction to the normal form, the distances of these lines from the origin are found to be

$$p_1 = \frac{-C}{\pm\sqrt{A^2 + B^2}}, \qquad p_2 = \frac{Ax_1 + By_1}{\pm\sqrt{A^2 + B^2}},$$

where, according to the rule of § 45, the sign must be taken the same as the sign of B (or, if $B = 0$, the same as the sign of A). But the distance from the line to the point is

$$d = QP = RS = OS - OR = p_2 - p_1.$$

The directed distance from the line

$$Ax + By + C = 0$$

to the point (x_1, y_1) *is*

$$\boldsymbol{d} = \frac{\boldsymbol{Ax_1 + By_1 + C}}{\pm\sqrt{\boldsymbol{A^2 + B^2}}}, \tag{3}$$

where the ambiguous sign is taken like the sign of B, or if B = 0, like the sign of A.

The distance as given by the formula is *positive* or *negative* according as P lies *above* or *below* the given line — i.e., according as the y-intercept of the line (2) is algebraically greater or less than that of the line (1). If the given line is parallel to the y-axis, the distance is positive or negative according as P is to right or left of the line.

Example (*a*): The distance from the line

$$3x - 2y = 5$$

to the point (3, 4) is

$$d = \frac{3 \cdot 3 - 2 \cdot 4 - 5}{-\sqrt{13}} = \frac{4}{\sqrt{13}}.$$

Example (*b*): Find the locus of points at the distance $\sqrt{5}$ from the line [Example (b), § 45]

$$x - 2y - 2 = 0. \tag{4}$$

Formula (3) gives the answer almost instantly. As usual, let $P:(x, y)$ be a point on the required locus, and set the distance from P to the line (4), as given by the formula, equal to the given distance $\sqrt{5}$, using both signs since P may be above or below the given line:

$$\frac{x - 2y - 2}{-\sqrt{5}} = \pm\sqrt{5}, \text{ etc.} \tag{5}$$

Problems such as (*b*) are sometimes confusing to the student, since we replace x and y by x and y, and therefore seem at first sight to be making no change at all. The change is in the *meaning* of the variables. In (4), x and y are the coordinates of a point on the line L, Fig. 63; in (5), x and y are the coordinates of a point on one or the other of the two parallels.

Example (*c*): Find the equations of the lines bisecting the angles between the lines L_1, L_2 in Fig. 65,

$$x + y = 2, \qquad x - 7y + 2 = 0.$$

The bisector of the angle between two lines is the locus of points *equidistant from the two lines*. Assume a point $P:(x, y)$ in one angle bisector, as shown in the figure. Then the distances M_1P and M_2P are numerically equal; they are also algebraically equal, since P is below both lines, or in the position P' it is above both lines. By (3),

$$M_1P = \frac{x + y - 2}{\sqrt{2}}, \qquad M_2P = \frac{-x + 7y - 2}{5\sqrt{2}},$$

so that the equation of the locus of P is

$$\frac{x + y - 2}{\sqrt{2}} = \frac{-x + 7y - 2}{5\sqrt{2}},$$

or

$$3x - y = 4.$$

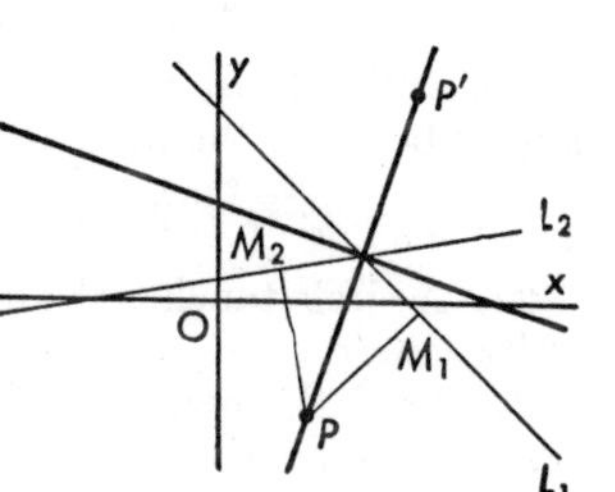

Figure 65

In a similar way the equation of the other angle bisector is found to be

$$\frac{x + y - 2}{\sqrt{2}} = -\left(\frac{-x + 7y - 2}{5\sqrt{2}}\right),$$

or

$$x + 3y = 3.$$

Exercises

In Exs. 1–8, find the distance from the line to the point. Check the sign by plotting.

1. $x = 3y - 12$, $(5, -1)$. *Ans.* $-2\sqrt{10}$.

2. $x + y = -7$, $(4, -1)$. *Ans.* $5\sqrt{2}$.

3. $3x + 4y = 11$, $(2, 5)$.

4. $4x - 3y = 11$, $(1, 1)$.

5. $3x - y = 25$, $(2, 1)$.

6. $2x + y = 16$, $(-3, 2)$.

7. $y = -4x$, $(8, 2)$.

8. $y = x$, $(3, 7)$.

9. Plot the points $(0, -10)$, $(8, 1)$, $(15, 10)$. How far is the second point from the line joining the other two?

10. Solve Ex. 9 for the points $(-10, -20)$, $(17, 0)$, $(30, 10)$.

11. A circle of radius 3 touches the line $12x - 5y = 7$. Find the locus of its center.

12. A circle of radius 4 touches the line $3x + 4y = 8$. Find the locus of its center.

13. A moving point remains equidistant from the point $(a, 0)$ and the line $y = x$. Find the equation of its locus.
Ans. $x^2 + 2xy + y^2 - 4ax + 2a^2 = 0$.

14. The distance of a point from the origin is twice its distance from the line $x - y = a$. Find the equation of its locus.
Ans. $x^2 - 4xy + y^2 - 4ax + 4ay + 2a^2 = 0$.

In Exs. 15–20, find the bisectors of the angles between the lines.

15. $x + 7y = 8$, $x - y = 10$. *Ans.* $2x - 6y = 21$, $3x + y = 29$.
16. $2x - y = 1$, $11x + 2y = 5$. *Ans.* $x + 7y = 0$, $21x - 3y = 10$.

17. $x = 2$, $y = 7$.
18. $4x = 3y$, $y = 1$.
19. $3x + 4y = 7$, $3x - 4y = 11$.
20. $3x + 4y = 7$, $4x - 3y = 11$.

In Exs. 21–22, use the method of Example (c) to find the bisectors of the angles between the lines, and interpret your results.

21. $y = 2x - 7$, $2x - y = 3$.
22. $y = 11 - 3x$, $3x + y = 7$.

23. A circle touches the lines $4x + y = 7$, $x + 4y = 13$. Find the locus of its center. *Ans.* $x + y = 4$, $x - y = -2$.

24. Find the equations of the lines through $(7, -4)$ passing at a distance 1 from the point $(2, 1)$. *Ans.* $3x + 4y = 5$; $4x + 3y = 16$.

25. Find the equations of the lines through $(3, 0)$ passing at a distance $\sqrt{5}$ from $(4, -3)$. *Ans.* $x + 2y = 3$; $2x - y = 6$.

26. A point moves so that the ratio of its distances from two intersecting lines is constant. Prove that the locus of the moving point is two straight lines.

27. A point moves so that the product of its distances from two parallel lines is constant. Prove that the locus consists of two lines parallel to the given lines.

28. Prove that the bisectors of the interior angles of a triangle are concurrent. Choose your notation so that the equations of the sides of the triangle appear as $y = 0$, $x \cos \beta_1 + y \sin \beta_1 = 0$, $x \cos \beta_2 + y \sin \beta_2 = p$.

29. In polar coordinates, show that the distance from the line $r \cos (\beta - \theta) = p$ to the point (r_1, θ_1) is given by the formula

$$D = r_1 \cos (\beta - \theta_1) - p.$$

47. Linear functions. A function (§ 22) of the first degree is called a *linear function:*

$$y = mx + b.$$

The graph is a straight line, or a segment of the line if the variable is restricted in range.

48. Rate of change of a linear function. If y is a function of x, then as x changes from any given value x_1 to a new value x_2, y will change from its original value y_1 to a new value y_2. As this happens, the point tracing the graph (Fig. 66 or Fig. 67) moves from $P_1:(x_1, y_1)$ to $P_2:(x_2, y_2)$. The net changes are

$$x_2 - x_1 = \Delta x, \qquad y_2 - y_1 = \Delta y.$$

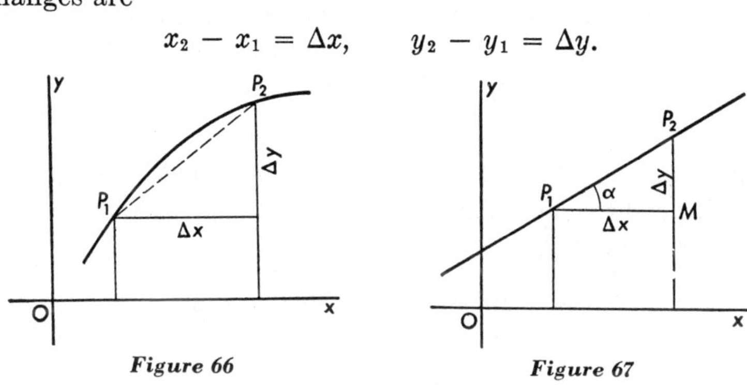

Figure 66 *Figure 67*

The above remarks apply to any function whatever (Fig. 66). But if y is a *linear function* (Fig. 67), it is apparent that when x changes by any amount, *the change in y is proportional to the change in x.* For, as we pass from any point P_1 to any other point P_2 on the graph, the change in x is P_1M, the change in y is MP_2, and

$$\frac{\Delta y}{\Delta x} = \frac{MP_2}{P_1M} = \tan \alpha = m,$$

which is *constant* — that is, will have the same value no matter what two points P_1, P_2 are chosen.

In Fig. 66, on the other hand, the value of $\dfrac{\Delta y}{\Delta x}$ (slope of the chord P_1P_2) is variable, depending on which points P_1, P_2 are chosen. Thus it must be firmly fixed in mind that the following definition of "rate of change," and the resulting theorem, apply to *linear functions only.*

The ratio of the change in y to the change in x is the *rate of change* of the linear function.

THEOREM: *The rate of change of a linear function is constant, and equal to the slope of its graph.*

Further, a positive rate (slope) means that the function is *increasing;* negative rate, *decreasing.* See Figs. 68–70.

When the rate of change of a linear function is given, together with the value of the function corresponding to any one value of the variable, the formula can be written down: for this is the same, analytically, as to give the slope and one point of a straight line.

Example (*a*): A car drives from A to B, 200 miles distant, at 40 mi. per hr. The distance from A in miles after time t in hours is (Fig. 68)

$$s = 40t, \qquad 0 \leqq t \leqq 5.$$

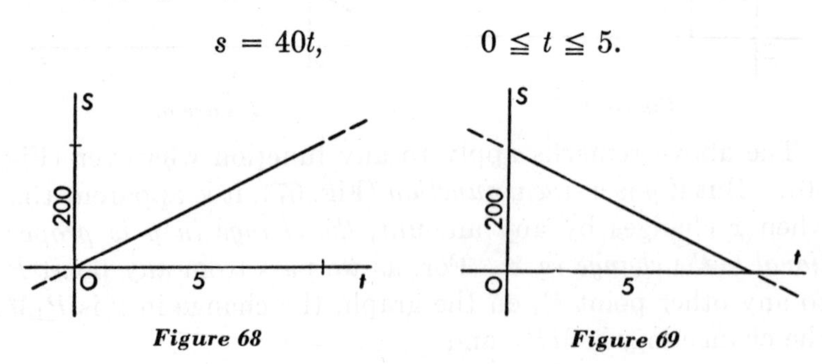

Figure 68 **Figure 69**

Example (*b*): A car drives from B to A, 200 miles, at 40 mi. per hr. The distance of the car from A, at time t, is (Fig. 69)

$$s = 200 - 40t, \qquad 0 \leqq t \leqq 5.$$

Example (*c*): A car drives from A to B, 200 miles, at 40 mi. per hr., stops 2 hours at B, then returns to A at 50 mi. per hr. The distance from A is (Fig. 70)

$$\begin{aligned} s &= 40t, & 0 \leqq t \leqq 5; \\ &= 200, & 5 \leqq t \leqq 7; \\ &= 200 - 50(t - 7) & \\ &= 550 - 50t, & 7 \leqq t \leqq 11. \end{aligned}$$

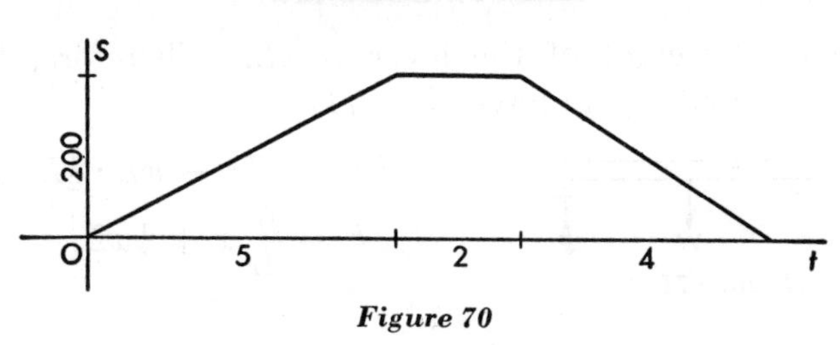

Figure 70

The reader could have solved these examples without ever having heard of analytic geometry; in fact, linear rate problems are solved every day, and very easily, by people with no mathematical training whatever. Of course the justification for our formal presentation is that, with it, we can go on to less simple problems.*

49. Parallel forces. As an application involving linear functions, let us consider a problem in mechanics.

The moment of a force with respect to a point is defined as the product of the force by its distance (i.e., the distance of its line of action) from that point. The moment is positive if the force tends to turn counterclockwise about the point.

Now, let a system of vertical forces (loads and supports) act on a horizontal beam. If the beam is to remain in equilibrium, the sum of the moments about any point in the line of the beam must equal zero — i.e., the counterclockwise and clockwise moments must balance.

In taking moments, a uniformly distributed load (for example, the weight of the beam itself) may be replaced by a concentrated load of the same magnitude, acting at the midpoint.

Example: A lever of length L weighing w lbs. per ft., with the fulcrum at one end, bears a load W at distance x from the fulcrum. Express the supporting force F as a function of x.

* Extension of the idea of rate to nonlinear functions accounts for a large proportion of the applications of calculus and more advanced mathematics.

The total weight of the lever is wL. Therefore, taking moments about A, we have (Fig. 71)

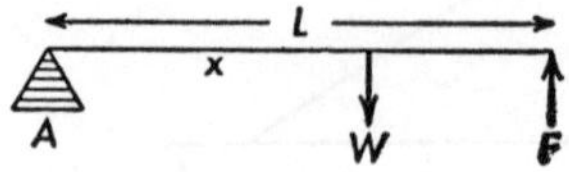

Figure 71

$$FL = Wx + wL \cdot \tfrac{1}{2}L,$$
$$F = \frac{W}{L}x + \tfrac{1}{2}wL.$$

Exercises

1. The radius of a circle increases 1 ft. How much does the circumference increase if the original radius is (*a*) 1 inch? (*b*) 1 mile?

2. Solve Ex. 1 with "circumference" replaced by "area."

3. The radius of a cylinder is 4 ft.; the altitude increases 3 ft. How much does the total surface area, including both bases, increase if the original altitude is (*a*) 6 ft.? (*b*) 8 ft.?

4. Solve Ex. 3 with the words "radius" and "altitude" interchanged.

5. How is the slope of a line OP affected if P moves (*a*) from (1, 1) to (1, 3)? (*b*) From (1, 5) to (1, 7)?

6. Solve Ex. 5 if P moves (*a*) from (1, 1) to (3, 1); (*b*) from (5, 1) to (7, 1).

7. A car, starting at noon, drives east at 30 mi. per hr. A second car, starting from the same point at 3:00 P.M., drives east at 50 mi. per hr. Find analytically and graphically when and where the cars will be together.

8. A car, starting at noon, drives east from A at 50 mi. per hr. A second car, starting at 3:00 P.M., drives east from B at 25 mi. per hr. If B is 200 mi. east of A, find analytically and graphically when and where they will be together.

9. A car, starting at noon, drives east from A at 50 mi. per hr. A second car, starting at 1:00, drives west from B at 40 mi. per hr. If B is 200 mi. east of A, find analytically and graphically when and where they will meet.

10. A car, starting at noon, drives east from A at 40 mi. per hr. A second car, starting at 1:00, drives west from B at 30 mi. per hr. A third car, starting at 2:00 drives west from B at 50 mi. per hr. If B is 200 mi. east of A, determine analytically and graphically the period during which the first car will be between the other two.

11. A manufacturer can turn out 100 units of his product for $450, 400 units for $600. Assuming that the cost C is a linear function of the number of units n, determine the function and draw the graph. What is the unit cost? What does the C-intercept mean? *Ans.* $C = \frac{1}{2}n + 400$.

12. A railroad can just break even on a certain run, carrying 50 passengers at $4.50 or 100 at $2.50. Assuming that the cost is a linear function of the number of passengers, determine the function and explain the meaning of the constants. *Ans.* $C = \frac{1}{2}n + 200$.

13. A ship strikes a rock at midnight, and water runs in at a uniform rate thereafter. At 4:00 A.M. a pump starts removing the water three-fourths as fast as it comes in. If, without pumping, the ship would have sunk at 10:00 A.M., determine analytically and graphically how long the pump can keep the ship afloat.

14. In Ex. 13, a second pump of the same capacity goes to work at noon. Determine analytically and graphically when the hold will be dry.

15. In the example of § 49, with $L = 6$ ft., suppose it is found that when $x = 2$, $F = 10$; when $x = 4$, $F = 18$. Determine W and w, and draw the graph. *Ans.* $F = 4x + 2,\ 0 \leqq x \leqq 6.$

16. For the lever of Fig. 72, show that

$$Fa = WL + \tfrac{1}{2}wL^2$$

where w is the weight of the lever per foot. Let $w = 2$, and suppose that when $W = 2$, $F = 33$; when $W = 4$, $F = 39$. Determine a and L.

Ans. $F = 3W + 27.$

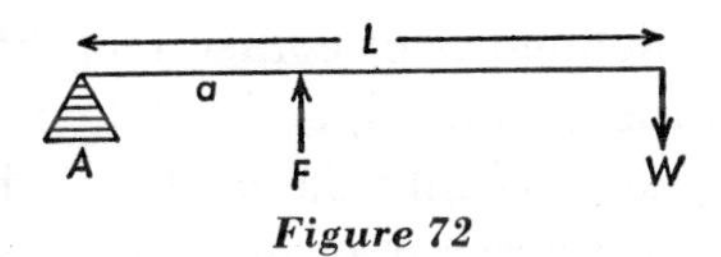

Figure 72

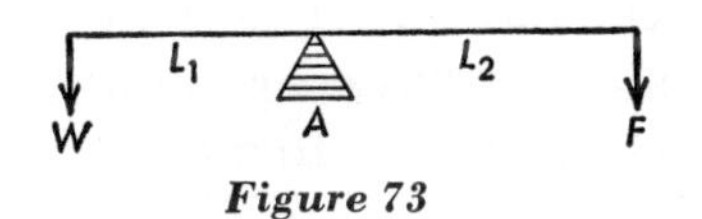

Figure 73

17. For the lever of Fig. 73, show that

$$FL_2 = WL_1 + \tfrac{1}{2}w(L_1^2 - L_2^2),$$

where w is the weight of the lever per foot.

18. A plank weighing 4 lbs. per ft., supported as in Fig. 74, bears a load W at the free end. A force F, just sufficient to hold the plank in place, is applied at A. Graph F as a function of W. *Ans.* $F = 0,\ W \leqq 80;$
$F = \tfrac{1}{3}(W - 80),\ W \geqq 80.$

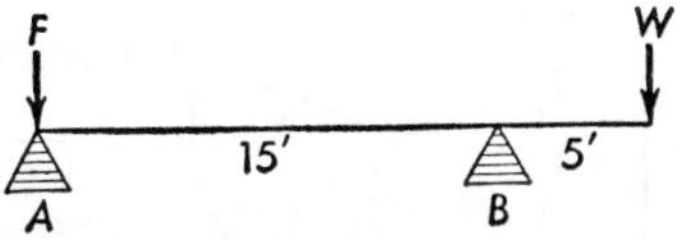

Figure 74

19. In Ex. 18, a child weighing 60 lbs. walks from A toward B. If x denotes the distance he can go before the plank tips, graph x as a function of W. *Ans.* $x = 20,\ 0 \leqq W \leqq 20;$
$x = \tfrac{1}{12}(260 - W),\ 20 \leqq W \leqq 260;\ x = 0,\ W \geqq 260.$

20. In Fig. 74, with $W = 60$, let w denote the least weight per foot of plank that will hold the plank in place. Graph w as a function of F.

Ans. $w = \tfrac{3}{20}(20 - F),\ 0 \leqq F \leqq 20;\ w = 0,\ F \geqq 20.$

CHAPTER 6 *The Circle*

50. Definitions; standard forms. A *circle* is the locus of a point that moves at a constant distance from a fixed point. The fixed point is the *center*, and the constant distance is the *radius*. The radius as thus defined is of course merely a number of linear units; the term is also used, as in elementary geometry, to mean a line-segment joining the center and a point of the curve. A *diameter* of a circle may mean either a straight line through the center, or the segment of such a line lying inside the curve.

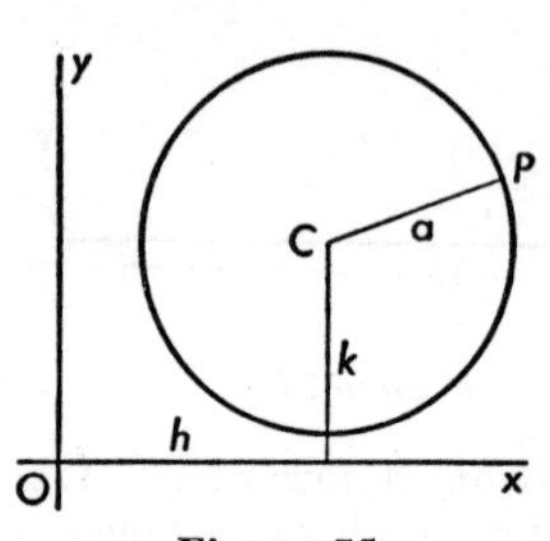

Figure 75

Given a circle of radius a with center at C: (h, k), assume a point P: (x, y) on the curve. Then

$$CP = \sqrt{(x-h)^2 + (y-k)^2} = a;$$

if the center is at the origin,

$$\sqrt{x^2 + y^2} = a.$$

Hence:

The equation of a circle of radius a is:

if the center is the origin,

$$x^2 + y^2 = a^2; \tag{1}$$

if the center is the point (h, k),

$$(x-h)^2 + (y-k)^2 = a^2. \tag{2}$$

51. General equation. It follows from (2), § 50, that the equation of a circle is always of the second degree.

The most general equation of the second degree in x and y may contain, at most, terms in x^2, xy, y^2, x, y, and a constant: i.e., it may be written in the form

$$Ax^2 + Bxy + Cy^2 + Dx + Ey + F = 0.$$

Consider now the special case in which $A = C$ and $B = 0$:

$$\text{(1)} \qquad \mathbf{Ax^2 + Ay^2 + Dx + Ey + F = 0} \qquad (A \neq 0).$$

We may always divide this equation through by A, transpose the constant term to the right member, and complete the squares in x and y (see the example below). The equation then has the form

$$\text{(2)} \qquad (x - h)^2 + (y - k)^2 = a^2,$$

and consequently represents a circle whenever the right member is positive.

Conversely, it appears from § 50 that the equation of every circle may be put in the form (1).

In some problems it is convenient to take, as the general form, the equation

$$\text{(3)} \qquad \mathbf{x^2 + y^2 + Dx + Ey + F = 0.}$$

That this is allowable follows from the fact that, in (1), the coefficient of the square terms can always be reduced to unity by dividing through by A.

When an equation of form (1) is reduced to form (2), it may happen that the right member becomes 0:

$$(x - h)^2 + (y - k)^2 = 0.$$

Since this equation holds only when $x = h$ and $y = k$, the locus is the single point (h, k) — a so-called "point-circle."

Finally, it may happen that the right member of (2) is negative: in this case there is clearly no locus.

THEOREM: *An equation of the second degree in which x^2 and y^2 have equal coefficients and the xy-term is missing represents a circle (exceptionally, a single point, or no locus).*

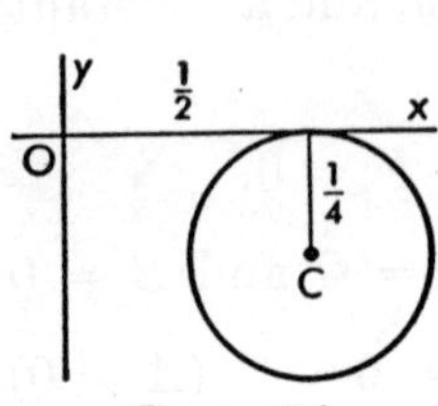

Figure 76

Example: Find the center and radius of the circle

$$4x^2 + 4y^2 - 4x + 2y + 1 = 0.$$

Transpose the constant term to the right member and divide by 4:

$$x^2 + y^2 - x + \tfrac{1}{2}y = -\tfrac{1}{4}.$$

Then complete the squares in x and y:

$$x^2 - x + \tfrac{1}{4} + y^2 + \tfrac{1}{2}y + \tfrac{1}{16} = -\tfrac{1}{4} + \tfrac{1}{4} + \tfrac{1}{16},$$

or

$$(x - \tfrac{1}{2})^2 + (y + \tfrac{1}{4})^2 = \tfrac{1}{16}.$$

The center is the point $C: (\frac{1}{2}, -\frac{1}{4})$, and the radius is $\frac{1}{4}$.

Exercises

In Exs. 1–13, write the equation of the circle.

1. With center at $(2, -3)$, radius 5.
2. With center at $(-4, 1)$, radius 6.
3. With radius a and touching both axes.
4. With center at $(2a, a)$ and touching the y-axis.
5. With center at (a, b) and passing through the origin.
6. With center at $(-1, -3)$ and passing through $(-2, 0)$.
7. With center at $(-4, 2)$ and passing through $(0, 5)$.
8. With the points $(2, 5)$, $(6, -1)$ as ends of a diameter.
9. With the points $(0, 2a)$, $(2b, 0)$ as ends of a diameter.
10. With center at $(0, 0)$ and touching the line $3x + 4y = 10$.
11. With center at $(0, 0)$ and touching the line $5x - 12y = 52$.
12. With center at $(-1, -2)$ and touching the line $x - 2y = -7$.
13. With center at $(4, 3)$ and touching the line $3x + y = -15$.
14. Using equation (2), § 50, find the condition that a circle shall
    (a) touch the x-axis;
    (b) pass through $(0, 0)$;
    (c) have its center on Oy;
    (d) touch both axes;
    (e) have its center on the line $3x + 4y = 12$.

In Exs. 15–24, draw the circle.

15. $x^2 + y^2 - 4x - 6y = 12$.
16. $x^2 + y^2 - 8x + 2y = 8$.
17. $x^2 + y^2 = 6x - 8y$.
18. $x^2 + y^2 = 2ay$.
19. $3x^2 + 3y^2 = y + 2$.
20. $2x^2 + 2y^2 = 6 - x$.
21. $4x^2 + 4y^2 + 4x = 12y - 1$.
22. $3x^2 + 3y^2 + 8x + 6y + 7 = 0$.
23. $2x^2 + 2y^2 = 2x + 2y - 1$.
24. $4x^2 + 4y^2 = 4y - 8x - 5$.

In Exs. 25–28, show that the circles are tangent to each other, and draw the figure.

25. $x^2 + y^2 - 6x + 1 = 0$, $x^2 + y^2 - 2y + 8x - 1 = 0$.
26. $x^2 + y^2 - 2x - 3 = 0$, $x^2 + y^2 + 4x - 8y + 11 = 0$.
27. $x^2 + y^2 = 2(y - x + 1)$, $x^2 + y^2 - 4x + 6y = 36$.
28. $x^2 + y^2 = 10x + 11$, $x^2 + y^2 - 2x - 6y + 9 = 0$.

29. Prove that equation (1), § 51, represents a point-circle if and only if $D^2 + E^2 - 4AF = 0$.

30. Prove that (1), § 51, has no locus if and only if $D^2 + E^2 - 4AF < 0$.

31. A point moves so that the sum of the squares of its distances from the points $(a, 0)$, $(-a, 0)$ is constant (equal to k^2). Find the equation of its locus, and draw the curve for the cases $k^2 = 6a^2$, $k^2 = 4a^2$, $k^2 = 3a^2$, $k^2 = 2a^2$.

32. A point moves so that the square of its distance from a fixed point is proportional to its distance from a fixed line. Find the equation of its locus, and draw the curve for various cases.

33. A cone is inscribed in a sphere of radius a. Express the radius of the cone as a function of its altitude, and draw the graph.

Ans. $r = \sqrt{2ah - h^2}$.

34. Prove that an angle inscribed in a semicircle is a right angle.

35. On the line $x + 2y = 3$, find points at a distance 5 from $(1, 6)$.

Ans. $(1, 1)$, $(-3, 3)$.

36. On the circle $x^2 + y^2 + 2x - 8y + 7 = 0$, find points at a distance 5 from $(2, -2)$.

Ans. $(2, 3)$, $(-2, 1)$.

37. Find the points of intersection of the circles

$$x^2 + y^2 - 18x - 4y + 35 = 0, \quad x^2 + y^2 + 2x + 6y - 15 = 0.$$

Draw the figure.

Ans. $(2, 1)$, $(4, -3)$.

38. Find the points of intersection of the circles

$$x^2 + y^2 - 6x - 8y - 24 = 0, \quad x^2 + y^2 = 6x + 8y.$$

Interpret the result.

39. Find the points of intersection of the circles

$$x^2 + y^2 - 4x - 2y + 1 = 0, \quad x^2 + y^2 + 4x + 3 = 0.$$

Interpret the result.

52. Circle determined by three conditions. We know from elementary geometry that a circle is determined by three points (not in a straight line). This is proved analytically by the fact that either (2), § 50, or (3), § 51, contains three constants.

More generally, a circle may be made to satisfy any three conditions which when expressed analytically lead to three consistent equations for determining the constants. Thus three tangents may be given, or two points and the radius, etc. It may not, however, be uniquely determined; there may be two or more circles satisfying the given conditions.

Example: Find the equation of the circle through the points $P_1:(1, 1)$, $P_2:(2, -1)$, and $P_3:(2, 3)$.

The direct general method, available in any similar problem no matter what kind of curve is involved, is to assume the equation of the circle in a suitable form and impose the conditions that the coordinates of each of the given points satisfy the equation of the circle. The algebra will be simple if we take as general equation of the circle

$$x^2 + y^2 + Dx + Ey + F = 0,$$

in which D, E, F are to be determined. Since the coordinates of the given points are to satisfy this equation, we get the following three equations for the unknowns D, E, F:

$$\begin{aligned} (1, 1)&: \quad 2 + D + E + F = 0, \\ (2, -1)&: \quad 5 + 2D - E + F = 0, \\ (2, 3)&: \quad 13 + 2D + 3E + F = 0. \end{aligned}$$

These equations give $D = -7$, $E = -2$, $F = 7$, whence the equation of the circle is

$$x^2 + y^2 - 7x - 2y + 7 = 0. \tag{1}$$

The following method, although it has the disadvantage of applying only to the circle, has the compensating advantage of a more direct geometric interpretation. The center lies on the perpendicular bisector of P_1P_2, whose equation is

$$\sqrt{(x-1)^2+(y-1)^2}=\sqrt{(x-2)^2+(y+1)^2},$$

or

$$2x-4y=3;$$

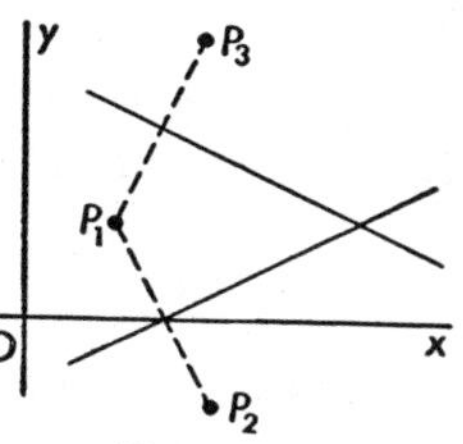

Figure 77

it also lies on the perpendicular bisector of P_1P_3, whose equation is

$$2x+4y=11.$$

The center is the point of intersection of these lines, $(\frac{7}{2}, 1)$. The radius is the distance from $(\frac{7}{2}, 1)$ to any one of the given points, which is found to be $\frac{5}{2}$. By formula (2), § 50, the equation of the circle is

$$(x-\tfrac{7}{2})^2+(y-1)^2=\tfrac{25}{4},$$

which reduces to (1). Evidently this method merely carries out analytically the ruler-and-compass construction for the center. For a third method, see Ex. 26 below.

53. Circles of Appolonius. Nearly 2200 years ago, the Greek mathematician Appolonius of Perga completed the solution of the following locus problem:

Given two fixed points, find the locus of a variable point such that the ratio of its distances from the two fixed points remains constant.

The solution of this problem has turned out to be of interest both in engineering and in advanced pure mathematics.

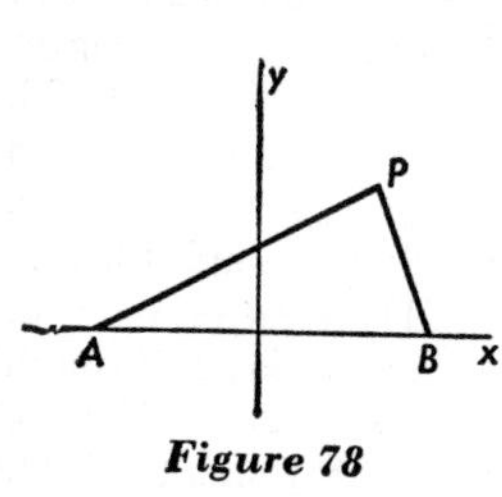

Figure 78

Let the fixed points be A, B, the variable point P, and the ratio of the distances $\dfrac{AP}{BP}$ be k. Call the distance between the fixed points $2c$ and place the axes as shown in Fig. 78, so that the coordinates of A, B, P are respectively $(-c, 0)$, $(c, 0)$, (x, y). Then

$$AP=k\cdot BP$$

yields

$$(1) \qquad \sqrt{(x + c)^2 + y^2} = k\sqrt{(x - c)^2 + y^2}.$$

Squaring both sides of equation (1) and collecting like terms, we find the equation of the locus to be

$$(2) \quad (1 - k^2)x^2 + (1 - k^2)y^2 + 2c(1 + k^2)x + c^2(1 - k^2) = 0.$$

It is a simple matter to show that, if $k \neq 1$, then (2) is the equation of a circle with center at $\left(c\dfrac{k^2 + 1}{k^2 - 1}, 0\right)$ and radius equal to the numerical value of $\dfrac{2ck}{k^2 - 1}$. If $k = 1$, equation (2) degenerates into the perpendicular bisector, $x = 0$, of the line segment joining A and B.

Note that, disassociating the result from the tools (coordinate system, distance formula, etc.) used to obtain those results, we have shown that the locus in question is a circle with its center on the line joining the fixed points.

Exercises

In Exs. 1–25, find the equation of the circle.

1. Through (2, 3), (3, 4), (−1, 2). Solve by two methods.

2. Through (3, 1), (5, 3), (−3, −1). Solve by two methods.

3. Circumscribing the triangle with vertices (−1, −4), (3, −2), (5, 2).

4. Circumscribing the triangle with vertices $(a, 0)$, (a, a), $(0, 2a)$.
Ans. $x^2 + y^2 + ax - ay - 2a^2 = 0$.

5. Passing through the points (−1, −3), (−5, 3), and having its center on the line $x - 2y + 2 = 0$. *Ans.* $(x + 6)^2 + (y + 2)^2 = 26$.

6. Passing through the points (−4, −2), (2, 0), and having its center on the line $5x - 2y = 19$. *Ans.* $x^2 + y^2 - 2x + 14y = 0$.

7. Touching the line $x + 2y = 8$ at (0, 4) and passing through (3, 7).
Ans. $(x - 1)^2 + (y - 6)^2 = 5$.

8. Touching the line $4x - 3y = 28$ at (4, −4) and passing through (−3, −5). *Ans.* $x^2 + y^2 + 2y - 24 = 0$.

9. Touching the line $3x - 2y = 5$ at (3, 2) and passing through (−2, 1).

10. Touching the line $x + y = 8$ at (2, 6) and passing through (4, 0).

11. Touching the x-axis and passing through the points (3, 1), (10, 8).
Ans. Centers: (−2, 13), (6, 5).

12. Touching the y-axis and passing through the points $(1, 5)$, $(8, 12)$.

13. Touching the line $x + y = 4$ at $(1, 3)$, and having a radius $\sqrt{2}$. Solve in two ways. *Ans.* Centers: $(0, 2)$, $(2, 4)$.

14. Touching the line $x - 2y = 3$ at $(-1, -2)$, and having a radius $\sqrt{5}$. Solve in two ways. *Ans.* Centers: $(0, -4)$, $(-2, 0)$.

15. Touching the lines $3x + 4y = 12$, $4x + 3y = 9$, and having its center on the line $3x + y = 7$. *Ans.* Centers: $(2, 1)$, $(1, 4)$.

16. Touching the lines $x - y = 3$, $7x + y = 5$, and having its center on the line $2x + y = 10$. *Ans.* Centers: $(3, 4)$, $(7, -4)$.

17. Having a radius $\sqrt{85}$, through $(5, 9)$, $(1, -7)$. Solve in two ways.

18. Having a radius $\sqrt{10}$, through $(4, 1)$, $(6, 3)$. Solve in two ways.

19. Touching the lines $x + 2y = 4$, $x + 2y = 2$, $y = 2x - 5$.

20. Touching the circle $x^2 + y^2 = 100$ at $(6, -8)$ and having a radius 15. *Ans.* Centers: $(-3, 4)$, $(15, -20)$.

21. Touching the circle $x^2 + y^2 + 2x - 6y + 5 = 0$ at $(1, 2)$ and passing through $(4, -1)$. *Ans.* Center: $(3, 1)$.

22. With radius $\sqrt{2}$, tangent to the line $x + y = 3$, and having its center on the line $y = 4x$. *Ans.* Centers: $(1, 4)$, $(\frac{1}{5}, \frac{4}{5})$.

23. With radius 1, tangent to the line $3x + 4y = 5$, and having its center on the line $x + 2y = 0$. *Ans.* Centers: $(0, 0)$, $(10, -5)$.

24. With radius 2, tangent to the x-axis, and passing through $(1, -1)$. *Ans.* Centers: $(1 \pm \sqrt{3}, -2)$.

25. With radius $2\sqrt{5}$, tangent to the line $y = 2x$, and passing through $(3, -4)$. *Ans.* Centers: $(1, -8)$, $(5, 0)$.

26. Prove that the equation

$$\begin{vmatrix} (x^2 + y^2) & x & y & 1 \\ (x_1^2 + y_1^2) & x_1 & y_1 & 1 \\ (x_2^2 + y_2^2) & x_2 & y_2 & 1 \\ (x_3^2 + y_3^2) & x_3 & y_3 & 1 \end{vmatrix} = 0$$

represents the circle through (x_1, y_1), (x_2, y_2), (x_3, y_3).

Solve Exs. 27–30 by the formula of Ex. 26.

27. Ex. 1. **28.** Ex. 2. **29.** Ex. 3. **30.** Ex. 4.

31. When does the equation of Ex. 26 reduce to an equation of first degree? Interpret geometrically.

32. In solving a three-point problem by the first method used in the example of § 52, if it should turn out that the equations in D, E, and F have no solution, what would be the geometric interpretation?

33. Find the locus of a point which remains twice as far from $(3, 2)$ as it does from $(2, -1)$. *Ans.* $3x^2 + 3y^2 - 10x + 12y + 7 = 0$.

34. Find the locus of points three times as far from $(0, 4)$ as from $(2, 0)$. *Ans.* $2x^2 + 2y^2 - 9x + 2y + 5 = 0$.

54. Tangents to a given circle. Three types of problems concerning tangents to circles will be treated here. We shall find: (*a*) the tangent at a given point on the circle, (*b*) the tangents in a prescribed direction, (*c*) the tangents to a circle from a point outside the circle.

Example (*a*): Find the equation of the tangent line at $(-1, 4)$ on the circle $x^2 + y^2 - 4x - 21 = 0$.

First put the equation of the circle into standard form,

$$(x - 2)^2 + y^2 = 25 \tag{1}$$

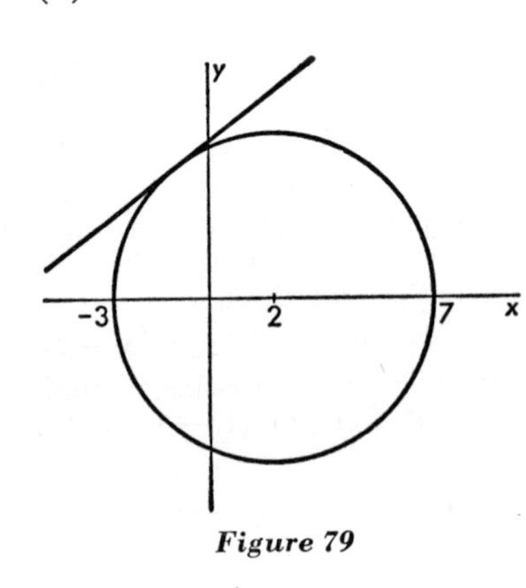

Figure 79

and thus locate its center, $(2, 0)$. Note that the coordinates of the given point $(-1, 4)$ satisfy equation (1), thus verifying that $(-1, 4)$ is actually on the circle.

Since the tangent at $(-1, 4)$ must be perpendicular to the radius through that point, we determine the slope of that radius, $m_r = -\frac{4}{3}$. The slope of the tangent, m_t, is the negative reciprocal of m_r. Hence the tangent line passes through $(-1, 4)$ and has slope $m_t = \frac{3}{4}$. Its equation is

$$y - 4 = \tfrac{3}{4}(x + 1),$$

or

$$3x - 4y = -19.$$

Example (*b*): Find lines tangent to the circle

$$x^2 + y^2 - 2x + 4y + 1 = 0 \tag{2}$$

and parallel to the line $5x - 12y = 7$.

The equation (2) may be put in standard form,

$$(x - 1)^2 + (y + 2)^2 = 4,$$

so that the circle is seen to have its center at (1, −2), its radius 2. The equation of a line through the center and parallel to the given line is

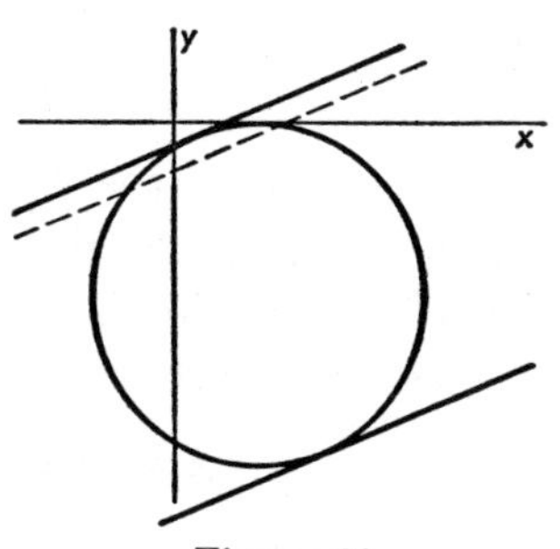

Figure 80

$$5x - 12y = 29.$$

The required tangents are parallel to this line at a distance 2 (the radius) from it. Hence, by § 46, their equations are

$$\frac{5x - 12y}{-13} = \frac{29}{-13} \pm 2,$$

or

$$5x - 12y = 3, \qquad 5x - 12y = 55.$$

Example (*c*): Find the tangents to the circle

$$(3) \qquad x^2 + (y - 3)^2 = 2$$

through the point (3, 4).

Write the equation of an arbitrary line through (3, 4),

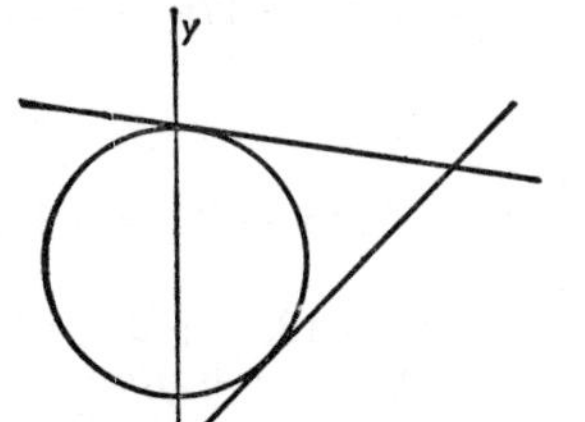

$$y - 4 = m(x - 3),$$

or

$$(4) \qquad mx - y - 3m + 4 = 0.$$

Figure 81

This line will be tangent to the circle if it passes at a distance $\sqrt{2}$ (the radius) from the center (0, 3). Therefore, find the distance from the line (4) to the center, and equate this distance to the radius:

$$\frac{0 - 3 - 3m + 4}{-\sqrt{1 + m^2}} = \pm\sqrt{2},$$

or

$$7m^2 - 6m - 1 = 0.$$

Thus we find $m_1 = 1$, $m_2 = -\frac{1}{7}$, the slopes of the desired tangent lines. Each tangent must pass through (3, 4), hence its equation is readily found using the point-slope form. The tangents are

$$x - y = -1, \qquad x + 7y = 31.$$

Exercises

In Exs. 1–20, find the tangent lines to the given circle as directed.

1. $x^2 + y^2 = 34$ at $(-3, 5)$. *Ans.* $3x - 5y = -34$.
2. $x^2 + y^2 = 26$ at $(5, 1)$. *Ans.* $5x + y = 26$.
3. $x^2 + y^2 = 5y$ at $(-2, 1)$. *Ans.* $4x + 3y = -5$.
4. $2x^2 + 2y^2 = 5x$ at $(2, -1)$. *Ans.* $3x - 4y = 10$.
5. $x^2 + y^2 + 2x - y - 17 = 0$ at $(3, -1)$.
6. $x^2 + y^2 - 5x + 2y + 3 = 0$ at $(2, -3)$.
7. $x^2 + y^2 = 17$ parallel to $4x - y = 5$. *Ans.* $4x - y = \pm 17$.
8. $x^2 + y^2 = 29$ parallel to $2x + 5y = 2$. *Ans.* $2x + 5y = \pm 29$.
9. $x^2 + y^2 + 10x - 6y - 2 = 0$ parallel to $y = 2x$.
Ans. $2x - y = -13 \pm 6\sqrt{5}$.
10. $x^2 + y^2 - 2x - 8y + 15 = 0$ parallel to $x - y = 9$.
Ans. $x - y = -1$, $x - y = -5$.
11. $x^2 + y^2 - 4x - 6y + 5 = 0$ perpendicular to $x - y = 7$.
12. $x^2 + y^2 + 6x - 2y + 5 = 0$ perpendicular to $2x + y = 8$.
13. $x^2 + y^2 = 10$ through $(5, 5)$. *Ans.* $3x - y = 10$, $x - 3y = -10$.
14. $x^2 + y^2 = 2$ through $(6, 4)$. *Ans.* $x - y = 2$, $7x - 17y = -26$.
15. $x^2 + y^2 = 25$ through $(7, 1)$.
16. $x^2 + y^2 = 5$ through $(4, 3)$.
17. $x^2 + y^2 + 4x - 9 = 0$ through $(-1, 5)$.
Ans. $3x + 2y = 7$, $2x - 3y = -17$.
18. $x^2 + y^2 - 2x + 6y + 5 = 0$ through $(6, -3)$.
Ans. $x - 2y = 12$, $x + 2y = 0$.
19. $x^2 + y^2 = a^2$ at (x_1, y_1) on the circle. *Ans.* $x_1x + y_1y = a^2$.
20. $x^2 + y^2 = a^2$ with slope m. *Ans.* $y = mx \pm a\sqrt{1 + m^2}$.

21. In finding tangents by the method of Example (*c*) above, what conclusions can be drawn if the values of m turn out to be imaginary? To be real and equal?

22. Given two points P_1, P_2 on a circle, prove analytically that the distance from P_1 to the tangent at P_2 equals the distance from P_2 to the tangent at P_1. Use the formula of Ex. 19.

55. Radical axis. If the equations of two nonconcentric circles are each put in the form

$$(1) \qquad x^2 + y^2 + Dx + Ey + F = 0,$$

then the quadratic terms can be eliminated by subtraction. The resultant linear equation represents a straight line which is called the *radical axis* of the two circles.

Example (*a*): Find the radical axis of the circles

$$(x - 1)^2 + y^2 = 3,$$
$$(x + 2)^2 + (y - 4)^2 = 8.$$

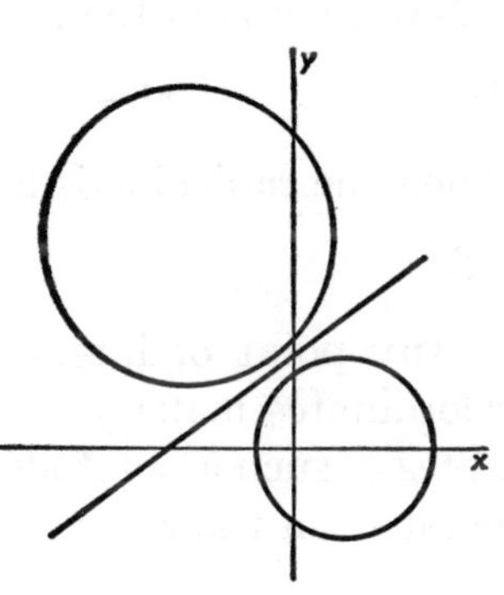

Figure 82

Write the equations of the circles in the form

$$x^2 + y^2 - 2x - 2 = 0,$$
$$x^2 + y^2 + 4x - 8y + 12 = 0,$$

and subtract to obtain the radical axis,

$$3x - 4y = -7.$$

Example (*b*): Prove that the radical axis of two circles is perpendicular to the line joining their centers.

Let us choose axes with origin at the center of one circle, with the x-axis passing through the center of the other circle. Then the equations of the circles may be written as

$$(2) \qquad x^2 + y^2 = a^2$$

$$(3) \qquad (x - h)^2 + y^2 = r^2,$$

with $h \neq 0$.

From equations (2) and (3) it follows that the radical axis is the vertical line

$$2hx = a^2 - r^2 + h^2.$$

Since the line of centers is horizontal, the proof is completed.

56. Common chord. If two circles intersect in two distinct points, the radical axis of the circles passes through their points of intersection, and is then called the *common chord*. See also Ex. 7 below.

To see that the radical axis does pass through any point of intersection, think of the equations of the circles as

$$u = 0 \qquad \text{and} \qquad v = 0,$$

where each is in the form

$$x^2 + y^2 + Dx + Ey + F = 0. \tag{1}$$

Then the radical axis has the equation

$$u - v = 0. \tag{2}$$

Any point of intersection of the circles is a point whose coordinates make $u = 0$ and make $v = 0$. Then the coordinates of such a point also satisfy (2). Hence any intersection point is on the radical axis.

57. The circle in polar coordinates. We already know (§ 16) a formula for the distance D between points with polar coordinates (r_1, θ_1) and (r_2, θ_2),

$$D^2 = r_1{}^2 + r_2{}^2 - 2r_1r_2 \cos (\theta_2 - \theta_1). \tag{1}$$

From (1) we obtain easily the polar equation of any circle (Ex. 15, p. 47). Let (r, θ) be a variable point on the circle, (r_1, θ_1) the center, and a the radius. Then, by (1),

$$r^2 + r_1{}^2 - 2r_1r \cos (\theta - \theta_1) = a^2. \tag{2}$$

Exercises

In Exs. 1–6, find the radical axis of the given circles. Draw the figure.

1. $x^2 + y^2 - 4x + 3 = 0,\ x^2 + y^2 - 2x - 4y + 19 = 0.$
2. $x^2 + y^2 - 6x + 6y + 14 = 0,\ x^2 + y^2 = 1.$
3. $x^2 + y^2 = 16a^2,\ (x - 2a)^2 + y^2 = a^2.$
4. $(x - a)^2 + y^2 = a^2,\ x^2 + (y - b)^2 = b^2.$

5. $2x^2 + 2y^2 = 2x + 2y + 1$, $x^2 + y^2 = 2x + 3$.

6. $2x^2 + 2y^2 = 6x - 2y + 3$, $4x^2 + 4y^2 = 12y - 5$.

7. Prove that if two circles are tangent, then their radical axis is the common tangent at the point of contact.

8. Prove that the radical axis of two circles bisects the line segment joining the centers if, and only if, the radii are equal. Choose axes so that the centers are at $(a, 0)$, $(-a, 0)$.

9. Find the radical axis of two point circles (zero radius). Solve directly and also by using Ex. 8.

10. Devise a method for finding the equation of the tangent at a point on a circle by considering the given point as a point circle. (Ex. 7.)

In Exs. 11–15, use the method of Ex. 10 to find the tangent to the circle at the given point on it.

11. $x^2 + y^2 - 3x + 5y = 12$ at $(1, 2)$.

12. $x^2 + y^2 + 6x - 3y = 31$ at $(3, -1)$.

13. $3x^2 + 3y^2 - 5x + 7y + 4 = 0$ at $(1, -2)$.

14. $2x^2 + 2y^2 + 3x + y = 20$ at $(2, -2)$.

15. $x^2 + y^2 = a^2$ at (x_1, y_1).

16. Show that, in general, the radical axes of three circles taken in pairs meet in a point: this is called the *radical center*. In what case do three circles, no two concentric, fail to have a radical center?

In Exs. 17–20, find the radical center and draw the figure.

17. $(x - 3)^2 + y^2 = 9$, $(x + 4)^2 + y^2 = 2$, $x^2 + (y - 3)^2 = 1$.
Ans. $(-1, \frac{1}{3})$.

18. $x^2 + y^2 = 2x$, $x^2 + y^2 = 4y$, $x^2 + y^2 + 2x + 8y = 32$. *Ans.* $(4, 2)$.

19. $x^2 + y^2 = 4$, $x^2 + y^2 = 4y$, $x^2 + y^2 - 6x + 8y + 24 = 0$.
Ans. $(6, 1)$.

20. $x^2 + y^2 = 1$, $x^2 + y^2 + 3x - 2y = 4$, $x^2 + y^2 - 2x - y = 6$.
Ans. $(-1, -3)$.

21. Write the polar equation of a circle of radius a through the pole, if a diameter through O makes an angle α with Ox.
Ans. $r = 2a \cos (\theta - \alpha)$.

22. Find the polar equation of the tangent line to the circle $r = a$ at the point (a, β). *Ans.* $r \cos (\theta - \beta) = a$.

23. Find the polar equation of the tangent line at the point (r_2, θ_2) on the circle with center at (r_1, θ_1).
Ans. $rr_2 \cos (\theta - \theta_2) - rr_1 \cos (\theta - \theta_1) + r_1r_2 \cos (\theta_1 - \theta_2) = r_2^2$.

CHAPTER 7 *Conic Sections. The Parabola*

58. Definitions. The path of a point which moves so that *its distance from a fixed point is in a constant ratio to its distance from a fixed line* is called a *conic section,* or simply a *conic.*

The fixed point is called the *focus* of the conic, the fixed line the *directrix,* and the constant ratio the *eccentricity.* In Fig. 83, if F is the focus, DD' the directrix, and P a point on the conic, then

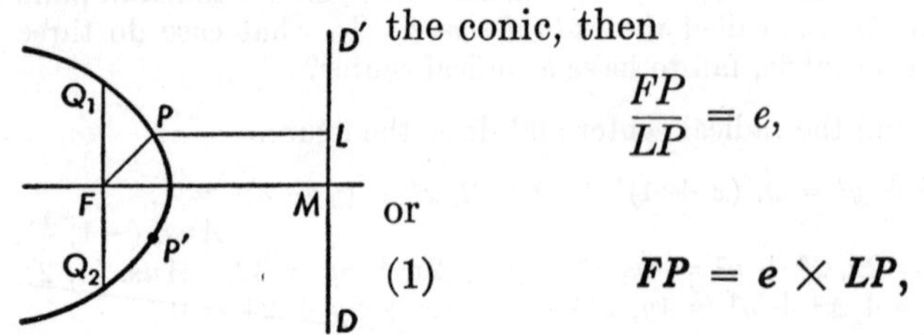

Figure 83

$$\frac{FP}{LP} = e,$$

or

$$FP = e \times LP, \tag{1}$$

where e denotes the eccentricity.

Equation (1) would still be true if the point P were in the position P', symmetric to P with respect to the line FM. Hence the line through the focus perpendicular to the directrix is a line of symmetry for the curve. It follows that the line through the focus parallel to the directrix intersects the curve in two points: the chord Q_1Q_2 joining these points is called the *latus rectum,* or *right chord.*

The conic sections fall into three classes, as follows (Fig. 84):

if $e < 1$, the conic is an *ellipse;*
if $e = 1$, the conic is a *parabola;*
if $e > 1$, the conic is a *hyperbola.*

Alternative denitions of the ellipse and hyperbola will appear in §§ 67, 73.

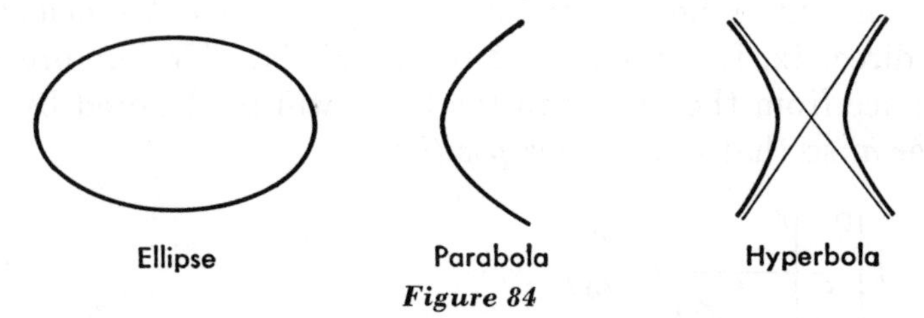

Figure 84

59. The circle; degenerate conics. In addition to the three typical conics mentioned above, it will be convenient to include under that term various other loci.

First, when $e = 0$ the definition fails. But when e *approaches zero*, the ellipse approaches a circle: we will therefore agree to consider the circle as a special case of the ellipse.

Second, we will agree to consider the "point-ellipse" (§ 69), two parallel or coincident lines (§ 64), and two intersecting lines (§ 75) as conic sections of exceptional type. These loci are called *degenerate conics*.

Every plane section of a circular cone is a curve of the class that we have called "conic sections" (provided the forms just discussed are included); this of course is the reason for the name. The student may amuse himself by discovering, intuitively, how the cutting plane must be passed in order to obtain the various sections.*

We will also establish in § 78 the important

THEOREM: *An equation of the second degree represents a conic section (exceptionally, no locus); and conversely.*

60. The parabola: first standard forms. The parabola has been defined in § 58 as *the conic whose eccentricity is* 1:

The parabola is the locus of points which are equidistant from a fixed point and a fixed line.

* Two parallel lines cannot be cut from a cone. They can, however, be cut from a cylinder, which is the form approached by a cone with a fixed right section as the vertex recedes indefinitely.

The line through the focus perpendicular to the directrix is called the *axis* of the curve. The point where the axis intersects the curve, i.e., the point midway between the focus and the directrix, is the *vertex* of the parabola. The undirected distance from the vertex to the focus will be denoted by the letter a, so that a is *always positive.*

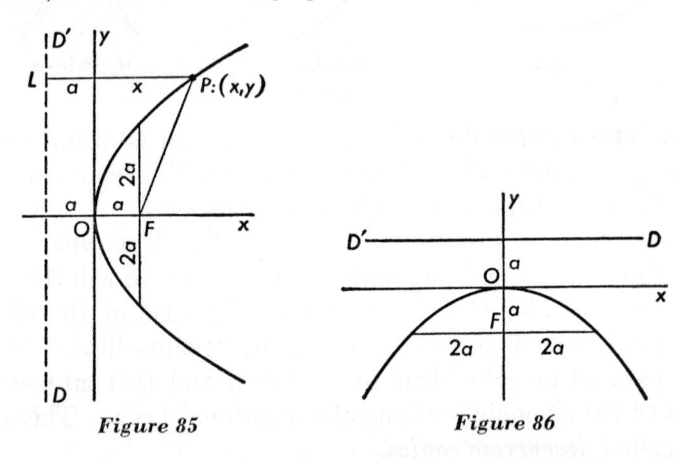

Figure 85 *Figure 86*

Let us take the vertex of a parabola as the origin and the focus at $(a, 0)$, where a is any positive number. Then the axis of the curve is the x-axis and the directrix is the line $x = -a$. If $P: (x, y)$ is any point on the curve, then by the definition of the parabola (Fig. 85)

$$FP = LP,$$

or

$$\sqrt{(x - a)^2 + y^2} = a + x,$$

which reduces to

$$y^2 = 4ax. \tag{1}$$

From (1) it appears that the curve is symmetric with respect to the x-axis, that y is imaginary for negative x, and that the two values of y increase numerically without limit as x increases: thus the curve opens indefinitely to the right.

When $x = a$, $y = \pm 2a$. Hence *the length of the latus rectum is four times the distance between the vertex and the focus.*

If the figure is the same except that the curve opens to the left — focus at $(-a, 0)$, directrix $x = a$ — the equation is

$$y^2 = -4ax. \tag{2}$$

The corresponding equations when the axis is vertical are

$$x^2 = 4ay, \tag{3}$$

$$x^2 = -4ay. \tag{4}$$

Figure 86 shows the curve (4).

61. Translation of axes. We have had many illustrations of the fact that a problem may be greatly simplified by taking the axes in a convenient position. It may happen, however, that the position of the axes is predetermined by the statement of the problem; it is then desirable that we have methods for shifting the axes to a more suitable position. Such methods will be developed as need arises.

Consider now the *translation* of axes, in which the axes are moved parallel to their original positions. Let Ox, Oy be the original axes, O_1x_1, O_1y_1 the new, and suppose the new origin O_1 to be the point (h, k) referred to the old axes. If we denote by (x, y) the coordinates of any point P in the original system, by (x_1, y_1) the coordinates of the same point* in the new system, then the two sets of coordinates are connected by the formulas

$$\begin{cases} x = x_1 + h, \\ y = y_1 + k. \end{cases} \tag{1}$$

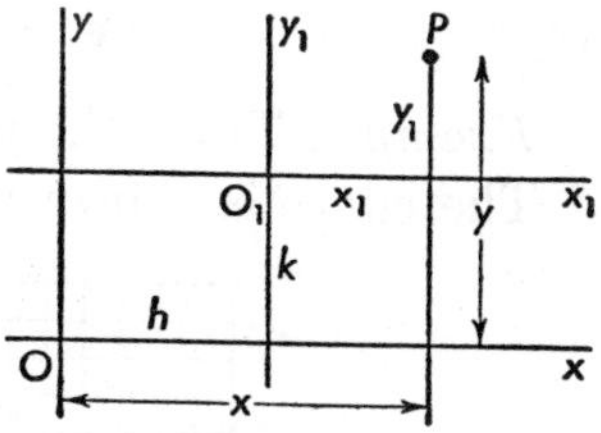

Figure 87

62. Other standard forms. To find the equation of a parabola with vertex at (h, k), axis

* Note that x_1 and y_1 are variables, so that we are departing from our usual convention (§ 18) whereby the coordinates (x_1, y_1) denote a fixed point.

horizontal, and opening to the right, let us introduce (temporarily) new axes O_1x_1, O_1y_1, where O_1 is the vertex and O_1x_1 the axis of the parabola. Then the equation, *referred to these axes*, may be written at once by § 60:

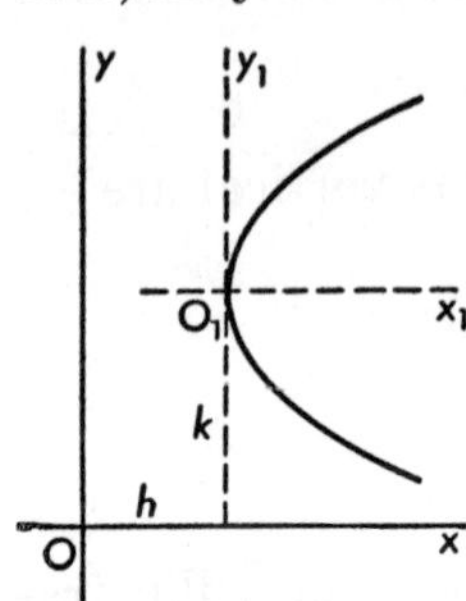

Figure 88

$$y_1{}^2 = 4ax_1.$$

Substituting the values of x_1 and y_1 as given by (1), § 61, we find that the equation referred to axes Ox, Oy is

$$(y - k)^2 = 4a(x - h).$$

By similar argument we may obtain the equation when the curve opens to the left, or upward or downward. In summary:

The equation of a parabola with vertex at (h, k) *is:*

if the axis is parallel to Ox *and the curve opens to the right,*

(1) $$(y - k)^2 = 4a(x - h);$$

if the axis is parallel to Ox *and the curve opens to the left,*

(2) $$(y - k)^2 = -4a(x - h);$$

if the axis is parallel to Oy *and the curve opens upward,*

(3) $$(x - h)^2 = 4a(y - k);$$

if the axis is parallel to Oy *and the curve opens downward,*

(4) $$(x - h)^2 = -4a(y - k).$$

Example: Trace the parabola $(x + \frac{1}{2})^2 = -\frac{3}{2}(y - \frac{1}{6})$. The curve is shown in Fig. 89 with 12 spaces as the unit.

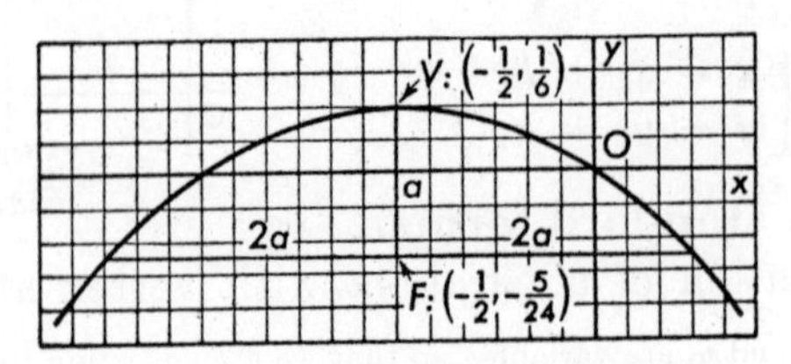

Figure 89

The equation is of form (4), so that the curve opens downward. The vertex is $V:(-\frac{1}{2}, \frac{1}{6})$. Since $4a = \frac{3}{2}$, $a = \frac{3}{8}$: the focus is at distance $a = \frac{3}{8}$ below the vertex. The ends of the latus rectum are at distance $2a = \frac{3}{4}$ right and left of the focus.

Unless for some reason we are interested in more distant regions, it is usually sufficient to draw only the part of the curve shown in Fig. 89 — i.e., extending a moderate distance beyond the ends of the latus rectum.

63. Geometric constructions. The work of elementary geometry falls into two parts: discovery of theorems pertaining to straight lines and circles, and geometric constructions based on the definitions and theorems — for instance, to bisect an angle, or to draw the common tangents to two circles.

In connection with the conics, the student will be asked from time to time to perform a variety of geometric constructions. For these the only tools allowed will be, just as in elementary geometry, the ruler and compass. No coordinate axes are to be drawn, and no algebra to be used.

In the hope of clearing up any uncertainty in the reader's mind, as to the role he is expected to play at a given time, let us say that an analyst, A, and a geometer, G, are two different persons. A lays down the necessary definitions and proceeds to establish (by analytic methods) a body of theorems and properties pertaining to the parabola. G studies this material; while he has never taken time to prove the theorems synthetically, he has full confidence in their truth. So, armed only with ruler and compass and A's definitions and theorems, G takes charge in Exs. 33–40 below.

The few problems given in this book represent only a small part of what can be done, with regard to the conics, by ruler and compass alone. As a matter of fact, the ancient Greek mathematicians knew a great deal about the conic sections; and their methods were purely synthetic, since analytic geometry was not invented until many centuries later.

Exercises

In Exs. 1–16, determine at sight the direction in which the curve opens; locate the vertex, focus, and ends of the latus rectum; draw the curve.

1. $y^2 = 7x$.
2. $y^2 = -20x$.
3. $x^2 + 8y = 0$.
4. $x^2 = 10y$.
5. $4y^2 = -3x$.
6. $8x^2 + 5y = 0$.
7. $(y - 3)^2 = 12(x + 2)$.
8. $(x - 1)^2 = -8(y + \frac{1}{2})$.
9. $(x + 4)^2 = -40y$.
10. $(y + 2)^2 = 8(x + 1)$.
11. $(x - 15)^2 = -100(y + 10)$.
12. $(y + 4)^2 = 8(x - 6)$.
13. $y^2 = -32(x + 3)$.
14. $x^2 = 12(y - 2)$.
15. $(x + \frac{1}{2})^2 - \frac{3}{2}y = 0$.
16. $(y - \frac{1}{2})^2 = -\frac{1}{2}x$.

In Exs. 17–30, find the equation of the parabola.

17. With vertex at O, axis Ox, and passing through $(3, -6)$.
Ans. $y^2 = 12x$.

18. With vertex at O, axis Oy, and passing through $(2, -\frac{1}{2})$.
Ans. $x^2 = -8y$.

19. With vertex $(-2, 3)$ and focus $(-4, 3)$.
20. With vertex $(5, 1)$ and focus $(5, -2)$.
21. With vertex $(2, -3)$ and directrix $y = -7$.
22. With vertex $(2, 4)$ and directrix $x = -3$.
23. With latus rectum joining $(-4, 1)$, $(2, 1)$.
Ans. $x^2 + 2x - 6y - 2 = 0$; $x^2 + 2x + 6y - 14 = 0$.

24. With latus rectum joining $(2, 5)$, $(2, -3)$.
25. With directrix $y = 3$, axis $x = 0$, and latus rectum 4. (Two answers.)
26. With directrix $x + 1 = 0$, axis $y = 2$, and latus rectum 2. (Two answers.)
27. With vertex on Oy, axis parallel to Ox, and passing through $(2, 2)$, $(8, -1)$. *Ans.* $(y - 1)^2 = \frac{1}{2}x$; $(y - 5)^2 = \frac{9}{2}x$.
28. With vertex on Ox, axis parallel to Oy, and passing through $(1, -4)$, $(-1, -1)$. *Ans.* $(x + 3)^2 = -4y$; $(x + \frac{1}{3})^2 = -\frac{4}{9}y$.
29. With vertex on the line $y = 2$, axis parallel to Oy, latus rectum 6, and passing through $(2, 8)$.
Ans. $(x - 8)^2 = 6(y - 2)$; $(x + 4)^2 = 6(y - 2)$.
30. With axis parallel to Ox, latus rectum 1, and passing through $(3, 1)$, $(-5, 5)$. *Ans.* $(y - 4)^2 = x + 6$; $(y - 2)^2 = -(x - 4)$.
31. Show that the following definition is equivalent to that of § 60: A *parabola* is the locus of a point moving so that the square of its distance from one of two fixed perpendicular lines is proportional to its distance from the other.
32. If P is a point on a parabola, show that the distance of P from the axis is a mean proportional between the latus rectum and the distance of P from the tangent at the vertex.

Solve Exs. 33–40 synthetically (i.e., by ruler and compass).

33. Show in detail how to construct points of a parabola if the focus and directrix are given.

34. Given a parabola with its vertex marked, construct the axis.

35. Given a parabola with its focus, construct the axis and directrix.

36. Given the directrix and two points of a parabola, find the focus. How many solutions are there, in general? Discuss exceptional cases.

37. Given the focus and two points of a parabola, find the directrix.

38. Given the directrix, the tangent at the vertex (vertex not marked), and one point of a parabola, find the focus.

39. Given a parabola with its axis, find the focus.

40. Given the axis, vertex, and one point of a parabola, find the focus. (Ex. 32.)

64. Reduction to standard form. The equation

$$Cy^2 + Dx + Ey + F = 0 \qquad (C \neq 0) \tag{1}$$

can be reduced to form (1) or (2), § 62, by dividing by C and completing the square in y; similarly the equation

$$Ax^2 + Dx + Ey + F = 0 \qquad (A \neq 0) \tag{2}$$

can be reduced to form (3) or (4), § 62. Exception arises in (1) when $D = 0$. The equation then involves only y. If the roots are real and distinct, the locus is (§ 26) two parallel lines; if the roots are real and equal, two coincident lines; if the roots are imaginary, there is no locus. A similar situation arises in (2) when $E = 0$.

THEOREM: *An equation of the second degree in which the xy-term is missing and only one square term is present represents a parabola with its axis parallel to a coordinate axis (exceptionally, two parallel or coincident lines, or no locus).*

Example: Reduce to standard form the equation

$$2x^2 + 2x + 3y = 0.$$

Complete the square in x:

$$x^2 + x + \tfrac{1}{4} = -\tfrac{3}{2}y + \tfrac{1}{4},$$
$$(x + \tfrac{1}{2})^2 = -\tfrac{3}{2}(y - \tfrac{1}{6}).$$

Thus it appears that the curve is the one shown in Fig. 89.

Exercises

In Exs. 1–16, reduce the equation to a standard form; plot the vertex, focus, and ends of the latus rectum; trace the curve.

1. $y^2 - 12x + 24 = 0$.
2. $y^2 + 8x + 16 = 0$.
3. $x^2 + 2y + 2 = 0$.
4. $x^2 - 4y + 12 = 0$.
5. $x^2 = 4(x + y)$.
6. $y^2 = 12(y - x)$.
7. $y^2 + 2x + 6y + 17 = 0$.
8. $x^2 - 2x + 2y + 7 = 0$.
9. $y^2 - x + y = 0$.
10. $y^2 + x + y = 0$.
11. $2x^2 - 2x + y - 1 = 0$.
12. $2x^2 + 2x + y - 1 = 0$.
13. $2y^2 - x - 8y + 8 = 0$.
14. $2y^2 + x + 12y + 18 = 0$.
15. $4x^2 + 6x - y + 2 = 0$.
16. $4x^2 - 6x + y + 1 = 0$.

17. Find the equation of a parabola with axis parallel to Ox and passing through $(5, 4)$, $(11, -2)$, $(21, -4)$. *Ans.* $y^2 - 2x - 4y + 10 = 0$.

18. Find the equation of a parabola with axis parallel to Oy and passing through $(1, 9)$, $(-2, 9)$, $(-1, 1)$. *Ans.* $4x^2 + 4x - y + 1 = 0$.

19. Find the locus of the center of a circle which passes through $(-2, 3)$ and touches the line $x = 6$. *Ans.* $y^2 + 16x - 6y - 23 = 0$.

20. Find the locus of the center of a circle which passes through $(2, 5)$ and touches the line $y = -7$. *Ans.* $x^2 - 4x - 24y - 20 = 0$.

21. Find the equation of a circle through $(0, 5)$, $(3, 4)$, touching the line $y + 5 = 0$. *Ans.* $x^2 + y^2 = 25$; $(x - 60)^2 + (y - 180)^2 = (185)^2$.

22. Find the locus of the center of a circle which touches the circle $x^2 + y^2 = 4$ and the line $x = 3$. *Ans.* $y^2 = 25 - 10x$; $y^2 = 1 - 2x$.

23. A circle touches the line $y = 2$ and the circle $x^2 + y^2 = 16$. Find the locus of its center. *Ans.* $x^2 + 12y = 36$; $x^2 - 4y = 4$.

24. Find the equation of a parabola with vertex on the line $y = x$, axis parallel to Ox, and passing through $(6, -2)$, $(3, 4)$.
Ans. $y^2 - 4x - 4y + 12 = 0$; $5y^2 - 36x - 28y + 140 = 0$.

25. Find the equation of a parabola with vertex on the line $y = x + 2$, axis parallel to Oy, latus rectum 6, and passing through $(-3, -1)$.
Ans. $(x + 3)^2 = \pm 6(y + 1)$; $(x + 9)^2 = 6(y + 7)$;
$(x - 3)^2 = -6(y - 5)$.

In Exs. 26–33, find the points of intersection of the given curves and draw the figure.

26. $x^2 = 4y$, $x^2 = y + 3$. *Ans.* $(2, 1)$, $(-2, 1)$.
27. $x^2 = 3y$, $x^2 = y - 2$.
28. $y^2 = 4x$, $x^2 - 3x + y = 0$. *Ans.* $(0, 0)$, $(1, 2)$ twice, $(4, -4)$.
29. $y^2 + y = x$, $x^2 + 12y = 8x$. *Ans.* $(0, 0)$, $(2, 1)$ twice, $(12, -4)$.
30. $(y - 1)^2 = x + 7$, $x^2 + (y + 5)^2 = 85$.
Ans. $(2, 4)$, $(9, -3)$, $(-6, 2)$, $(-7, 1)$.
31. $y^2 + x - 2y + 2 = 0$, $x^2 + 9x + 6y + 2 = 0$.
Ans. $(-1, 1)$, $(-2, 2)$, $(-5, 3)$, $(-10, -2)$.

32. $x^2 + y^2 = 2x + 5y - 7$, $x^2 = 2x + y - 3$. *Ans.* (1, 2) four times.

33. $x^2 = 4y$, $y^2 = 6y - 4x + 3$. *Ans.* (2, 1) three times, $(-6, 9)$.

34. Show that if the axes of two parabolas are parallel, the parabolas cannot intersect in more than two points.

35. Show that if a line is parallel to the axis of a parabola, it intersects the curve in one and only one point; and conversely.

36. If a parabola with horizontal axis passes through the points (x_1, y_1), (x_2, y_2), (x_3, y_3), prove that its equation is

$$\begin{vmatrix} y^2 & x & y & 1 \\ y_1^2 & x_1 & y_1 & 1 \\ y_2^2 & x_2 & y_2 & 1 \\ y_3^2 & x_3 & y_3 & 1 \end{vmatrix} = 0.$$

37. In the equation of Ex. 36, show that the coefficient of x vanishes if any two of y_1, y_2, y_3 are equal. How do we know in advance that this must be so? What becomes of the curve under these conditions?

38. In Ex. 36, investigate the case $x_1 = x_2 = x_3$.

39. In Ex. 36, under what condition does the equation degenerate to a linear equation? Interpret the condition geometrically.

40. Derive an equation analogous to that of Ex. 36 for the parabola with vertical axis.

41. In the equation of Ex. 36, show that the coefficient of x vanishes only if two of the numbers y_1, y_2, y_3 are equal.

65. Quadratic functions. A function of the form

$$y = ax^2 + bx + c \qquad (a \neq 0)$$

is called a *quadratic function.* Such functions are of frequent occurrence in the applications of mathematics.

Since the above equation is of form (2), § 64, it follows that *the graph of every quadratic function is a parabola with vertical axis.* Of course the constant a as used here does not mean the distance from vertex to focus.

As x increases, many functions increase to a *maximum* value, decreasing thereafter, or decrease to a *minimum* and then begin to increase. In such cases, it is frequently a problem of prime importance to determine the maximum or minimum value. To solve this problem for functions in general, differential calculus is required; for the quadratic function it may evidently be solved by merely finding the vertex of the parabola, since the ordinate of the vertex is the (algebraically) greatest or least of all the ordinates, according as the parabola opens downward or upward.

Example: Find the dimensions of the largest rectangular lot that can be inclosed by 100 yd. of fencing.

Let x and y denote the sides of the lot. Then the area

$$A = xy$$

is to be given its maximum value. The first step is to express *A in terms of a single variable* — either x or y. The total perimeter is 100 yd.: i.e.,

$$2x + 2y = 100, \qquad x + y = 50,$$

whence

$$A = x(50 - x) = 50x - x^2.$$

Standardize the equation:

$$x^2 - 50x + 625 = 625 - A,$$
$$(x - 25)^2 = -(A - 625).$$

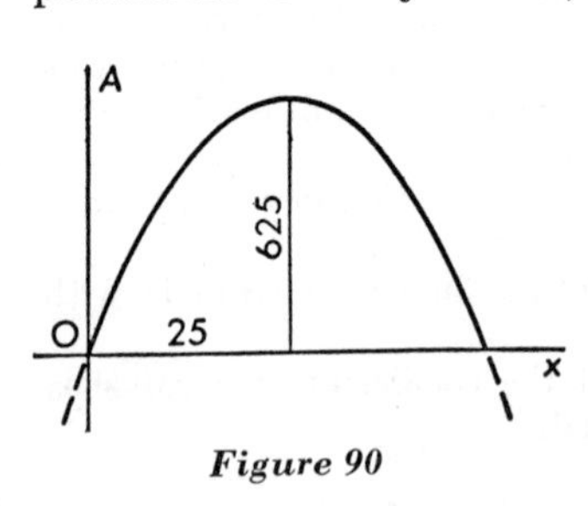

Figure 90

Thus $A_{\text{max.}} = 625$ sq. yd., occurring when $x = y = 25$ yd.

Exercises

In each exercise, draw the pertinent graph indicating that portion which has meaning.

1. Express the total surface area, including both ends, of a circular cylinder of altitude 4, as a function of the radius.

2. Express the total surface area of a circular cone of slant height 6, as a function of the radius.

3. What number exceeds its square by the maximum amount? *Ans.* $\frac{1}{2}$.

4. Two numbers differ by 4. What are their values when their product is (*a*) algebraically least? (*b*) Numerically least? *Ans.* (*a*) 2, −2.

5. What is the shape of a rectangle of given perimeter when its area is a maximum? *Ans.* A square.

6. Express $\cos 2\theta$ as a function of $\cos \theta$.

7. Find the radius of the circular cross-section cut from a sphere of radius r by a plane passing at a distance h from the center. Taking $r = 1$, express the area of the section as a function of h.

8. When the load is uniformly distributed horizontally, a suspension-bridge cable hangs in a parabolic arc. If the bridge is 200 ft. long, the towers 40 ft. high, and the cable 15 ft. above the floor of the bridge at the midpoint, find the equation of the parabola with the midpoint of the bridge as origin; also the height 50 ft. from the middle.

9. A circular cylinder is inscribed in a circular cone of radius 4 in., altitude 12 in. Find the radius of the cylinder if its convex surface area is a maximum. *Ans.* 2 in.

10. Solve Ex. 9 if the total surface area of the cylinder is a maximum.

11. A rectangular lot is to be fenced off along the bank of a river. If no fence is needed along the river, find the dimensions of the largest lot that can be inclosed with 100 yd. of fencing. *Ans.* 25×50 yd.

12. A rectangular field is to be inclosed, and divided into three lots by parallels to one of the sides. Find the dimensions of the largest field that can be inclosed with 1000 yd. of fencing. *Ans.* 250×125 yd.

13. A triangular lot has 60 ft. frontage on one street, 80 ft. on another street at right angles to the first. Find the dimensions of the largest rectangular building that can be erected facing one of the streets.

Ans. 30×40 ft.

14. A rectangular lot is to be fenced off along a highway. If the fence on the highway costs $1.50 per yd., on the other sides $1 per yd., find the dimensions of the largest lot that can be inclosed for $100.

Ans. 20×25 yd.

15. Solve Ex. 14 if there is a 30 ft. opening in front. *Ans.* $23 \times 28\frac{3}{4}$ yd.

16. Graph the difference in area of a square of side x and a square of side 1.

17. Graph the difference in volume V between two boxes of dimensions c, c, 1 and c, 6, 1. What values of c make V a minimum?

18. In Ex. 17, find c if $V = 8$. *Ans.* 2, 4, 7.12.

19. When a body moves in a vertical line under gravity, the distance from the starting point after time t, with the positive direction *upward*, is

$$s = v_0 t - \tfrac{1}{2}gt^2,$$

where v_0 is the initial velocity. Graph s for $v_0 > 0$, $v_0 = 0$, $v_0 < 0$.

20. In Ex. 19, the velocity at any time t is

$$v = v_0 - gt.$$

Draw the graph of v for the three cases, $v_0 > 0$, $v_0 = 0$, $v_0 < 0$.

21. Given the formulas of Exs. 19–20, obtain the relation between v and s, and graph v as a function of s. *Ans.* $v^2 = v_0^2 - 2gs$.

22. For the quadratic function $y = Ax^2 + Bx + C$, with $A > 0$, show that y takes on its minimum value when $x = -\dfrac{B}{2A}$.

23. For the quadratic function $y = Ax^2 + Bx + C$, with $A < 0$, show that y takes on its maximum value when $x = -\dfrac{B}{2A}$.

CHAPTER 8 *The Central Conics*

66. Ellipse referred to its axes. By § 58, the ellipse is *the conic section for which* $e < 1$. Let F be the focus and DD' the directrix. The line FM through the focus perpendicular to the directrix intersects the curve in two points, say V_1 and V_2. (Two points, because the segment FM can be divided both internally and externally in the ratio e.) These points are the *vertices*, and C, the midpoint of V_1V_2, is the *center* (Fig. 91).

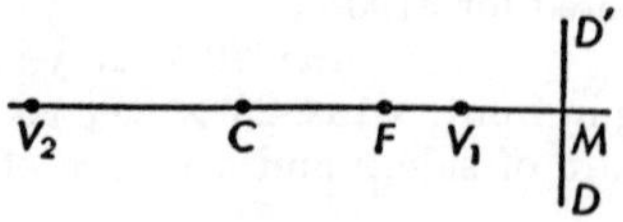

Figure 91

Let us set

$$CV_1 = a$$
$$CF = c$$
$$CM = d.$$

Then, applying (1), § 58, to the points V_1 and V_2, we have

$$a - c = e(d - a), \qquad a + c = e(d + a).$$

By subtraction and addition we find

$$2c = 2CF = 2ae, \qquad c = ae,$$

$$2ed = 2e \cdot CM = 2a, \qquad d = \frac{a}{e}.$$

Let us now (Fig. 92) choose our origin at the center C, so that the focus F becomes $(ae, 0)$ and the equation of the directrix DD' is

$$x = \frac{a}{e}. \tag{1}$$

Then, if $P{:}\,(x, y)$ is a point on the curve, we have

$$\sqrt{(x - ae)^2 + y^2} = e\left(\frac{a}{e} - x\right) = a - ex,$$

$$x^2 - 2aex + a^2e^2 + y^2 = a^2 - 2aex + e^2x^2,$$

$$x^2(1 - e^2) + y^2 = a^2(1 - e^2),$$

$$\frac{x^2}{a^2} + \frac{y^2}{a^2(1 - e^2)} = 1. \tag{2}$$

For simplicity, let us put

$$\mathbf{b^2 = a^2(1 - e^2),} \tag{3}$$

where b is real because $e < 1$, $a^2(1 - e^2) > 0$. This reduces (2) to the *standard form*

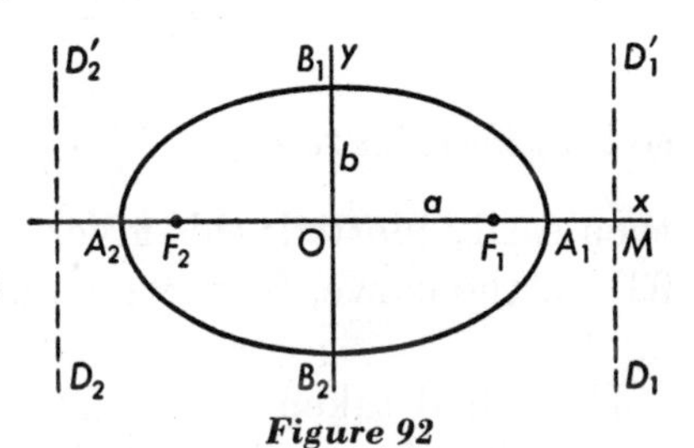

Figure 92

$$\mathbf{\frac{x^2}{a^2} + \frac{y^2}{b^2} = 1} \qquad (a > b). \tag{4}$$

From (4) we see that:

(*a*) The curve is symmetric with respect to Ox and Oy. From the latter statement it follows that there is *a second focus, the point* $(-\boldsymbol{ae}, \mathbf{0})$, and *a second directrix, the line* $\boldsymbol{x = -\frac{a}{e}}$.

(*b*) The intercepts on the axes are $(\pm \boldsymbol{a}, \mathbf{0})$, $(\mathbf{0}, \pm \boldsymbol{b})$.

(*c*) The equation when solved for y has the form

$$y = \pm \frac{b}{a}\sqrt{a^2 - x^2}: \tag{5}$$

thus y is imaginary when x is numerically greater than a. Similarly, x is imaginary when y is numerically greater than b.

It is convenient occasionally to speak of the two lines of symmetry as the *axes* of the curve, but usually we consider the axes to be the *segments* of these lines included within the curve: the segment A_2A_1 is the *major axis*, the segment B_2B_1 the *minor axis*. It follows from (3) that, for the ellipse, *a is always greater than b*—hence the terms "major" and "minor."

From equation (4), the ends of the axes may be plotted immediately. By (3),

$$ae = \sqrt{a^2 - b^2}:$$

i.e., *the distance from center to foci is* $\sqrt{a^2 - b^2}$. By (5), when

$$x = ae, \qquad y = \pm \frac{b}{a}\sqrt{a^2 - a^2e^2} = \pm \frac{b^2}{a},$$

so that *the latus rectum is* $\frac{2b^2}{a}$. The ends of the latera recta are then easily plotted; this gives a total of eight points (see Fig. 93) on the curve, from which a fairly accurate sketch can be made.

If we had taken the focus on the y-axis and the directrix parallel to the x-axis, the above discussion would have been the same except that x and y would have been interchanged. Hence, the equation

$$(6) \qquad \frac{y^2}{a^2} + \frac{x^2}{b^2} = 1 \qquad (a > b)$$

represents *the ellipse with center at the origin and major axis on Oy.*

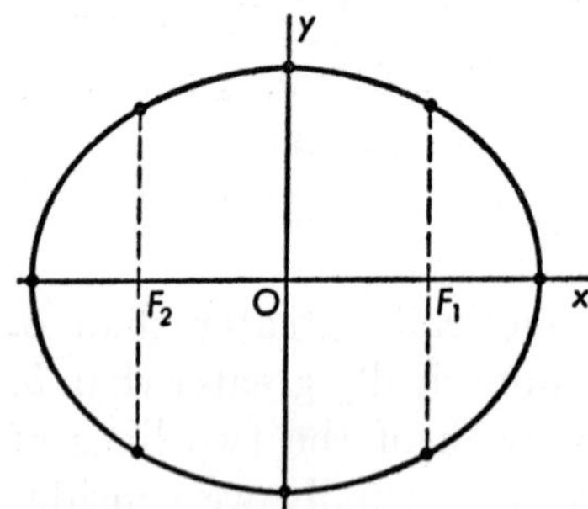

Figure 93

Example: Trace the ellipse

$$4x^2 + 6y^2 = 3.$$

Divide by 3, to reduce the right member to unity:

$$\frac{4x^2}{3} + 2y^2 = 1.$$

Divide numerator and denominator in the first term by 4, in the second by 2, to reduce the numerator-coefficients to unity:

$$(7) \qquad \frac{x^2}{\frac{3}{4}} + \frac{y^2}{\frac{1}{2}} = 1.$$

Since the denominator of x^2 is the larger, we have equation (4) rather than (6): an ellipse with its major axis on Ox. From (7) we read off $a = \frac{1}{2}\sqrt{3} = 0.87$, $b = \frac{1}{2}\sqrt{2} = 0.71$; the foci are at distance $\sqrt{a^2 - b^2} = \frac{1}{2}$ to right and left of the center; the ends of the latera recta are at distance

$$\frac{b^2}{a} = \frac{\frac{1}{2}}{\frac{1}{2}\sqrt{3}} = \frac{1}{3}\sqrt{3} = 0.57$$

above and below the foci. The eccentricity is

$$e = \frac{\sqrt{a^2 - b^2}}{a} = \frac{\frac{1}{2}}{\frac{1}{2}\sqrt{3}} = \frac{1}{3}\sqrt{3};$$

the directrices are the lines $x = \pm\frac{a}{e} = \pm\frac{\frac{1}{2}\sqrt{3}}{\frac{1}{3}\sqrt{3}} = \pm\frac{3}{2}$.

67. Another definition of the ellipse. The following property of the ellipse is often used as a *definition:*

An ellipse is the locus of a point which moves so that the sum of its distances from two fixed points is constant. The fixed points are the *foci;* the constant sum is the *length of the major axis.*

To prove that this curve is the same as the one defined in § 58, we will show first that the curve of § 58 possesses the property just stated. Let the foci be $F_1: (ae, 0)$, $F_2: (-ae, 0)$, the directrices $x = \pm\frac{a}{e}$, and $P: (x, y)$ any point of an ellipse.

By (1), § 58,

$$F_1P = e\left(\frac{a}{e} - x\right) = a - ex,$$

$$F_2P = e\left(\frac{a}{e} + x\right) = a + ex,$$

so that

$$(1) \qquad F_1P + F_2P = 2a.$$

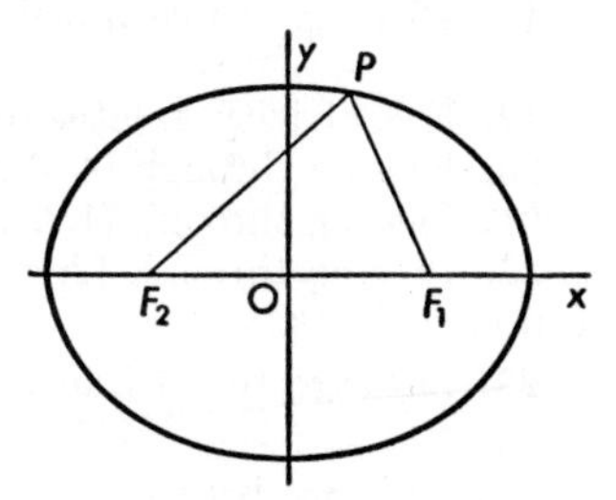

Figure 94

To complete the proof of equivalence of the two definitions, it must be shown that, when a point moves so that the sum of its distances from two fixed points is constant, its locus is an ellipse as defined in § 58. This part of the proof may be supplied by the student. (Ex. 32 following.)

Exercises

In Exs. 1–10, find the center and foci, plot the vertices and the ends of the latera recta, and draw the curve. Find the eccentricity and the equations of the directrices.

1. $\frac{x^2}{25} + \frac{y^2}{16} = 1.$ **2.** $\frac{x^2}{25} + \frac{y^2}{9} = 1.$

3. $\frac{x^2}{144} + \frac{y^2}{169} = 1.$ **4.** $\frac{x^2}{36} + \frac{y^2}{100} = 1.$

5. $2x^2 + y^2 = 8.$ **6.** $3x^2 + 4y^2 = 12.$

7. $3x^2 + 5y^2 = 15.$ **8.** $2x^2 + 3y^2 = 6.$

9. $144x^2 + 225y^2 = 400.$ **10.** $4x^2 + y^2 = 1.$

In Exs. 11–24, find the equation of the ellipse, assuming form (4), § 66.

11. Eccentricity $\frac{1}{3}$, distance between foci 2. *Ans.* $8x^2 + 9y^2 = 72.$

12. Eccentricity $\frac{1}{2}$, distance between directrices 24. *Ans.* $3x^2 + 4y^2 = 108.$

13. Major axis 8, distance between foci 6. *Ans.* $7x^2 + 16y^2 = 112.$

14. Minor axis 10, distance between foci 24. *Ans.* $\frac{x^2}{169} + \frac{y^2}{25} = 1.$

15. Distance between foci 2, between directrices 8. *Ans.* $3x^2 + 4y^2 = 12.$

16. Distance between foci 4, between directrices 36. *Ans.* $8x^2 + 9y^2 = 288.$

17. Latus rectum 4, distance between foci $4\sqrt{2}$. *Ans.* $x^2 + 2y^2 = 16.$

18. Distance between foci $8\sqrt{6}$, rectangle on the axes of area 80. *Ans.* $x^2 + 25y^2 = 100.$

19. Eccentricity $\frac{2}{3}$, latus rectum $\frac{2}{3}$. *Ans.* $25x^2 + 45y^2 = 9.$

20. Passing through (4, 3), (6, 2). *Ans.* $x^2 + 4y^2 = 52.$

21. Passing through (1, 2), (3, 1). *Ans.* $3x^2 + 8y^2 = 35.$

22. Passing through (2, 3), latus rectum three times the distance from center to focus. *Ans.* $3x^2 + 4y^2 = 48.$

23. Latus rectum $\frac{60}{19}$, distance between directrices $2\sqrt{19}$. *Ans.* $15x^2 + 19y^2 = 60$; $10x^2 + 19y^2 = 90.$

24. Distance between foci $\frac{4}{3}\sqrt{33}$, passing through (2, 1). *Ans.* $x^2 + 12y^2 = 16.$

25. Show that as e approaches zero with a kept fixed, the ellipse approaches a circle. Let $e \to 0$ in (2), § 66. As this happens, how do the foci and directrices move?

26. The orbits of the planets in our solar system are ellipses with the sun at one focus. The eccentricities are tabulated below. For each planet, as a measure of the nearly circular character of its orbit, determine to the nearest hundredth the ratio of the length of the minor axis to the length of the major axis.

Mercury:	$e = 0.206$.	*Ans.* 0.98.
Venus:	$e = 0.007$.	*Ans.* 1.00.
Earth:	$e = 0.017$.	*Ans.* 1.00.
Mars:	$e = 0.093$.	*Ans.* 1.00.
Jupiter:	$e = 0.048$.	*Ans.* 1.00.
Saturn:	$e = 0.056$.	*Ans.* 1.00.
Uranus:	$e = 0.047$.	*Ans.* 1.00.
Neptune:	$e = 0.008$.	*Ans.* 1.00.
Pluto:	$e = 0.249$.	*Ans.* 0.97.

27. In the asteroid belt between the orbits of Mars and Jupiter some asteroids are known to have orbits with eccentricities as high as $\frac{2}{3}$. For such an elliptic orbit, find the ratio of minor to major axis. *Ans.* 0.75.

28. Show that the ordinates of the ellipse $\dfrac{x^2}{a^2} + \dfrac{y^2}{b^2} = 1$ are $\dfrac{b}{a}$ times those of the circle $x^2 + y^2 = a^2$.

29. The orbit of the earth is an ellipse with the sun at a focus; the semi-major axis is 93 million miles, the eccentricity $\frac{1}{60}$, nearly. Find the greatest and least distances of the earth from the sun.

30. In Fig. 91, show that $CF:CV_1 = CV_1:CM$.

31. A line segment of fixed length moves with its ends following two perpendicular lines. The line is divided by a point P into two segments of lengths a, b. Find the locus of P. *Ans.* An ellipse.

32. A point moves so that the sum of its distances from $(ae, 0)$, $(-ae, 0)$ is $2a$. Find the equation of its locus.

33. A point moves so that the sum of its distances from $(4, 0)$, $(-4, 0)$ is 9. Find the equation of its locus. (§ 67.)

34. A circle touches the circle $(x + 1)^2 + y^2 = 9$, and passes through $(1, 0)$. Find the locus of its center. *Ans.* $20x^2 + 36y^2 = 45$.

35. A moving circle is tangent to each of the fixed circles $(x - c)^2 + y^2 = c^2$, $(x + c)^2 + y^2 = 16c^2$. Find the locus of its center.

Ans. $84x^2 + 100y^2 = 525c^2$; $20x^2 + 36y^2 = 45c^2$.

Solve Exs. 36–41 synthetically (by ruler and compass).

36. Given an ellipse with its center, construct the axes.

37. Given an ellipse with its axes, construct the foci.

38. Given an ellipse with its axes, construct the directrices.

39. Given the foci and major axis of an ellipse, construct the directrices. (Ex. 30.)

40. Given the foci and major axis of an ellipse, show how to construct points of the curve.

41. Given the foci and one point of an ellipse, construct the axes.

68. Other standard forms. By a method analogous to that of § 62, we may establish the following:

The equation of an ellipse with center at (h, k) *is:*

if the major axis is parallel to Ox,

$$(1) \qquad \frac{(x-h)^2}{a^2} + \frac{(y-k)^2}{b^2} = 1 \qquad (a > b);$$

if the major axis is parallel to Oy,

$$(2) \qquad \frac{(y-k)^2}{a^2} + \frac{(x-h)^2}{b^2} = 1 \qquad (a > b).$$

69. Reduction to standard form. The equation

$$(1) \qquad Ax^2 + Cy^2 + Dx + Ey + F = 0,$$

where A and C have *the same sign*, can in general be reduced by completing the squares in x and y, to one of the standard forms of § 68. (See the example below.) There are two exceptional cases. When the left member is written as the sum of two squares, the right member may be zero, or it may be negative. In the former case the locus is evidently the single point (h, k) — the so-called "point-ellipse"; in the latter case there is no locus.

THEOREM: *An equation of the second degree in which the xy-term is missing and the coefficients of* x^2 *and* y^2 *have the same sign represents an ellipse with axes parallel to the coordinate axes (exceptionally, a single point, or no locus).*

Example: Trace the curve

$$9x^2 + 4y^2 - 36x + 8y + 31 = 0.$$

Transpose the constant and complete the squares:

$$9(x^2 - 4x + 4) + 4(y^2 + 2y + 1) = -31 + 36 + 4,$$
$$9(x - 2)^2 + 4(y + 1)^2 = 9,$$
$$\frac{(x-2)^2}{1} + \frac{(y+1)^2}{\frac{9}{4}} = 1.$$

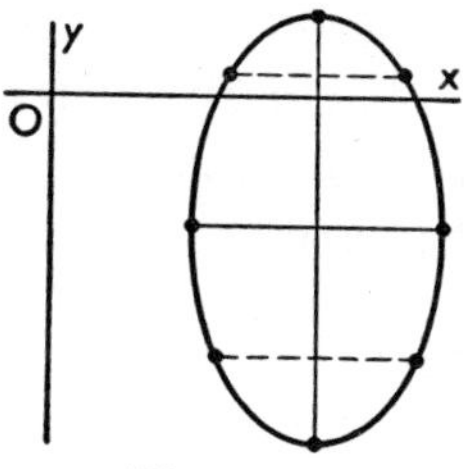

Figure 95

The center is at $(2, -1)$, major axis parallel to Oy, semi-axes $a = \frac{3}{2}$, $b = 1$, distance from center to foci $\sqrt{a^2 - b^2} = \sqrt{\frac{5}{4}} = \frac{1}{2}\sqrt{5}$, latus rectum $= \frac{4}{3}$.

Exercises

In Exs. 1–18, find the center and foci, plot the ends of the axes and of the latera recta, and draw the curve.

1. $\frac{(x-2)^2}{25} + \frac{(y-1)^2}{9} = 1.$ **2.** $\frac{(x-4)^2}{4} + \frac{(y+2)^2}{9} = 1.$

3. $\frac{(x+1)^2}{1} + \frac{(y+3)^2}{9} = 1.$ **4.** $\frac{(x+5)^2}{1} + \frac{(y-4)^2}{\frac{1}{4}} = 1.$

5. $\frac{(x+2a)^2}{a^2} + \frac{y^2}{2a^2} = 1.$ **6.** $\frac{(x+a)^2}{4a^2} + \frac{(y-2a)^2}{a^2} = 1.$

7. $x^2 + 2y^2 - 6x + 5 = 0.$ **8.** $x^2 + 9y^2 + 18y = 0.$

9. $x^2 + 9y^2 + 3x - 18y + 9 = 0.$ **10.** $x^2 + 2y^2 + 4x - 12y + 16 = 0.$

11. $2x^2 + y^2 - 4x - 2y = 1.$

12. $3x^2 + 2y^2 - 36x - 16y + 134 = 0.$

13. $3x^2 + y^2 = 6cx.$

14. $x^2 + 4y^2 = 4cy.$

15. $16x^2 + 12y^2 - 12y - 9 = 0.$

16. $7x^2 + 2y^2 + 14x - 8y + 1 = 0.$

17. $3x^2 + 4y^2 - 18x + 8y + 19 = 0.$

18. $9x^2 + 10y^2 - 90x - 40y + 175 = 0.$

In Exs. 19–24, find the points of intersection and trace the curves.

19. $x^2 = 3y$, $3x^2 + y^2 - 24y + 36 = 0.$ *Ans.* $(\pm 3, 3)$, $(\pm 6, 12)$.

20. $y^2 = 4x$, $4x^2 + 9y^2 = 8x + 72y - 112.$ *Ans.* $(4, 4)$, $(1, 2)$.

21. $x^2 = y + 10$, $y^2 + 5(x - 1)^2 = 81.$
Ans. $(-3, -1)$, $(-2, -6)$, $(1, -9)$, $(4, 6)$.

22. $4x^2 + y^2 = 100$, $4(x + 3)^2 + 7y^2 = 448.$ *Ans.* $(-3, \pm 8)$, $(4, \pm 6)$.

23. $x^2 + y^2 + 4ax = 0$, $2x^2 + y^2 = 4ax.$ *Ans.* $(0, 0)$ twice.

24. $x^2 + y^2 - 4ax = 0$, $2x^2 + y^2 = 4ax.$ *Ans.* $(0, 0)$ four times.

25. A circle is tangent to the circles $x^2 + y^2 = 4$, $x^2 + y^2 = 12y + 64$. Find the locus of its center and draw the figure.
Ans. $4x^2 + 3y^2 - 18y - 81 = 0$; $16x^2 + 7y^2 - 42y - 49 = 0$.

70. Hyperbola referred to its axes. By definition, the hyperbola is *the conic section for which* $e > 1$.

The derivation of equation (2), § 66, does not depend upon the fact that $e < 1$; we therefore conclude at once that, with the axes chosen as in § 66, the equation of the hyperbola is

$$\frac{x^2}{a^2} + \frac{y^2}{a^2(1 - e^2)} = 1. \tag{1}$$

But, since now $e > 1$, it follows that $\frac{a}{e} < a < ae$; hence the directrices pass between the vertices, and the foci lie in the axis produced each way. From the fact that $e > 1$ it also follows that the quantity $a^2(1 - e^2)$ is negative; therefore to make b real, we set

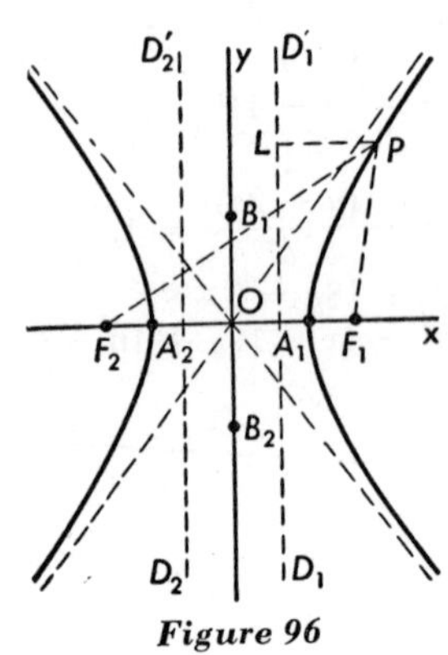

Figure 96

$$b^2 = a^2(e^2 - 1), \tag{2}$$

and write (1) in the form

$$\frac{x^2}{a^2} - \frac{y^2}{b^2} = 1. \tag{3}$$

This equation shows that:

(*a*) The curve is symmetric with respect to both axes. Hence there are two foci, the points $(\pm ae, 0)$, and two directrices, the lines $x = \pm \frac{a}{e}$.

(*b*) The intercepts on Ox are $(\pm a, 0)$; on Oy, imaginary.

(*c*) The equation when solved for y has the form

$$y = \pm \frac{b}{a} \sqrt{x^2 - a^2}: \tag{4}$$

thus y is imaginary if and only if $x^2 < a^2$: i.e., if $-a < x < a$. The curve consists of two disconnected branches, one to the right of the line $x = a$, the other to the left of the line $x = -a$.

(*d*) *The latus rectum is* $\frac{2b^2}{a}$.

As in the case of the ellipse, the lines of symmetry are sometimes spoken of as the *axes* of the curve, but unless the contrary is indicated the axes will be considered to be the *segments* A_2A_1 of length $2a$ and B_2B_1 of length $2b$: the former is the *transverse axis*, the latter the *conjugate axis*. The conjugate axis does not intersect the curve, but plays an important part in its theory. The ends A_2, A_1 of the transverse axis are the *vertices;* the point of intersection of the axes is the *center*.

It appears from (2) that *the distance from center to foci* is

$$ae = \sqrt{a^2 + b^2}. \tag{5}$$

It should also be noted that in the case of the hyperbola b may be *greater than*, *equal to*, or *less than* a, according to the value of e.

With the foci on the y-axis and directrices parallel to the x-axis, the analysis is the same except that x and y are interchanged; hence, the equation

$$\frac{y^2}{a^2} - \frac{x^2}{b^2} = 1$$

represents a *hyperbola with transverse axis along Oy.*

It is customary to call the ellipse and hyperbola the *central conics*, since each has a center of symmetry.

71. Asymptotes. Given the standard equation of the hyperbola

$$\frac{x^2}{a^2} - \frac{y^2}{b^2} = 1, \tag{1}$$

let us clear of fractions:

$$b^2x^2 - a^2y^2 = a^2b^2. \tag{2}$$

When this curve and the straight line

$$bx - ay = 0 \tag{3}$$

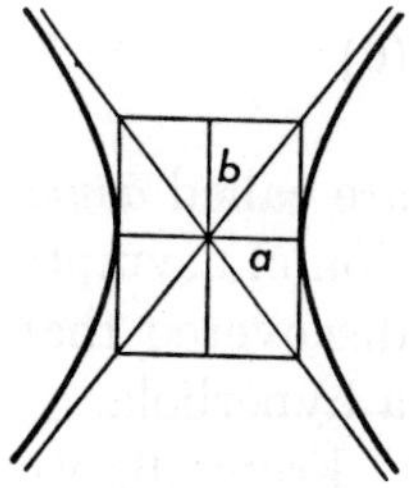

are drawn on the same axes, the figure indicates that the

b
a

Figure 97

hyperbola approaches the straight line more and more closely as the distance from the center increases, but without ever reaching the line. To prove this, let $P:(x_1, y_1)$ be a point on the hyperbola in the first or third quadrant. The distance of P from the line (3) is

$$d = \frac{bx_1 - ay_1}{\sqrt{a^2 + b^2}}. \tag{4}$$

Since P is on the curve, its coordinates satisfy (2):

$$b^2x_1^2 - a^2y_1^2 = a^2b^2,$$

whence

$$bx_1 - ay_1 = \frac{a^2b^2}{bx_1 + ay_1}.$$

Substituting this value of $bx_1 - ay_1$ in (4), we find

$$d = \frac{a^2b^2}{\sqrt{a^2 + b^2}} \cdot \frac{1}{bx_1 + ay_1}.$$

Evidently d can never equal zero, but as P recedes, so that x_1 and y_1 both increase indefinitely, d becomes smaller and smaller, approaching the limit zero.

A similar result is easily established for the line

$$bx + ay = 0 \tag{5}$$

when P lies in the second or fourth quadrant.

The lines (3) and (5), which together may be written as

$$y = \pm \frac{b}{a} x, \tag{6}$$

are called *asymptotes* of the hyperbola (1). A general definition of asymptote will be given in § 94. In practice it is soon discovered that the asymptotes are of great aid in sketching a hyperbola.

From (6) we derive the following theorem, which gives a convenient method for drawing the asymptotes of any hyperbola whose axes are given:

The asymptotes of a hyperbola are the diagonal lines of the rectangle whose center is the center of the curve and whose sides are parallel and equal to the axes of the curve.

To trace a hyperbola whose equation is in a standard form, we plot the vertices and the ends of the latera recta, and draw the asymptotes.

Example: Trace the curve

$$4y^2 = x^2 - 1.$$

First, reduce to standard form:

$$\frac{x^2}{1} - \frac{y^2}{\frac{1}{4}} = 1.$$

Since x^2 is in the *positive term*, this is form (3), § 70: the transverse axis is along the x-axis. We read off $a = 1, b = \frac{1}{2}$; the foci are at distance $\sqrt{a^2 + b^2} = \frac{1}{2}\sqrt{5} = 1.12$ to right and left of the center; the ends of the latera recta are at distance $\dfrac{b^2}{a} = \dfrac{1}{4}$ above and below the foci; the asymptotes are the diagonals of the rectangle of sides $2a$, $2b$. The eccentricity is $e = \dfrac{\sqrt{a^2 + b^2}}{a} = \dfrac{1}{2}\sqrt{5}$; the directrices are the lines

$$x = \pm\frac{a}{e} = \pm\frac{2}{5}\sqrt{5}.$$

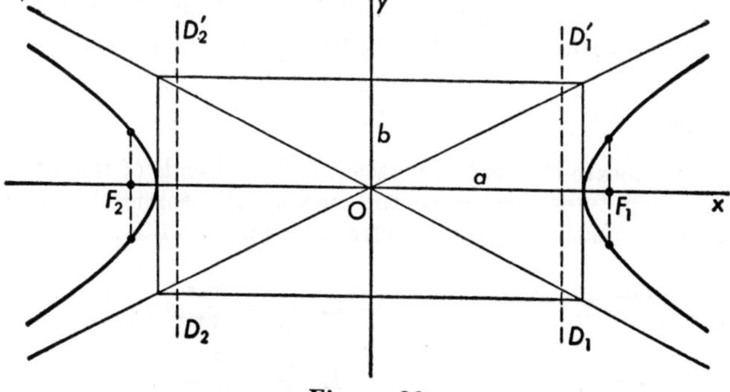

Figure 98

72. Equilateral, or rectangular, hyperbola. The hyperbola for which a and b are equal is called, on account of the equality of the semi-axes, the *equilateral* hyperbola. Since the rectangle of Fig. 97 is in this case a square, the asymptotes of the equilateral hyperbola are at right angles: for this reason it is also called the *rectangular* hyperbola. Since $ae = \sqrt{a^2 + b^2}$, the eccentricity of the equilateral hyperbola is $e = \sqrt{2}$.

When the hyperbola is equilateral, equation (3) of § 70 evidently assumes the form

$$x^2 - y^2 = a^2.$$

73. Another definition of the hyperbola. A property of the hyperbola frequently used as a *definition* of the curve is as follows (compare with § 67):

A hyperbola is the locus of a point which moves so that the difference of its distances from two fixed points is constant. The fixed points are the *foci* and the constant difference is *the length of the transverse axis.*

That is, in Fig. 96,

$$F_2P - F_1P = 2a, \tag{1}$$

or, for a point P' on the left-hand branch,

$$F_1P' - F_2P' = 2a.$$

The proof that this definition is equivalent to our former one is very similar to that of § 67. (Exs. 23–24 below.)

Exercises

In Exs. 1–10, locate the center, vertices, foci, and ends of the latera recta, draw the asymptotes, and trace the curve. Determine the eccentricity and write the equations of the directrices and asymptotes.

1. $\dfrac{x^2}{16} - \dfrac{y^2}{9} = 1.$ **2.** $\dfrac{x^2}{25} - \dfrac{y^2}{144} = 1.$

3. $\dfrac{y^2}{9} - \dfrac{x^2}{1} = 1.$ **4.** $\dfrac{y^2}{4} - \dfrac{x^2}{2} = 1.$

5. $2y^2 - 9x^2 = 18.$

6. $3x^2 - 4y^2 = 12.$

7. $5x^2 = 4y^2 + 20.$

8. $x^2 = 3y^2 - 27.$

9. $y^2 = x^2 - a^2.$

10. $y^2 = x^2 + a^2.$

In Exs. 11–22, find the equation of the hyperbola, assuming form (3), § 70. Draw the curve.

11. Distance between foci 18, between directrices 2. *Ans.* $8x^2 - y^2 = 72.$

12. Eccentricity 2, distance between foci $4\sqrt{2}$. *Ans.* $3x^2 - y^2 = 6.$

13. Latus rectum 1, slope of asymptotes $\pm\frac{1}{2}$. *Ans.* $x^2 - 4y^2 = 4.$

14. Latus rectum 6, distance between foci twice the distance between directrices. *Ans.* $x^2 - y^2 = 9.$

15. Eccentricity 3, latus rectum $\frac{8}{3}$. *Ans.* $72x^2 - 9y^2 = 2.$

16. Latus rectum $\frac{14}{3}$, distance between directrices $\frac{9}{2}$. *Ans.* $7x^2 - 9y^2 = 63.$

17. Distance between directrices unity, passing through (2, 3). *Ans.* $3x^2 - y^2 = 3$; $12x^2 - y^2 = 39.$

18. Latus rectum 2, passing through (6, 3). *Ans.* $x^2 - 3y^2 = 9.$

19. Latus rectum 18, distance between foci 12. *Ans.* $3x^2 - y^2 = 27.$

20. Eccentricity 2, distance between directrices $\sqrt{2}$. *Ans.* $3x^2 - y^2 = 6.$

21. Passing through (2, 1), (4, 3). *Ans.* $2x^2 - 3y^2 = 5.$

22. Foci $(\pm 4, 0)$, slope of asymptotes ± 3. *Ans.* $45x^2 - 5y^2 = 72.$

23. Prove equation (1), § 73. (Compare with § 67.)

24. A point moves so that the difference of its distances from $(ae, 0)$, $(-ae, 0)$ is $2a$. Find the equation of its locus.

25. A circle passes through a given point and touches a given circle. Find the locus of its center, for all possible cases. (§§ 67, 73.)

26. The sound of a gun and the ring of the ball on the target are heard simultaneously at P. Show that the locus of P is one branch of a hyperbola.

27. A point moves so that the difference of its distances from (4, 0), (−4, 0) is 6. Find the equation of its locus.

28. A point moves so that the difference of its distances from (0, 3), (0, −3) is 2. Find the equation of its locus.

29. A circle touches the circle $x^2 + y^2 + 2cx = 0$ and passes through $(c, 0)$. Find the locus of its center. *Ans.* $12x^2 - 4y^2 = 3c^2.$

30. Prove analytically that a line parallel to an asymptote of a hyperbola intersects the curve in one and only one point.

31. Show that the product of the distances of any point of a hyperbola from its asymptotes is constant.

32. Show that the following definition is equivalent to those already laid down: A *hyperbola* is the locus of a point moving so that the product of its distances from two intersecting lines is constant. (Take the lines $y = \pm mx$.) Draw the curve.

33. If the angle between the transverse axis and an asymptote of a hyperbola is denoted by α, show that $\sec \alpha = e$.

34. A moving circle is tangent externally to each of the two fixed circles $(x - 2c)^2 + y^2 = c^2$, $(x + 2c)^2 + y^2 = 4c^2$. Find the locus of its center.
Ans. $60x^2 - 4y^2 = 15c^2$.

Solve Exs. 35–39 by ruler and compass.

35. Given a hyperbola with its center, draw the transverse axis.
36. Given a hyperbola with its axis, find the foci and directrices.
37. Given the foci and transverse axis of a hyperbola, show how to construct points of the curve.
38. Given the foci and one point of a hyperbola, construct the axes.
39. Given the foci and asymptotes of a hyperbola, find the vertices.

74. Other standard forms. Formulas for the hyperbola analogous to those of § 68 for the ellipse are as follows:

The equation of a hyperbola with center at (h, k) is:

if the transverse axis is parallel to Ox,

$$(1) \qquad \frac{(x-h)^2}{a^2} - \frac{(y-k)^2}{b^2} = 1;$$

if the transverse axis is parallel to Oy,

$$(2) \qquad \frac{(y-k)^2}{a^2} - \frac{(x-h)^2}{b^2} = 1.$$

The proof is left to the student.

75. Reduction to standard form. The equation

$$(1) \qquad Ax^2 + Cy^2 + Dx + Ey + F = 0,$$

where A and C have *opposite* signs, can be reduced, by completing the squares in x and y, to one of the forms (1), (2) of § 74. The only exceptional case is the one in which, when the left member has been expressed as the difference of two squares, the right member reduces to zero:

$$\frac{(x-h)^2}{a^2} - \frac{(y-k)^2}{b^2} = 0.$$

This equation can be factored, and therefore represents two straight lines, intersecting at (h, k).

THEOREM: *An equation of the second degree in which the xy-term is missing and the coefficients of x^2 and y^2 have unlike signs represents a hyperbola with its axes parallel to the coordinate axes (exceptionally, two intersecting lines).*

Example: Trace the curve

$$x^2 - 2y^2 + 4x + 4y + 4 = 0.$$

Completing squares, we get

$$(x + 2)^2 - 2(y - 1)^2 = -2,$$

or, dividing by -2,

$$\frac{(y-1)^2}{1} - \frac{(x+2)^2}{2} = 1.$$

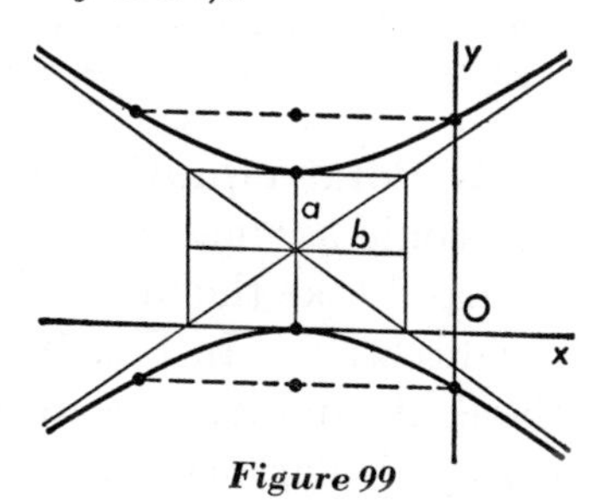

Figure 99

This is a hyperbola with center at $(-2, 1)$ and transverse axis parallel to Oy; the semi-axes are $a = 1, b = \sqrt{2}$. The vertices are at the distance 1, the foci at the distance $\sqrt{a^2 + b^2} = \sqrt{3}$, above and below the center. The ends of the latera recta are at distance 2 to right and left of the foci.

Exercises

In Exs. 1–14, locate the center, vertices, foci, and ends of the latera recta, draw the asymptotes, and trace the curve.

1. $\frac{(x-1)^2}{16} - \frac{(y-2)^2}{9} = 1.$ **2.** $\frac{(x-3)^2}{4} - \frac{(y+2)^2}{1} = 1.$

3. $\frac{(y+1)^2}{1} - \frac{(x+4)^2}{3} = 1.$ **4.** $\frac{(y+2)^2}{9} - \frac{(x-4)^2}{7} = 1.$

5. $\frac{y^2}{4} - \frac{(x+3)^2}{5} = 1.$ **6.** $\frac{x^2}{8} - \frac{(y-3)^2}{8} = 1.$

7. $3x^2 - y^2 + 12cx + 9c^2 = 0.$ **8.** $4x^2 - y^2 + 2cy + 3c^2 = 0.$

9. $5x^2 - 4y^2 = 20x + 24y + 36.$ **10.** $9x^2 - y^2 = 90x - 12y - 225.$

11. $x^2 - 3y^2 - 2x - 2 = 0.$ **12.** $x^2 - y^2 + 6x + 2y = -10.$

13. $16x^2 - 4y^2 = 12y - 16x + 1.$ **14.** $4x^2 - 3y^2 = 6(x + y).$

In Exs. 15–23, find the points of intersection and draw the curves.

15. $x^2 - 4y^2 + 2x + 8y + 1 = 0, 2y = x + 3.$ *Ans.* None.

16. $9x^2 - y^2 + 18x + 2y + 17 = 0, 3x + y + 2 = 0.$

17. $(x - 1)^2 = y + 1, 5x^2 - y^2 = 4x.$

Ans. $(0, 0), (1, -1), (4, 8), (-1, 3).$

18. $2x^2 - y^2 - 8x - y + 8 = 0, 4x^2 - 3y^2 - 16x - 3y + 18 = 0.$

Ans. $(1, 1), (3, 1), (1, -2), (3, -2).$

19. $x^2 - y^2 - 2ay = 0,\ x^2 + y^2 - 2ay = 0.$ *Ans.* (0, 0) four times.

20. $2x^2 + y^2 - 8x - 4y + 10 = 0,\ x^2 - 3y^2 + 10x + 12y - 23 = 0.$ *Ans.* (1, 2) four times.

21. $x^2 - y^2 = 4x + 4y - 1,\ x^2 - y^2 = 4x + 2y - 7.$ *Ans.* (2, −3) twice.

22. $x^2 - y^2 + 3x - y + 8 = 0,\ x^2 - y^2 + 4x - 4y + 16 = 0.$

23. $x^2 - y^2 + 5x + 3y + 2 = 0,\ x^2 - y^2 + 2x = 4.$ *Ans.* (−4, 2).

76. Polar equation of a conic. The polar equation of any conic assumes a simple form if we choose a focus as the pole and take the initial line perpendicular to the corresponding directrix. In Fig. 100, let $P:(r, \theta)$ be a random point on the conic and let Q be at one end of the latus rectum. Drop perpendiculars as indicated from P and Q to the initial line OA and the directrix AB. For convenience, let the length of the latus rectum be $2L$.

By the definition of eccentricity e (p. 94), we may write

$$PO = e \cdot PB, \qquad QO = e \cdot QM.$$

Therefore, $PB = \dfrac{r}{e}$ and $QM = \dfrac{L}{e}\cdot$ But also,

$$PB = OA - OC$$
$$= QM - OC,$$

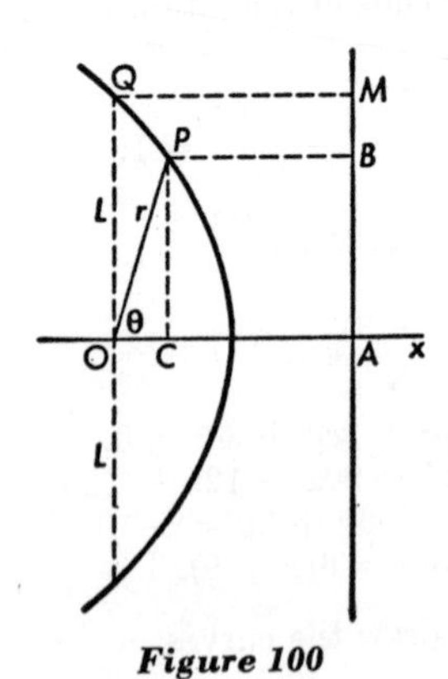

Figure 100

from which it follows that

$$\frac{r}{e} = \frac{L}{e} - r\cos\theta. \tag{1}$$

When solved for r, equation (1) takes the form

$$r = \frac{L}{1 + e\cos\theta}. \tag{2}$$

Other forms for the polar equation of a conic appear in the exercises.

Exercises

In Exs. 1–11, derive the polar equation of the conic with focus as pole, eccentricity and directrix as stipulated.

1. With $e = 2$ and directrix perpendicular to Ox through $(4, \pi)$.

Ans. $r = \dfrac{8}{1 - 2\cos\theta}$.

2. With $e = \frac{1}{2}$ and directrix perpendicular to Ox through $(6, \pi)$.

3. With $e = \frac{1}{3}$ and directrix parallel to Ox through $(2, \frac{1}{2}\pi)$.

Ans. $r = \dfrac{2}{3 + \sin\theta}$.

4. With $e = 2$ and directrix parallel to Ox through $(4, \frac{1}{2}\pi)$.

5. With $e = 6$ and directrix parallel to Ox through $(1, \frac{3}{2}\pi)$.

Ans. $r = \dfrac{6}{1 - 6\sin\theta}$.

6. With $e = 3$ and directrix parallel to Ox through $(2, \frac{3}{2}\pi)$.

7. With $e = 1$ and directrix perpendicular to Ox through $(4, 0)$.

8. With $e = \frac{1}{2}$ and directrix perpendicular to Ox through $(3, 0)$.

9. With eccentricity e and directrix perpendicular to Ox through $\left(\dfrac{L}{e}, \pi\right)$.

Ans. $r = \dfrac{L}{1 - e\cos\theta}$.

10. With eccentricity e and directrix parallel to Ox through $\left(\dfrac{L}{e}, \frac{1}{2}\pi\right)$.

Ans. $r = \dfrac{L}{1 + e\sin\theta}$.

11. With eccentricity e and directrix parallel to Ox through $\left(\dfrac{L}{e}, \frac{3}{2}\pi\right)$.

Ans. $r = \dfrac{L}{1 - e\sin\theta}$.

In Exs. 12–17, sketch the conic.

12. $r(3 + 2\sin\theta) = 6$.

13. $r(1 - \cos\theta) = 4$.

14. $r(1 - \sin\theta) = 2$.

15. $r(5 + \cos\theta) = 20$.

16. $r(2 + 3\sin\theta) = 6$.

17. $r(1 + 3\cos\theta) = 15$.

18. Find the polar equation of a central conic, with center at the pole and transverse axis along Ox, by transforming to polar coordinates equation (4) § 66.

19. Derive the polar equation of a parabola with vertex at the pole and focus at $(a, 0)$ directly, and also by transforming the equation $y^2 = 4ax$.

Ans. $r\sin\theta\tan\theta = 4a$.

20. Chords are drawn from the vertex of a parabola. Show that the locus of their midpoints is a parabola whose latus rectum is half as long as that of the original parabola. Use the answer to Ex. 19.

21. Find the locus of the midpoints of the focal radii of the conic (2) of this section.

$$\textit{Ans. } r = \frac{\frac{1}{2}L}{1 + e \cos \theta}.$$

22. Find the locus of the midpoints of chords drawn through a fixed point on a circle.

23. Given the axis, focus, and one point of a parabola, construct the directrix by ruler and compass, using the property (2) of this section.

CHAPTER 9 *The General Equation of Second Degree*

77. Rotation of axes. Before taking up the problem of this chapter, we must develop formulas for *rotation of axes.*

Let Ox, Oy and Ox_1, Oy_1 be two pairs of rectangular axes with the same origin, and denote by φ the angle through which the first pair must be rotated to come to coincidence with the second, the angle φ being considered *positive* when measured *counterclockwise.* Let P have the coordinates x, y in the first system and x_1, y_1 in the second, so that

$$OM = x, \qquad MP = y,$$
$$OM_1 = x_1, \qquad M_1P = y_1.$$

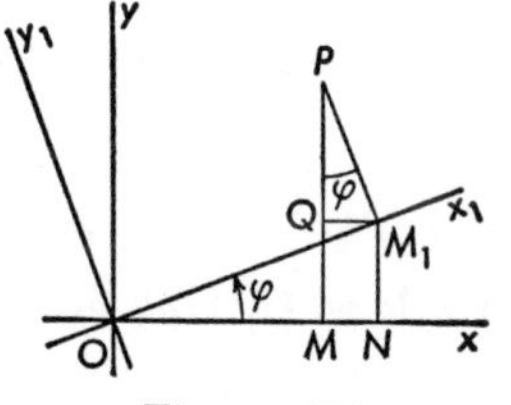

Figure 101

Now

$$OM = ON - MN = ON - QM_1;$$
$$MP = MQ + QP = NM_1 + QP.$$

We thus have the formulas

$$(1) \qquad \begin{cases} x = x_1 \cos \varphi - y_1 \sin \varphi, \\ y = x_1 \sin \varphi + y_1 \cos \varphi. \end{cases}$$

78. Removal of the product term. The most general equation of the second degree has the form

$$(1) \qquad Ax^2 + Bxy + Cy^2 + Dx + Ey + F = 0.$$

If $B = 0$, then (1) is the equation of a conic as treated in Chapters 7 and 8. If $B \neq 0$, we shall show that equation (1) can be transformed by rotation of axes into an equation of the form

$$A_1x_1^2 + C_1y_1^2 + D_1x_1 + E_1y_1 + F_1 = 0, \tag{2}$$

which has no x_1y_1 term and is therefore the equation of a conic.

Let us substitute for x and y in (1) the values given by (1), § 77:

$$\begin{aligned} &A(x_1 \cos \varphi - y_1 \sin \varphi)^2 \\ &+ B(x_1 \cos \varphi - y_1 \sin \varphi)(x_1 \sin \varphi + y_1 \cos \varphi) \\ &+ C(x_1 \sin \varphi + y_1 \cos \varphi)^2 + D(x_1 \cos \varphi - y_1 \sin \varphi) \\ &+ E(x_1 \sin \varphi + y_1 \cos \varphi) + F = 0. \end{aligned} \tag{3}$$

In equation (3), we wish to choose φ so that the term involving x_1y_1 will drop out. The x_1y_1 term in (3) is

$$\begin{aligned} [-2A \sin \varphi \cos \varphi + B \cos^2 \varphi - B \sin^2 \varphi \\ + 2C \sin \varphi \cos \varphi]x_1y_1. \end{aligned} \tag{4}$$

Now $2 \sin \varphi \cos \varphi = \sin 2\varphi$ and $\cos^2 \varphi - \sin^2 \varphi = \cos 2\varphi$, so that (4) becomes

$$[B \cos 2\varphi - (A - C) \sin 2\varphi]x_1y_1. \tag{5}$$

To make the x_1y_1 term drop out of equation (3), we must set

$$(A - C) \sin 2\varphi = B \cos 2\varphi. \tag{6}$$

If $A = C$, this equation gives*

$$\cos 2\varphi = 0, \qquad \varphi = 45°.$$

If $A \neq C$, equation (6) gives

$$\tan 2\varphi = \frac{B}{A - C}. \tag{7}$$

Now for every value of $\tan 2\varphi$, from $-\infty$ to $+\infty$, there is a value of 2φ between 0 and π, hence a value of φ between† 0

* Unless also $B = 0$: in that case (6) is true for all values of φ, which is to be expected. Why?

† There are of course infinitely many other values of φ, differing from this one by multiples of $\frac{1}{2}\pi$, but for definiteness we will agree to choose always the positive acute angle.

and $\frac{1}{2}\pi$, so that it is always possible to determine a positive acute angle φ satisfying (6). If the curve (1) be referred to axes making this angle with the original ones, the resulting equation will be free of the product term, and its locus must be a conic.

THEOREM: *Every equation of the second degree (if it has a locus) represents a conic section, whose axes are inclined to the coordinate axes at the positive acute angle φ given by formula* (7), *or, if $A = C$, $B \neq 0$, at an angle of* $45°$.

Thus to trace the locus of an equation of the second degree ($B \neq 0$), for instance as in Fig. 102, the first step is to determine $\tan 2\varphi$ by (7), and then use the formulas

$$(8)\qquad \begin{cases} \cos\varphi = \sqrt{\dfrac{1+\cos 2\varphi}{2}}, \\ \sin\varphi = \sqrt{\dfrac{1-\cos 2\varphi}{2}}. \end{cases}$$

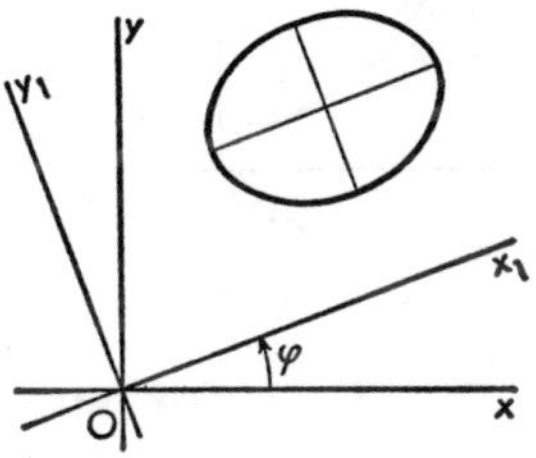

Figure 102

The values of $\cos\varphi$ and $\sin\varphi$, when substituted in (1), § 77, give us the expressions for x and y in terms of the new coordinates. Substituting these expressions in the original equation, we have the equation of the curve referred to the new axes. This equation will contain no x_1y_1-term, so that the curve may be traced by the methods of Chapters 7 and 8.

Example: Trace the curve

$$(9)\qquad 9x^2 - 24xy + 16y^2 - 18x - 101y + 19 = 0.$$

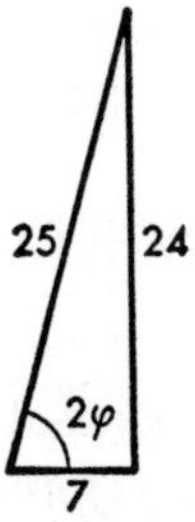

Here we have

$$\tan 2\varphi = \frac{-24}{9-16} = \frac{24}{7},$$

whence

$$\cos 2\varphi = \tfrac{7}{25},$$

and by (8),

$$\cos\varphi = \tfrac{4}{5}, \qquad \sin\varphi = \tfrac{3}{5}.$$

Figure 103

Thus by § 77,

$$x = \tfrac{1}{5}(4x_1 - 3y_1), \qquad y = \tfrac{1}{5}(3x_1 + 4y_1).$$

In terms of the new coordinates equation (9) reduces to

$$25y_1^2 - 75x_1 - 70y_1 + 19 = 0,$$

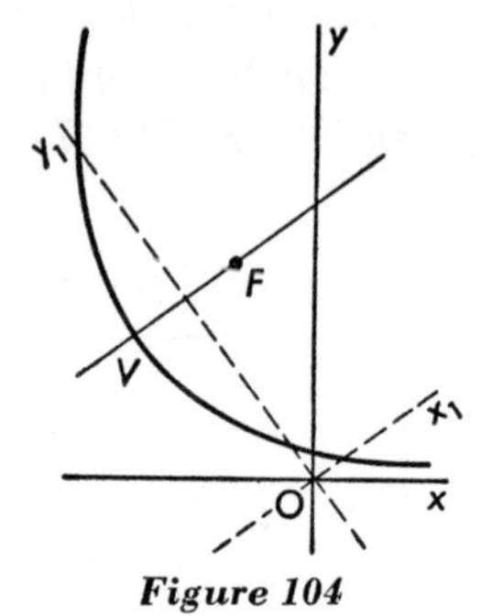

Figure 104

or, in the standard form,

$$(y_1 - \tfrac{7}{5})^2 = 3(x_1 + \tfrac{2}{5}).$$

This is a parabola with axis parallel to Ox_1, opening in the positive direction, with its vertex at the point $(-\frac{2}{5}, \frac{7}{5})$ referred to the new axes. The x_1-axis has a slope $\tan \varphi = \frac{3}{4}$; after the new axes are drawn the curve is traced as in Fig. 104.

79. Test for species of a conic. We sometimes wish merely to determine what kind of conic is represented by a given equation, without going to the trouble of removing the xy-term. For this purpose the following test may be used.

An equation of the second degree represents:

(*a*) *if* $B^2 - 4AC < 0$, *an ellipse;*
(*b*) *if* $B^2 - 4AC = 0$, *a parabola;*
(*c*) *if* $B^2 - 4AC > 0$, *a hyperbola.*

The exceptional forms listed in §§ 69, 64, 75 are included under (*a*), (*b*), (*c*) respectively.

Proof of this test can be accomplished by doing Exs. 34–38 below.

The condition $B^2 - 4AC = 0$ means that $Ax^2 + Bxy + Cy^2$ is a perfect square.

THEOREM: *An equation of second degree represents a parabola (exceptionally, two parallel or coincident lines, or no locus) if and only if the terms of second degree form a perfect square.*

Exercises

In Exs. 1–21, determine the species of conic represented by the given equation; remove the product term by rotation of axes, reduce the resulting equation to a standard form, and trace the curve on the new axes.

1. $xy = 8$. **2.** $2xy + 1 = 0$.
3. $x^2 - 4xy + y^2 = -9$. **4.** $5x^2 - 8xy + 5y^2 = 9$.

5. $13x^2 - 10xy + 13y^2 = 72$. *Ans.* $4x_1^2 + 9y_1^2 = 36$.
6. $8x^2 - 12xy + 17y^2 = 80$. *Ans.* $x_1^2 + 4y_1^2 = 16$.
7. $2x^2 + 12xy - 7y^2 = 45$. *Ans.* $x_1^2 - 2y_1^2 = 9$.
8. $2x^2 - 3xy - 2y^2 = -10$. *Ans.* $x_1^2 - y_1^2 = 4$.
9. $11x^2 + 6xy + 19y^2 = 80$. *Ans.* $2x_1^2 + y_1^2 = 8$.
10. $x^2 + 8xy + 7y^2 = 1$. *Ans.* $9x_1^2 - y_1^2 = 1$.
11. $5x^2 + 12xy = 4$. *Ans.* $9x_1^2 - 4y_1^2 = 4$.
12. $4xy + 3y^2 = 1$. *Ans.* $4x_1^2 - y_1^2 = 1$.
13. $5x^2 + 24xy - 5y^2 = -13$. *Ans.* $y_1^2 - x_1^2 = 1$.
14. $x^2 - 4xy - 2y^2 = 6$. *Ans.* $2y_1^2 - 3x_1^2 = 6$.
15. $13x^2 - 8xy + 7y^2 = 45$. *Ans.* $x_1^2 + 3y_1^2 = 9$.
16. $3x^2 + 12xy - 13y^2 = -120$. *Ans.* $3y_1^2 - x_1^2 = 24$.
17. $17x^2 + 18xy - 7y^2 = 80$. *Ans.* $2x_1^2 - y_1^2 = 8$.
18. $16x^2 - 24xy + 9y^2 = 30x + 40y$. *Ans.* $y_1^2 = 2x_1$.
19. $16x^2 - 24xy + 9y^2 - 60x - 80y + 100 = 0$. *Ans.* $y_1^2 = 4(x_1 - 1)$.
20. $11x^2 - 24xy + 4y^2 + 6x + 8y = -15$. *Ans.* $(x_1 - 1)^2 - 4y_1^2 = 4$.
21. $3x^2 + 2\sqrt{3}\,xy + y^2 = 12x - 12\sqrt{3}\,y + 24$.
Ans. $x_1^2 = -6(y_1 - 1)$.

22. By rationalizing each equation, show that the three equations $x^{\frac{1}{2}} + y^{\frac{1}{2}} = a^{\frac{1}{2}}$, $x^{\frac{1}{2}} - y^{\frac{1}{2}} = a^{\frac{1}{2}}$, $y^{\frac{1}{2}} - x^{\frac{1}{2}} = a^{\frac{1}{2}}$ collectively represent a parabola, and trace the curve. What part of the curve is represented by each equation?

23. Show that if the terms of first degree are missing, the equation of second degree cannot represent a parabola.

24. Prove that if A and C have opposite signs the equation of the second degree always represents a hyperbola. Is the converse true?

25. Prove that if the xy-term is present and either or both square terms are missing the equation of second degree always represents a hyperbola.

26. It follows from the theorem of § 79 that the equation of every parabola can be put in the form $(ax + by)^2 + Dx + Ey + F = 0$. Show that the axis of the parabola is parallel to the line $ax + by = 0$. (Ex. 35, p. 103.)

27. Show that if the terms of first degree are missing, an equation of second degree represents a conic with center at the origin; and conversely.

28. Prove that a straight line can intersect a conic in not more than two points.

29. Prove that two conics can intersect in not more than four points.

30. Show that a conic is determined by five points.

31. Show that a parabola is determined by four points. (Note the equation in Ex. 26.)

32. Show that if the equations of a hyperbola and a straight line, when solved as simultaneous, have no solution real or imaginary, the line must be an asymptote to the curve. (Use the equations $\frac{x^2}{a^2} - \frac{y^2}{b^2} = 1$, $y = mx + c$.)

33. If in the equation of a hyperbola the terms of first degree are missing, show that the equations of the asymptotes may be found by equating to 0 the group of terms of second degree and factoring. [Take the equation in the form $(a_1x + b_1y)(a_2x + b_2y) + c = 0$; see Ex. 32.]

34. In § 78, equation (1) was transformed by rotation of axes into equation (3). Let the coefficients in (3) be designated by A_1, B_1, etc., so that (3) may be written

$$A_1x_1^2 + B_1x_1y_1 + C_1y_1^2 + D_1x_1 + E_1y_1 + F_1 = 0. \tag{10}$$

By identifying (3) and (10), show that

$$\begin{aligned} A_1 &= A \cos^2 \varphi + B \cos \varphi \sin \varphi + C \sin^2 \varphi, \\ B_1 &= B \cos 2\varphi - (A - C) \sin 2\varphi, \\ C_1 &= A \sin^2 \varphi - B \cos \varphi \sin \varphi + C \cos^2 \varphi. \end{aligned} \tag{11}$$

35. Use the formulas (11) of Ex. 34 to show that

$$A_1 + C_1 = A + C, \tag{12}$$

which is expressed in advanced mathematics by saying that $(A + C)$ is *invariant* (unchanged) under the transformation (rotation) employed.

36. Formulas (11) Ex. 34 yield

$$A_1 - C_1 = B \sin 2\varphi + (A - C) \cos 2\varphi. \tag{13}$$

From (13) and the expression for B_1 in terms of A, B, C, show that

$$B_1^2 + (A_1 - C_1)^2 = B^2 + (A - C)^2.$$

37. Since $B^2 - 4AC = B^2 + (A - C)^2 - (A + C)^2$, conclude from the results of Exs. 35–36 that

$$B_1^2 - 4A_1C_1 = B^2 - 4AC, \tag{14}$$

another invariant under rotation of axes.

38. Let φ be chosen as in § 78 so that $B_1 = 0$, and use (14) of Ex. 37 to prove the validity of the test in § 79.

80. Rectangular hyperbola. The equation

$$Bxy + Dx + Ey + F = 0, \qquad (B \neq 0), \tag{1}$$

if not factorable, represents a hyperbola, since

$$B^2 - 4AC = B^2 > 0.$$

Let us solve the equation for y and for x in turn:

$$y = -\frac{Dx + F}{Bx + E}, \qquad x = -\frac{Ey + F}{By + D}.$$

From the above equations we see that if $(Bx + E) = 0$, y does not exist; if $(By + D) = 0$, x does not exist. As

$$(Bx + E) \to 0,\ y \to \pm\infty;\ \text{as } (By + D) \to 0,\ x \to \pm\infty.$$

This shows that the asymptotes are the lines

$$Bx + E = 0, \qquad By + D = 0.$$

Thus equation (1) represents *a rectangular hyperbola with its asymptotes parallel to the coordinate axes.*

An important special case of (1) is the following:

The equation

$$(2) \qquad 2xy = a^2$$

represents a rectangular hyperbola whose asymptotes are the coordinate axes.

Since x and y must have the same sign, the curve lies in the first and third quadrants. The constant a is the semi-axis: this may be shown by rotation of axes, or by noting that the curve intersects the line $y = x$ at $(\pm\frac{1}{2}\sqrt{2}a, \pm\frac{1}{2}\sqrt{2}a)$.

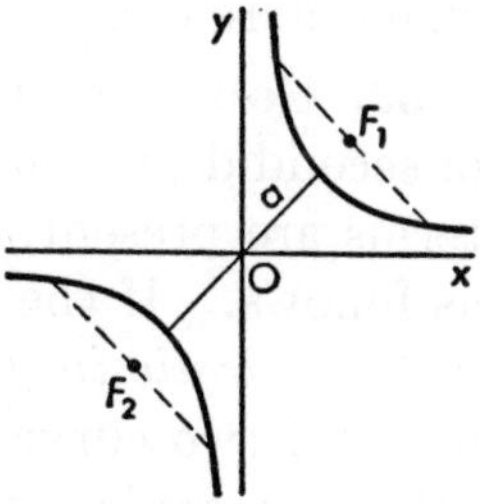

Figure 105

Of course the equation

$$(3) \qquad 2xy = -a^2$$

represents the hyperbola lying in the second and fourth quadrants (since x and y must now have opposite signs).

Both (2) and (3) are evidently included in the single formula

$$(4) \qquad xy = k, \qquad (k \neq 0),$$

where k is unrestricted as to sign. This equation, written in the form

$$y = \frac{k}{x},$$

shows that y is *inversely proportional to* x. This type of variation occurs frequently in science — as examples, see Exs. 13–14 below.

81. Composition of ordinates. Given two functions of x

(1) $$y = u(x), \qquad y = v(x),$$

let it be required to trace the curve

(2) $$y = u(x) + v(x).$$

It may happen that the curve (2) is difficult to trace by our usual methods, but that the curves (1) are simple and easily handled. In such a case we may make use of the obvious fact that, for any given value of x, the ordinate of (2) is the *sum of the ordinates* of the separate curves (1). If these two curves are traced on the same axes, any number of points of (2) are easily plotted.

That we may compound abscissas instead of ordinates, when more convenient, should be clear without argument.

82. Conics traced by composition. Given an equation of second degree in which the xy-term and one or both square terms are present, the conic may be traced by composition* as follows. If the term in y^2 is present, we *solve the equation as a quadratic in* y, which exhibits y as the sum of the ordinates of two curves, one of which is a straight line, while the other is a conic with axes parallel to the coordinate axes, and thus amenable to the methods of Chapters 7 and 8.

Of course, if the term in y^2 is missing, we solve for x instead. When both square terms are present, so that a choice is possible, there is sometimes a definite advantage in one alternative over the other (Exs. 36–37 below).

* If the product term is missing, the methods of Chapters 7 and 8 are applicable. If both square terms are missing, we may use § 80. Thus in those cases the method of composition, though available, is not indicated.

Example: Trace the parabola

$$x^2 - 2xy + y^2 + x - 1 = 0.$$

Solving for y, we find

$$\begin{aligned} y &= x \pm \tfrac{1}{2}\sqrt{4x^2 - 4x^2 - 4x + 4} \\ &= x \pm \sqrt{1 - x}. \end{aligned}$$

The ordinates of this curve are found by adding the ordinates of the straight line

$$(1) \qquad y = x$$

and the curve

$$y = \pm\sqrt{1 - x}.$$

This equation reduces to

$$(2) \qquad y^2 = -(x - 1):$$

a parabola with vertex at (1, 0), opening to the left. We now trace the curves (1) and (2) and locate points on the required curve by adding (algebraically) the ordinates of these two curves. For a given value of x there are two points on the required curve, with ordinates

$$\begin{aligned} y &= x + \sqrt{1 - x}, \\ y &= x - \sqrt{1 - x}. \end{aligned}$$

For instance, when $x = OM$, the two ordinates are

$$\begin{aligned} MP_1 &= MR + MQ_1, \\ MP_2 &= MR + MQ_2. \end{aligned}$$

Due regard must of course be paid to the sign of each segment.

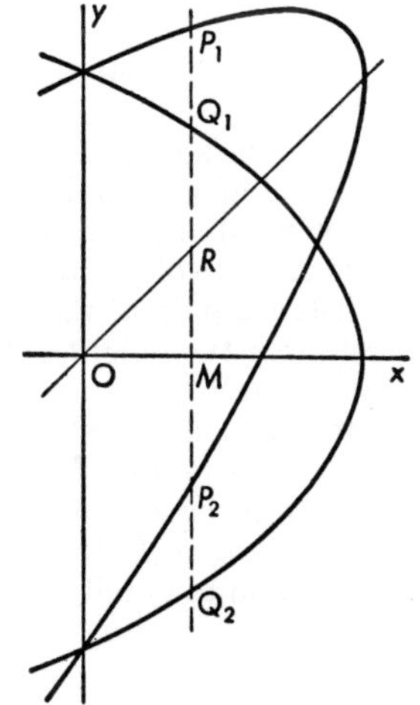

Figure 106

Although this curve can (of course) be traced by use of § 78, the present method is definitely easier.

Exercises

In Exs. 1–4, find the asymptotes and the intercepts, plot a few other points, and draw the curve.

1. $xy + 4x - 2y + 6 = 0$.
2. $xy - 3x + 2y = 0$.
3. $xy - 5y + 5 = 0$.
4. $6xy - 3x - 2y + 3 = 0$.

In Exs. 5–8, locate the vertices, plot a few points, and draw the curve.

5. $xy = 18$.
6. $xy = -2$.
7. $2xy = -3$.
8. $6xy = 1$.

9. As the number of terms increases indefinitely, the sum of a geometric progression of first term a and ratio r approaches the limiting value

$$S = \frac{a}{1 - r}, \qquad -1 < r < 1.$$

Graph S as a function of r.

10. Represent $\tan(\theta + 45°)$ as a function of $\tan\theta$. [Expand by the addition formula; put $\tan\theta = x$, $\tan(\theta + 45°) = y$.]

11. A man wishes to drive a specified distance at constant speed. Draw a curve from which he can read off the time needed to cover the distance at any given speed; show how to read from the same curve the speed required to cover the distance in a given time.

12. Two cars drive from A to B, 200 mi. The second car starts one hour later than the first, but drives 10 mi. per hr. faster. If both reach B at the same time, find analytically and graphically the time and average speed for each car.

13. Boyle's Law for "perfect gases" states that volume is inversely proportional to pressure. Draw the graph (a) of volume as a function of pressure; (b) of pressure as a function of volume.

14. A lever of length l with the fulcrum at one end is to lift a weight W at distance a from the fulcrum, by means of a force F at the free end. Neglecting the weight of the lever, graph F as a function of l.

15. Graph the current I from a battery as a function of the external resistance R, if the internal resistance r and electromotive force E are constant. (Current equals electromotive force divided by total resistance.)

16. In Atwood's Machine two masses m_1, m_2 are joined by a cord hung over a pulley. If the masses of the cord and pulley are negligible, the acceleration of gravity is reduced in the ratio

$$r = \frac{m_1 - m_2}{m_1 + m_2}. \qquad (m_1 > m_2.)$$

If originally $m_1 = 7$, $m_2 = 5$, and each is changed by an amount x, express r as a function of x, and draw the curve. *Ans.* $r = \dfrac{1}{x + 6}$, $x > -5$.

In Exs. 17–32, trace the curve by composition, first determining the species of the conic.

17. $5x^2 - 4xy + y^2 = 1.$
18. $2x^2 - 2xy + y^2 = 4.$
19. $x^2 - 2xy + y^2 - x = 0.$
20. $x^2 - 2xy + y^2 = 4x - 4.$
21. $xy - y^2 - 4 = 0.$
22. $4xy - y^2 + 8 = 0.$
23. $2x^2 - 2xy + 3y = 0.$
24. $2x^2 - xy = 29 - 4y.$

25. $x^2 - 2xy + y^2 + 4x - 2y - 1 = 0.$
26. $x^2 - 6xy + 8y^2 + 2x - 6y = 0.$
27. $5x^2 - 2xy + y^2 = 4.$
28. $x^2 - 2xy + 10y^2 - 4x + 4y - 5 = 0.$
29. $x^2 - 6xy + 10y^2 + 8x - 23y + 10 = 0.$
30. $x^2 - 4xy + 5y^2 - 20x + 42y + 100 = 0.$
31. $x^2 - 2xy + 2y^2 - 2x + 2y + 1 = 0.$ *Ans.* The point (1, 0).
32. $4x^2 + 4xy + 5y^2 + 8x + 4y + 5 = 0.$ *Ans.* No locus.
33. Trace the curve of Ex. 22, p. 129, by composition of ordinates.

34. Show that the conic $Ax^2 + Bxy + Dx + F = 0$ can be traced by compounding ordinates of a straight line and a rectangular hyperbola referred to its asymptotes [equation (4), § 80]. Obtain an analogous result for the conic $Bxy + Cy^2 + Ey + F = 0$.

35. The following statements (§ 82) have not been proved: (*a*) when the equation of a conic is solved for y, y appears as the sum of the ordinates of a straight line and a conic; (*b*) the axes of this conic are parallel to the coordinate axes; (*c*) the equations of the component line and conic have rational coefficients whenever the original coefficients are rational. By solving the general equation of second degree for y, prove these statements ($C \neq 0$).

36. By solving the general equation of second degree for y ($C \neq 0$), show that when $B^2 - 4AC = -4C^2$, the ordinates of the conic are the sum of ordinates of a straight line and a circle.

37. Obtain a theorem like that of Ex. 36, when abscissas are compounded.

CHAPTER 10 *Tangents and Normals to Conics*

83. Tangents to plane curves. In high school geometry a tangent to a circle is often defined as a line which intersects the circle in only one point. Such a definition is highly undesirable for the study of curves not so simple as the circle. The following concept of a tangent line is valid for other curves as well as for circles.

Let P be a fixed point of a plane curve, and P' a neighboring point. If P' be made to approach P along the curve, the secant PP' evidently approaches, in general, a definite limiting position, the line PQ in Fig. 107. The line thus approached is called the *tangent to the curve at* P, or is said to *touch the curve at* P. The point P is the *point of contact*.

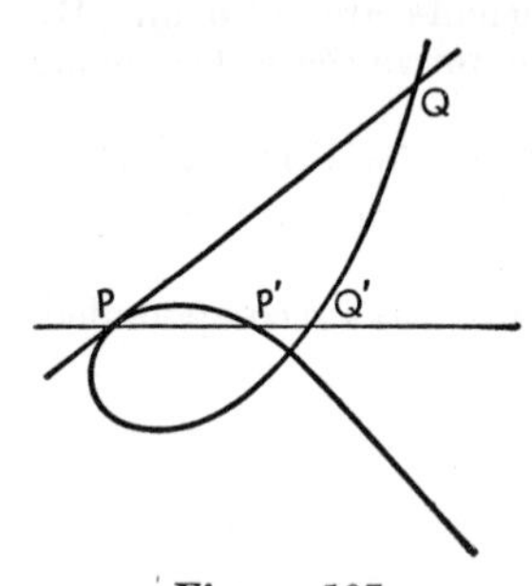

Figure 107

The *slope of a curve* at any point is defined as the *slope of the tangent* at that point. Two curves intersecting in a point P are said to be *tangent at* P if they have the same slope.

Notice that the position of the tangent as described above is a local property of the curve; it depends only on the shape of the curve near P. Whether the tangent intersects the curve at other, distant points, as at Q in Fig. 107, has nothing to do with the existence and position of the tangent at P.

We know by the theorem of § 29 that a straight line may intersect a curve of the nth degree in not more than n points. Since a tangent is the limiting position of a secant when two points of intersection come to coincidence in the manner described above, it follows that the point of tangency counts as two intersections.

THEOREM: *A tangent to a curve of nth degree may intersect the curve in, at most,* $n - 2$ *points besides the point of tangency.*

84. Tangent at a given point. The definition of tangent line given in § 83 leads to an analytic method for finding the slope of the tangent to a curve at a specified point on the curve. The method will be applied here only to conics. A systematic exploitation of this method forms a basic part of any course in calculus.

Example (a): Find the tangent to the parabola

$$y^2 - 3x - y + 1 = 0 \tag{1}$$

at the point P:(1, 2). (Fig. 108.)

Choose a point P' on the curve and near P. Denote the distances PR, RP' by Δx, Δy respectively, so that the coordinates of P' are $(1 + \Delta x, 2 + \Delta y)$. Since P' is on the curve, its coordinates must satisfy equation (1). Thus we obtain

$$(2 + \Delta y)^2 - 3(1 + \Delta x) - (2 + \Delta y) + 1 = 0,$$

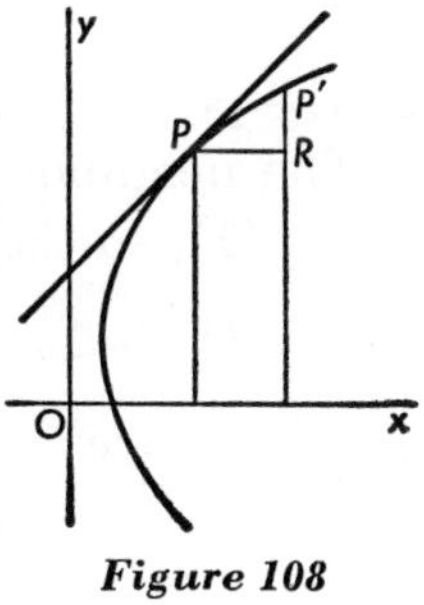

Figure 108

which reduces to

$$3\,\Delta y + \overline{\Delta y}^2 - 3\,\Delta x = 0.$$

Dividing throughout by Δx, we get

$$3\,\frac{\Delta y}{\Delta x} + \Delta y\,\frac{\Delta y}{\Delta x} - 3 = 0. \tag{2}$$

The quantity $\frac{\Delta y}{\Delta x}$ is the slope of the secant PP'. Now let $\Delta x \to 0$, so that $P' \to P$ along the curve. Then also $\Delta y \to 0$ and, by § 83, the ratio $\frac{\Delta y}{\Delta x}$ will approach as its limit the slope m of the curve at P. Hence, when $\Delta x \to 0$, equation (2) yields

$$3m + 0 \cdot m - 3 = 0, \tag{3}$$

or $m = 1$.

The desired tangent is, therefore, a line with slope 1 and passing through P: (1, 2). The equation of the tangent line is

$$x - y + 1 = 0.$$

If the desired tangent happens to be vertical, the above method must break down somewhere because the slope of the tangent does not exist. The following example shows how to recognize this situation and how to find the required tangent.

Example (*b*): Find the tangent to the parabola

$$y^2 - 3x - y + 1 = 0 \tag{1}$$

at the point $(\frac{1}{4}, \frac{1}{2})$.

The neighboring point $(\frac{1}{4} + \Delta x, \frac{1}{2} + \Delta y)$ is to be on the curve (1), so the coordinates of the point satisfy (1):

$$(\tfrac{1}{2} + \Delta y)^2 - 3(\tfrac{1}{4} + \Delta x) - (\tfrac{1}{2} + \Delta y) + 1 = 0,$$

which reduces to

$$\overline{\Delta y}^2 - 3\,\Delta x = 0. \tag{4}$$

Divide by Δx throughout to obtain

$$\Delta y \frac{\Delta y}{\Delta x} - 3 = 0,$$

and let $\Delta x \to 0$, $\frac{\Delta y}{\Delta x} \to m$. We thus get

$$0 \cdot m - 3 = 0, \tag{5}$$

which is impossible. There is no such m.

Since m does not exist, we suspect the presence of a vertical tangent. The vertical line through the given point $(\frac{1}{4}, \frac{1}{2})$ is

$$x = \tfrac{1}{4}. \tag{6}$$

If we solve (6) together with the equation of the curve (1) we obtain

$$y^2 - y + \tfrac{1}{4} = 0,$$

which has equal roots, $y = \frac{1}{2}, \frac{1}{2}$. Hence the line (6) is tangent to the parabola (1) at $(\frac{1}{4}, \frac{1}{2})$.

The method exemplified above can be extended to yield the tangent to any curve at any point where a tangent line exists. For conics the method may be summarized as follows:

To find the slope m of a curve at $P:(x_1, y_1)$ on the curve, choose $P':(x_1 + \Delta x, y_1 + \Delta y)$ a point on the curve, substitute the coordinates of P' in the equation of the curve, and simplify. Divide each term of the equation by Δx. Let $\Delta x \to 0$, $\dfrac{\Delta y}{\Delta x}$ at the same time approaching m, and solve for m. If the process does not yield a value for m, the tangent is vertical and its equation is $x = x_1$.

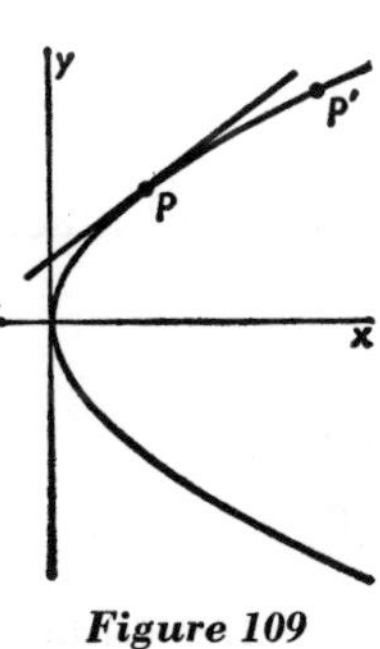

Figure 109

85. Tangent at a given point of a standard conic. Let $P:(x_1, y_1)$ be a point on the parabola

$$y^2 = 4ax, \tag{1}$$

and $P':(x_1 + \Delta x, y_1 + \Delta y)$ be a neighboring point on the curve. Since P' is on the parabola (1), we get

$$y_1{}^2 + 2y_1\,\Delta y + \overline{\Delta y}^2 = 4ax_1 + 4a\,\Delta x. \tag{2}$$

The point (x_1, y_1) is on the curve, so we must have

(3) $$y_1^2 = 4ax_1,$$

whence (2) becomes

$$2y_1\,\Delta y + \overline{\Delta y}^2 = 4a\,\Delta x,$$

or

(4) $$2y_1\frac{\Delta y}{\Delta x} + \Delta y\frac{\Delta y}{\Delta x} = 4a.$$

Let $\Delta x \to 0$ in (4). Then $\Delta y \to 0$, $\dfrac{\Delta y}{\Delta x} \to m$, and we find that

(5) $$2y_1m = 4a.$$

If $y_1 = 0$, equation (5) does not yield a value for m, but then $x_1 = 0$ also, the point P becomes the vertex, and the tangent is the vertical line $x = 0$.

If $y_1 \neq 0$, then by (5)

$$m = \frac{2a}{y_1},$$

and the tangent is the line

$$y - y_1 = \frac{2a}{y_1}(x - x_1)$$

or

(6) $$y_1y - y_1^2 = 2ax - 2ax_1.$$

Using equation (3) to simplify (6), we obtain the result: The equation of the tangent to the parabola

$$y^2 = 4ax$$

at the point (x_1, y_1) on the curve is

(7) $$\boldsymbol{y_1y = 2ax + 2ax_1}.$$

In a similar way we can prove that: The equation of the tangent to the central conic

(8) $$\frac{x^2}{a^2} \pm \frac{y^2}{b^2} = 1$$

at the point (x_1, y_1) on the curve is

(9) $$\frac{x_1x}{a^2} \pm \frac{y_1y}{b^2} = 1.$$

86. Normal; subtangent; subnormal. A line perpendicular to the tangent at a point on the curve (*PN* in Fig. 110)

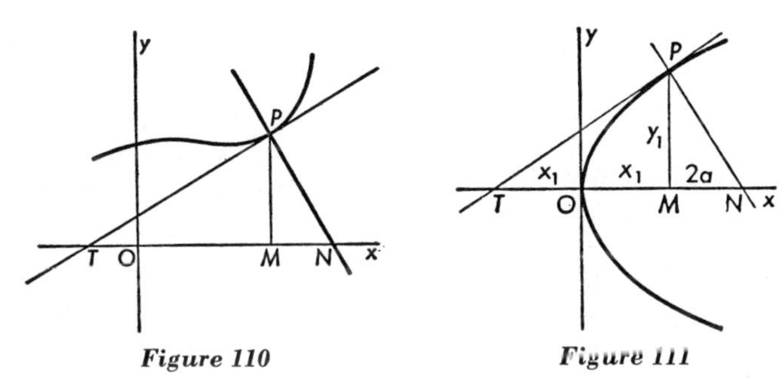

Figure 110 *Figure 111*

is called the normal to the curve at that point. The equation of the normal can be written at once when that of the tangent is known.

If the tangent and normal at P (Fig. 110) cross the x-axis at T and N respectively, and if OM is the abscissa of P, then the distances TM and MN are called the subtangent and the subnormal corresponding to the point P.

For the standard parabola

(1) $$y^2 = 4ax$$

we know the tangent line at $P:(x_1, y_1)$ on the curve to be

(2) $$y_1y = 2ax + 2ax_1.$$

The x-intercept of this line is $OT = -x_1$; hence the subtangent is $TM = 2x_1$. That is, for the parabola $y^2 = 4ax$, the subtangent is bisected at the vertex.

The normal at P is perpendicular to the tangent (2) at $P:(x_1, y_1)$. The equation of the normal is thus found to be

(3) $$y_1x + 2ay = x_1y_1 + 2ay_1.$$

The x-intercept of the normal is $ON = x_1 + 2a$, so that the subnormal is

$$MN = ON - OM = 2a.$$

That is, for the parabola $y^2 = 4ax$, the subnormal is constant, and equal to the distance from the directrix to the focus.

Exercises

In Exs. 1–6, find the tangent to the conic at the given point, using the formulas of § 85.

1. $4y^2 + x = 0$ at $(-1, \frac{1}{2})$.
2. $y^2 = 8x$ at $(2, -4)$.
3. $3x^2 + 2y^2 = 35$ at $(1, 4)$.
4. $x^2 + 8y^2 = 3$ at $(-1, \frac{1}{2})$.
5. $9x^2 - y^2 = 11$ at $(-2, 5)$.
6. $4x^2 - y^2 = 12$ at $(-2, -2)$.

In Exs. 7–11, find the tangent at the given point on the conic.

7. $x^2 = 4ay$ at (x_1, y_1). *Ans.* $x_1x = 2a(y + y_1)$.
8. $x^2 + y^2 = a^2$ at (x_1, y_1). *Ans.* $x_1x + y_1y = a^2$.
9. $2xy = a^2$ at (x_1, y_1). *Ans.* $y_1x + x_1y = a^2$.
10. $x^2 + y^2 = 2ay$ at (x_1, y_1). *Ans.* $x_1x + y_1y = a(y + y_1)$.
11. $x^2 - y^2 = 2ax$ at (x_1, y_1). *Ans.* $x_1x - y_1y = a(x + x_1)$.
12. Derive formulas (9), § 85.
13. Prove that the tangents at the ends of the latus rectum of a parabola intersect on the directrix.
14. Prove the theorem of Ex. 13 for the central conics.
15. Prove that any tangent to a parabola intersects the directrix and the latus rectum (produced) at points equally distant from the focus.
16. Tangents are drawn to the ellipse $\frac{x^2}{a^2} + \frac{y^2}{b^2} = 1$ and to the circle $x^2 + y^2 = a^2$ at points having the same abscissa. Show that these tangents cross Ox at the same point.
17. Find the x-intercept of the normal to the hyperbola $x^2 - y^2 = a^2$ at (x_1, y_1) on the curve. *Ans.* $2x_1$.
18. Show that the tangents at the ends of the latera recta of a standard central conic have slopes $\pm e$.
19. Given two points P_1, P_2 on a circle, show that the distances from P_1 to the tangent at P_2 and from P_2 to the tangent at P_1 are equal. (Ex. 8.)
20. Prove that the perpendicular from a focus to any tangent to an ellipse, and the line joining the center to the point of contact, intersect on the directrix.
21. Prove the theorem of Ex. 20 for the hyperbola.
22. Prove that the product of the distances from the foci to any tangent to an ellipse is constant (equal to b^2 for ellipse in standard form).
23. Prove the theorem of Ex. 22 for the hyperbola.

Solve Exs. 24–28 by ruler and compass.

24. Given a parabola with its axis, draw the tangent at any point.

25. Given the axis, vertex, and one point of a parabola, construct the focus and directrix.

26. Given the axis and one point of a parabola, with the tangent at that point, construct the focus and directrix.

27. Given the axis, vertex, and one tangent to a parabola, construct the focus and directrix.

28. Given an equilateral hyperbola with its center, draw the tangent at any point. (Ex. 17.)

87. The discriminant condition for tangency. We already know that when the equation of a curve and that of a line tangent to it are solved simultaneously, two of the pairs of values of x and y will be the same: i.e., two of the points of intersection will coincide.

Let us now confine our attention to curves of the second degree. If the line and the conic are tangent, the quadratic in x (or y) obtained by substituting in the equation of the conic the value of y (or x) from the equation of the line will have equal roots. Conversely, if this quadratic has equal roots, the line is tangent to the curve, since the points of intersection then have equal abscissas (or ordinates) and must coincide.* Now we know from algebra that the quadratic equation

$$ax^2 + bx + c = 0$$

has equal roots if and only if

$$b^2 - 4ac = 0. \tag{1}$$

The condition (1) applied to a quadratic obtained in the way just described is called the *condition for tangency* of the line and the conic. If either the equation of the line or that of the curve contains an undetermined constant, the condition for tangency serves to determine that constant.

* If the line is parallel to the y-axis the points have equal abscissas without necessarily coinciding. But in that case the equation of the line is $x = k$, so that in the process of solving the simultaneous equations we are led to a quadratic in y, and if this has equal roots the line is a tangent.

88. Tangents having a given slope. Let us now apply the theory of § 87.

Example: Find the tangents to the circle,

$$x^2 + y^2 = 4 \tag{1}$$

parallel to the line

$$y = x - 1.$$

The equation of any line parallel to the given line is

$$y = x + k. \tag{2}$$

To determine k so that this line shall intersect the circle in two coincident points, substitute the value of y from (2) in (1):

$$x^2 + (x + k)^2 = 4,$$

or

$$2x^2 + 2kx + k^2 - 4 = 0.$$

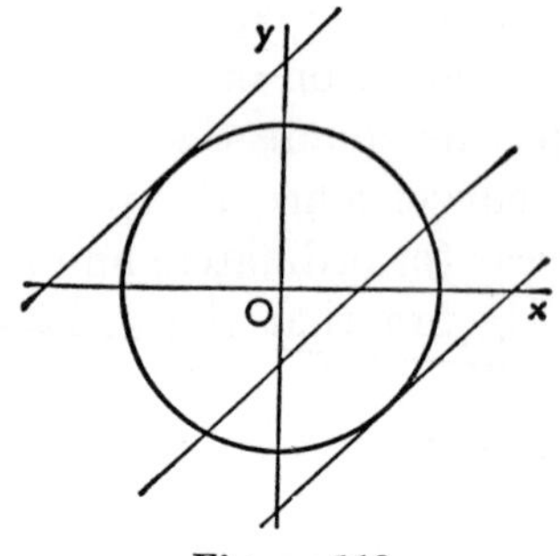

Figure 112

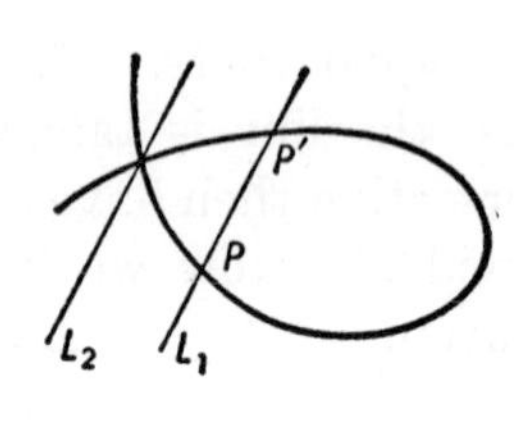

Figure 113

The roots of this quadratic in x are the abscissas of the points of intersection of the line (2) with the circle; if these roots are equal, the points will coincide. We therefore set $b^2 - 4ac = 0$; i.e.,

$$(2k)^2 - 4 \cdot 2(k^2 - 4) = 0, \qquad -k^2 + 8 = 0, \qquad k = \pm 2\sqrt{2},$$

and the required tangents are

$$y = x \pm 2\sqrt{2}.$$

It may be objected that the above process does not rest squarely upon the definition of tangent (§ 83), since the tangent here appears as the limiting position of a secant which moves parallel to its original position until the two points of intersection coincide. For the higher plane curves, this objection is sound: in Fig. 113, as the line L_1 approaches the parallel position L_2, the points P and P' approach coincidence, but the line L_2 is not a tangent. However, for the conic sections the same result is obtained by this manner of approaching the limit as by the method used in the definition. An analytic proof of this statement may be written out by use of calculus.

89. Tangents of given slope to the standard conics. Given a line

$$y = mx + k,$$

where m is supposed known, the method of § 88 enables us to determine k so that the line will be tangent to a given conic. For the standard conics. the results are as follows.

For all values of m, except as noted, the line

$$y = mx + \frac{a}{m} \qquad (m \neq 0) \tag{1}$$

is tangent to the parabola $y^2 = 4ax$; the lines

$$y = mx \pm \sqrt{a^2m^2 + b^2} \tag{2}$$

are tangent to the ellipse $\dfrac{x^2}{a^2} + \dfrac{y^2}{b^2} = 1$; the lines

$$y = mx \pm \sqrt{a^2m^2 - b^2} \quad (a^2m^2 - b^2 > 0) \tag{3}$$

are tangent to the hyperbola $\dfrac{x^2}{a^2} - \dfrac{y^2}{b^2} = 1$.

90. Two important properties of the parabola.

THEOREM I: *The tangent bisects the angle between the focal radius drawn to the point of contact and the line through that point parallel to the axis of the parabola.*

THEOREM II: *The foot of the perpendicular from the focus upon any tangent lies on the tangent through the vertex.*

That is, in Fig. 114, the line PT bisects the angle FPL, and the line FL intersects PT on the tangent through V.

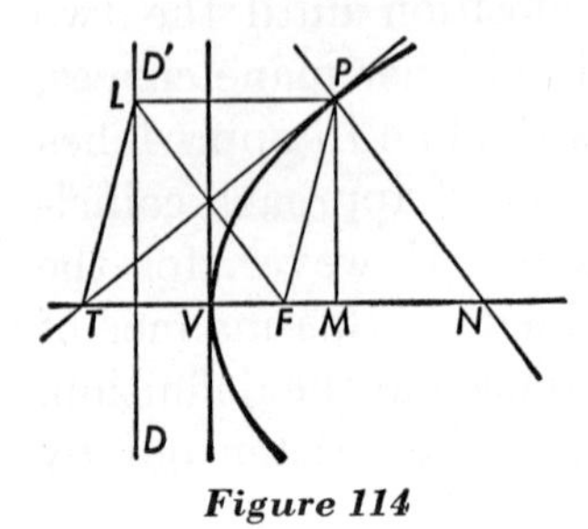

Figure 114

The property embodied in Theorem I is the underlying principle of the "parabolic reflector," used in searchlights, spotlights, etc. Rays issuing from a point-source at F and striking the polished interior of a parabolic surface (§ 185) are reflected parallel to the axis of the surface.

The proof of these theorems is left to the student.

Exercises

In Exs. 1–7, find the tangents as indicated.

1. To the parabola $y^2 = 3x - 2y - 1$ perpendicular to the line $2y = -3x$. *Ans.* $24y = 16x + 3$.

2. To the parabola $3x^2 + x - y = 10$ parallel to the line $y = 13x$. *Ans.* $y = 13x - 22$.

3. To the ellipse $4x^2 + y^2 = 4y - 6x$ parallel to the line $2y = 3x$.

4. To the ellipse $x^2 + 3y^2 + 2x + y = 0$ perpendicular to the line $4y = 2x + 1$. *Ans.* $y = -2x$; $6x + 3y + 13 = 0$.

5. To the hyperbola $y^2 - 4x^2 + 6x - 3y + 6 = 0$ perpendicular to the line $10y = x - 2$.

6. To the hyperbola $2x^2 - 3y^2 = 2$ parallel to the line $5x - 6y = 9$.

7. To the parabola $2x^2 - 4xy + 2y^2 = y$ perpendicular to the line $4y = -5x$.

8. Find the equations of the horizontal and vertical lines which box in the ellipse $x^2 + xy + y^2 + y = 0$. *Ans.* $y = 0$, $y = -\frac{4}{3}$; $x = 1$, $x = -\frac{1}{3}$.

9. Solve Ex. 8 for the ellipse $2x^2 - 2xy + y^2 - 4x + 2y + 1 = 0$. *Ans.* $y = \pm\sqrt{2}$; $x = 0$, $x = 2$.

In Exs. 10–15, draw the figure.

10. Make the parabola $y^2 = 4ax$ touch the line $2x + 3y = 4$. *Ans.* $a = -\frac{8}{9}$.

11. Make the hyperbola $y^2 - x^2 = a^2$ touch the line $x + 2y = 3$. *Ans.* $a = \sqrt{3}$.

12. Make the parabola $y = ax^2 + bx$ touch each of the lines $x - y = 1$, $y = 5x - 1$. Write out the condition for tangency of each line to the curve; solve for a and b.
Ans. $a = 1$, $b = 3$.

13. Make the conic $Ax^2 + By^2 = 1$ touch each of the lines $x - 5y + 4 = 0$, $2x + 5y + 2 = 0$.
Ans. $5y^2 - x^2 = 4$.

14. Find the common tangents to the curves $x^2 + y^2 = 2$, $y^2 = 8x$. Write the condition for tangency of the line $y = mx + k$ to each curve; solve for m and k.
Ans. $y = \pm(x + 2)$.

15. Find the common tangents to the curves $x^2 + y^2 + 10x = 20$, $y^2 = 20x$.
Ans. $2y = \pm(x + 20)$.

16. Derive formula (1), § 89.

17. Derive formulas (2) and (3), § 89.

18. Prove that the tangents drawn to a parabola from any point on the directrix are perpendicular.

19. In Fig. 114, prove that $FP = FT = FN$.

20. Prove Theorem I, § 90.

21. Prove Theorem II, § 90.

22. Find the tangent of slope m to the parabola $x^2 = 4ay$.
Ans. $y = mx - am^2$.

23. Find the tangents of slope m to the hyperbola $2xy = a^2$.
Ans. $y = mx \pm a\sqrt{-2m}$, $m < 0$.

91. Tangents through an external point. The discriminant condition for tangency can be used efficiently to find tangents to a conic from an external point.

Example: Find the tangents to the ellipse

$$x^2 + 4y^2 = 8 \tag{1}$$

through the point $(3, \frac{1}{2})$.

The equation of any line through $(3, \frac{1}{2})$ is

$$y - \tfrac{1}{2} = m(x - 3),$$

or

$$y = mx - 3m + \tfrac{1}{2}. \tag{2}$$

Substituting this expression for y in (1), we get

$$x^2 + 4(mx - 3m + \tfrac{1}{2})^2 = 8,$$

which reduces to

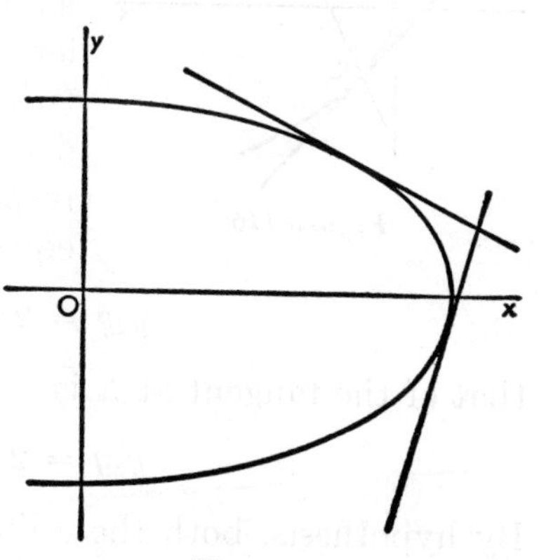

Figure 115

$$(1 + 4m^2)x^2 + 4m(1 - 6m)x + 36m^2 - 12m - 7 = 0.$$

This quadratic in x will have equal roots, and the line (2) will be tangent to the ellipse (1), if

$$16m^2(1 - 6m)^2 - 4(1 + 4m^2)(36m^2 - 12m - 7) = 0.$$

This reduces to

$$4m^2 - 12m - 7 = 0,$$

whence

$$m = -\tfrac{1}{2}, \tfrac{7}{2},$$

and the tangents are

$$x + 2y = 4, \qquad 7x - 2y = 20.$$

92. Chord of contact. If tangents are drawn to a conic through an external point P, the line through the points of contact is called the *chord of contact* of P, with reference to the given conic. The term is also used as in elementary geometry to mean the line segment joining the points of contact.

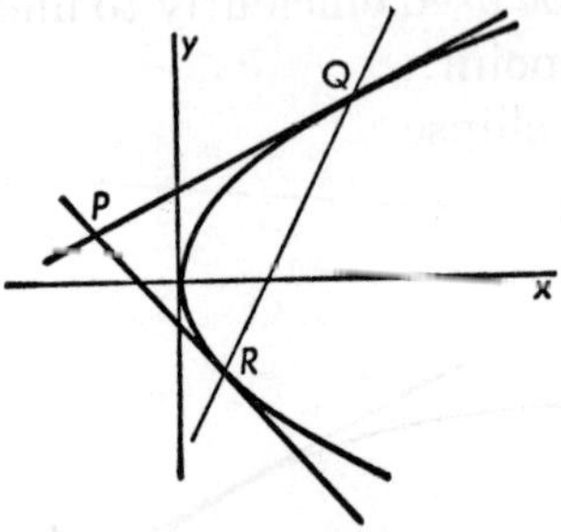

To find the chord of contact of the external point $P:(x_1, y_1)$ with respect to the parabola

$$y^2 = 4ax,$$

let QR be the required chord, and let the coordinates of Q and R (which are of course unknown) be denoted by (x_2, y_2) and (x_3, y_3) respectively. Then by (7), § 85, the equation of the tangent at Q is

Figure 116

$$y_2y = 2ax + 2ax_2,$$

that of the tangent at R is

$$y_3y = 2ax + 2ax_3.$$

By hypothesis, both these lines pass through P: substituting the coordinates of P for x and y in the two equations we obtain the relations

$$(1) \qquad y_2y_1 = 2ax_1 + 2ax_2,$$
$$(2) \qquad y_3y_1 = 2ax_1 + 2ax_3.$$

Consider now the straight line

$$(3) \qquad y_1y = 2ax + 2ax_1.$$

This line passes through Q, for if we substitute in (3) the coordinates (x_2, y_2) we obtain (1), which is known to be true. Similarly we see from (2) that the line (3) passes through R. This line is therefore the required chord of contact. Hence:

If the point $P: (x_1, y_1)$ is outside the parabola

$$y^2 = 4ax,$$

the equation

$$(4) \qquad \boldsymbol{y_1y = 2ax + 2ax_1}$$

represents the chord of contact corresponding to P.

In the same way we find:

The equation of the chord of contact corresponding to the external point $P: (x_1, y_1)$, with reference to the conic

$$\frac{x^2}{a^2} \pm \frac{y^2}{b^2} = 1,$$

is

$$(5) \qquad \boldsymbol{\frac{x_1x}{a^2} \pm \frac{y_1y}{b^2} = 1.}$$

The proof is left to the student.

Exercises

In Exs. 1–8, find the tangents to the conic through the given point, and draw the figure.

1. $y^2 = -3x$; $(-1, -2)$. *Ans.* $x - 2y = 3$, $3x - 2y = 1$.

2. $x^2 = 6y + 10$; $(7, 5)$. *Ans.* $3y = 4x - 13$, $3y = 10x - 55$.

3. $xy = 2$; $(6, -1)$. *Ans.* $x + 18y = -12$, $x + 2y = 4$.

4. $x^2 - 2y^2 = 7$; $(-1, 5)$. *Ans.* $3x + 2y = 7$, $6y = 19x + 49$.

5. $3x^2 + 4y^2 = 16$; $(4, -2)$. *Ans.* $3x + 2y = 8$, $y = -2$.

6. $x^2 - y^2 - 2x - 2y + 4 = 0$; (0, 1). *Ans.* $y = 1$, $4x + 5y = 5$.

7. $x^2 + y^2 = 2x$; (0, 6). *Ans.* $x = 0$, $35x + 12y = 72$.

8. $(x - y)^2 = x - 1$; (1, 2). *Ans.* $x = 1$, $4y = 5x + 3$.

9. In applying the method of § 91, if the values of m turn out to be equal, what conclusion can be drawn? What if m is imaginary?

10. In applying the method of § 91, if the equation to determine m turns out to be of degree one, what conclusion can be drawn? (Exs. 7, 8.)

11. By employing the chord of contact, devise a new method for finding tangents to the standard conics through an external point. Employ this method to solve the example of § 91.

In Exs. 12–15, use the method suggested in Ex. 11.

12. Ex. 1. **13.** Ex. 4. **14.** Ex. 5.

15. Find the tangents to the parabola $y^2 = 5x$ through (3, −4).

16. Prove formula (5), § 92.

17. Using the answer to Ex. 9, p. 142, find the equation of the chord of contact of an external point (x_1, y_1) with reference to the hyperbola $2xy = a^2$. *Ans.* $y_1x + x_1y = a^2$.

18. Solve Ex. 3 by a new method (Ex. 17).

19. For the parabola, prove that the chord of contact of any point on the directrix passes through the focus.

20. For the parabola, prove that if the chord of contact of the point P passes through the focus, then P lies on the directrix.

21. Prove the theorem of Ex. 19 for the central conics.

22. For the parabola, prove that the chord of contact of any point on the directrix is perpendicular to the line joining that point to the focus.

23. Prove the theorem of Ex. 22 for the central conics.

24. For the parabola, prove that if a point lies on the latus rectum (produced), its chord of contact passes through the point of intersection of the axis and the directrix.

25. Prove the theorem of Ex. 24 for the central conics.

26. The points P_1, P_2 being external to a parabola, prove that if P_1 lies on the chord of contact of P_2, then P_2 lies on the chord of contact of P_1.

27. Prove the theorem of Ex. 26 for the central conics.

28. If O is the center of a circle, P an external point, and λ the chord of contact of P, prove that OP is perpendicular to λ.

29. In Ex. 28, if K is the point of intersection of OP with λ, prove that $OP \cdot OK = a^2$, a being the radius of the circle.

30. Prove that if the chord of contact of P with respect to the circle $x^2 + y^2 = a^2$ is tangent to the circle $4x^2 + 4y^2 = a^2$, then P lies on the circle $x^2 + y^2 = 4a^2$.

CHAPTER 11 *Algebraic Curves*

93. Analysis of the equation. Curves of degree one (straight lines) and of degree two (conics) have been treated in some detail in previous chapters. All straight lines look alike. There are three species of nondegenerate conics. The number of shapes possible for curves of a given degree increases rapidly with the degree. We make no exhaustive study of curves of degree greater than two, but restrict ourselves to certain reasonably representative examples.

The method of plotting by points is useful if we wish to inspect an accurately drawn segment of a curve, but for most purposes it is a very dull tool. To develop the general appearance and properties of the curve, we *discover algebraic properties of the equation,* and then *interpret the results geometrically.*

A start on the process of analyzing an equation was made in § 21 with the introduction of tests for symmetry. Further tools are indicated in § 94–97 below. To avoid any danger of overstatement, it should be admitted to begin with that if a complicated equation were to be written down at random, the task of tracing the curve might be very difficult by any method or combination of methods. That is, if our forthcoming theory is to work well, the problems must be to some extent handpicked. Fortunately, however, a great many interesting and important curves yield very nicely to these methods.

94. Asymptotes. Being given any curve, imagine a circle drawn with center at the origin. If, no matter how large the radius be taken, there will always be points of the curve lying outside the circle, we say that the curve *has an infinite branch*, or *runs to infinity*.

If the curve has an infinite branch, it may happen that, as the tracing point recedes farther and farther, the curve approaches some fixed straight line, and the tangent to the curve approaches that line as a limiting position. When this happens, the line so approached is called an *asymptote* of the curve. In Fig. 117, p. 156, there is no asymptote; in Fig. 118, p. 158, there are three (the dotted lines); in Fig. 97, p. 115, there are two (the sloping lines).

THEOREM: *An asymptote to a curve of nth degree may intersect the curve in, at most, $n - 2$ points.*

Proof of this theorem will be omitted. Although an asymptote is not a tangent in the ordinary sense, it is the line approached by the tangent as the point of contact recedes to infinity, and behaves like a tangent in many ways.

95. Behavior in distant regions; horizontal and vertical asymptotes. When x increases indefinitely in either direction, y may behave in a variety of ways. The two most frequently occurring situations are:

(*a*) y increases numerically without limit; or

(*b*) y approaches a limit a.

These results are easily interpreted. For example, in (*a*), if y becomes large and positive as x becomes large and positive, we know that the curve recedes indefinitely from both axes, in the first quadrant.

In (*b*), the curve approaches the line $y = a$: this line is a *horizontal asymptote*.*

* In rare instances a curve approaches a line but the tangent approaches no limiting position, so that the line is not an asymptote. Such exceptions cannot occur among algebraic curves.

It may happen that y increases indefinitely as x approaches some value b. This means that the curve approaches more and more closely the line $x = b$: this line is a *vertical asymptote.*

Example (*a*): Determine the behavior, for large x, of the curve

$$y = (x - 1)^2(x^2 - 4).$$

When x is large and positive, y is large and positive; when x is large and negative, y is large and positive. Thus the curve rises indefinitely in the first and second quadrants. (Fig. 117, p. 156.)

Example (*b*): Test the curve

$$y = \frac{x^2}{x^2 - 4x + 3}$$

for horizontal and vertical asymptotes.

Dividing numerator and denominator of the fraction by x^2, we find

$$y = \frac{1}{1 - \frac{4}{x} + \frac{3}{x^2}}:$$

from this it appears that when x increases without limit (in either direction), y approaches 1. The line $y = 1$ is a horizontal asymptote.

As the denominator of the fraction approaches zero, y will increase numerically without limit. Therefore, set

$$x^2 - 4x + 3 = 0, \qquad (x - 1)(x - 3) = 0,$$
$$x = 1 \text{ or } 3.$$

The lines $x = 1$, $x = 3$ are vertical asymptotes (Fig. 118, p. 158).

96. Restriction to definite regions. It is frequently possible to determine certain definite portions of the plane within which the curve must lie. While no general directions can be given, in many cases we can *solve the equation for y* (or y^2) and note the *changes of sign* of the right member. When this step can be effectively carried out, it is one of the most helpful of all in tracing the curve.

Example: Test the curve

$$y = (x - 1)^2(x^2 - 4)$$

for changes of sign of y.

The equation may be written

$$y = (x - 1)^2(x - 2)(x + 2).$$

Think of x increasing continuously as we move along the curve from left to right (or decreasing continuously as we move from right to left). For a value of x slightly less than -2, the last factor is negative; x slightly greater than -2, the last factor is positive; the signs of the other factors are unaffected. Therefore, as x passes through -2, *y changes sign.* By similar argument, y changes sign as x goes through 2. As x goes through 1, y takes the value zero for an instant, but does not change sign because $(x - 1)^2$ cannot be negative.

The geometric interpretation of these results is easy, and highly illuminating. By Example (*a*), § 95, at the extreme left the curve is above Ox. Therefore, as we move in from the left, the curve will remain above Ox until the first change of sign is reached, at $x = -2$; it will then cross Ox, swing back to touch the axis (from below) at $x = 1$, bend down again, then up and across at $x = 2$, finishing up in the first quadrant as required by the previous example. (Fig. 117.)

97. Summary. The preceding remarks may now be brought together in the form of a definite sequence of steps, as follows:

1. *Test the curve for symmetry.*
2. *Find the intersections with the axes.*
3. *Determine the behavior of y for large values of x. Test for*

horizontal and vertical asymptotes.

4. *Determine as narrowly as possible those regions of the plane in which the curve lies.*

5. *Look for any further general information that may be obtainable; and if necessary, plot a few points.*

98. Polynomials. Consider first the case in which y is a *polynomial:*

$$y = a_0x^n + a_1x^{n-1} + \cdots + a_{n-1}x + a_n \quad (n \geqq 2). \tag{1}$$

If $n = 0$ or 1, the curve is a straight line; if $n = 2$, the curve is a parabola.

Before considering special examples, it will be well to apply our analysis step by step to the polynomial in general, thus deducing certain results applicable to all curves of this class.

1. The curve cannot be symmetric with respect to Ox. There may or may not be symmetry with respect to Oy or O.

2. The x-intercepts are the real zeros of the polynomial.

3. As x becomes large in either direction, y becomes large (though not necessarily of the same sign as x): at the extreme right and extreme left the curve recedes indefinitely from the x-axis, so that there is no horizontal asymptote. Since there is a value of y for every value of x, there can be no vertical asymptote.

4. A polynomial can change sign only by passing through the value zero, hence the curve can cross the x-axis only by intersecting it. (A "discontinuous" curve may cross by jumping — see Fig. 118.) The curve actually will cross the x-axis at every intersection, *except when the vanishing factor carries an even exponent,* in which case the curve will touch the axis and then turn back. (Review the example of § 96.)

5. General remarks: (a) For every value of x there is one and only one value of y. Hence the curve extends across the plane in an unbroken arc from left to right. (b) A curve of the nth degree can intersect a straight line in only n points. Hence (and this remark applies to all algebraic curves except

the straight line) no segment of the curve is straight, and no part should ever be put in with a ruler. For if the curve contained a segment of a straight line, it would have in common with that line an infinite number of points, which is impossible. (Theorem, § 29.)

Example: Trace the curve

$$y = (x - 1)^2(x^2 - 4).$$

1. There is no symmetry. (That is, no symmetry with respect to Ox, Oy, or O. Of course there might be a line or center of symmetry elsewhere.)

2. Intercepts: $(0, -4)$; $(1, 0)$, $(\pm 2, 0)$.

3. When x is large positive, y is large positive; x large negative, y large positive. The curve rises indefinitely in the first and second quadrants.

4. This step has been fully worked out. (Example, § 96.) The curve is restricted to the unshaded regions* in Fig. 117.

5. Plotting the additional point $(-1, -12)$, we draw the curve, with the x-scale four times the y-scale. [To find the *lowest* point on the curve, differential calculus is required: it must not be supposed that $(-1, -12)$ is that point.]

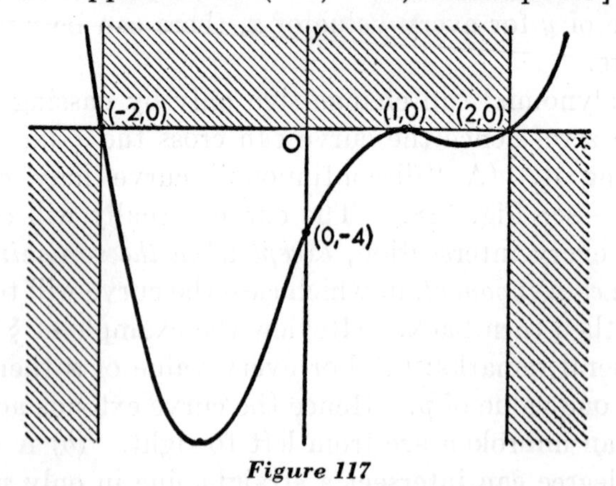

Figure 117

*** In the actual drawing of curves it is not desirable to shade the figure in this way. The device is adopted here to emphasize the meaning of step 4.**

Exercises

In Exs. 1–24, work out a full analysis, and trace the curve.

1. $y = x(x^2 - 1)$.
2. $y = x(9 - x^2)$.
3. $y = x(x^2 + 4x + 3)$.
4. $y = -x(x^2 - 5x + 4)$.
5. $y = x^2(x - 2)$.
6. $y = x^2(2 - x)$.
7. $y = (x - 1)^2(x - 2)$.
8. $y = (x - 1)(x - 2)^2$.
9. $y = (x + 2)^3$.
10. $y = (2 - x)^3$.
11. $y = x^3 - 2x^2 - 5x + 6$.
12. $y = 2x^3 + 9x^2 + 7x - 6$.
13. $y = x^2(1 - x^2)$.
14. $y = x^2(x^2 + 7x + 6)$.
15. $y = x^4 - 5x^3 + 6x^2$.
16. $y = 16 - x^4$.
17. $y = -x^4 + 5x^2 - 4$.
18. $y = x^4 - 3x^3 + 3x^2 - x$.
19. $y = 4x^3 + x^4$.
20. $y = 2x^3 - x^4$.
21. $y = (x + 1)^2(x^3 + 2x^2)$.
22. $y = x(1 - x^2)(x - 2)^2$.
23. $y = (1 - x^2)^3$.
24. $y = x^2(1 - x^2)^2$.

In Exs. 25–31, draw the graph of the function, indicating that portion of the graph which has physical meaning.

25. The difference in volume between a cube of edge c and one of edge 3.

26. The volume remaining when a slab of thickness 2 is cut from one face of a cube of edge c.

27. The volume remaining when slabs of thickness x, $2x$, $3x$ are cut from three mutually perpendicular faces of a cube of edge 2.

28. The volume of a box made by cutting squares of side x out of the corners of a piece of cardboard 8 in. square and turning up the sides. Find x if the volume is to be 12 cu. in. *Ans.* $x = 3, 0.21$.

29. Ex. 28 if the cardboard is 4×8 in. *Ans.* $x = 1, 0.70$.

30. The volume of a circular cylinder of radius r inscribed in a circular cone of radius 2 and height 8.

31. The volume of a circular cone of altitude y inscribed in a sphere of radius 3. *Ans.* $V = \frac{1}{3}\pi y^2(6 - y)$.

99. Rational fractions. Consider next the case in which y is a *rational fraction* — i.e., the quotient of two polynomials:

$$y = \frac{N}{D}, \tag{1}$$

where

$$N = a_0x^n + a_1x^{n-1} + \cdots + a_{n-1}x + a_n \qquad (n \geqq 0),$$
$$D = b_0x^m + b_1x^{m-1} + \cdots + b_{m-1}x + b_m \qquad (m \geqq 1).$$

It will be assumed that our fraction is in its lowest terms — i.e., that N and D contain no common factor.

Example: Trace the curve

$$y = \frac{x^2}{(x-1)(x-3)}. \tag{2}$$

1. There is no symmetry.
2. The only intersection with the axes is (0, 0).
3. As $x \to \pm\infty$, $y \to 1$: the line $y = 1$ is an asymptote. [Review Example (*b*), § 95.] For large positive x, the denominator is less than the numerator and $y > 1$; for large negative x, the denominator is greater than the numerator and $y < 1$. Thus the curve approaches the asymptote from above at the extreme right and from below at the extreme left. Putting $y = 1$ in (2), we find $x = \frac{3}{4}$: the curve crosses the asymptote at $(\frac{3}{4}, 1)$. (Theorem, § 94.)

As $x \to 1$, and as $x \to 3$, y increases indefinitely: the lines $x = 1$, $x = 3$ are asymptotes. (§ 95.)

4. The numerator vanishes at $x = 0$, but does not change sign because of the even exponent; the denominator, and hence the function, changes sign as x goes through 1, 3. This combined with step 3 limits the curve as shown.

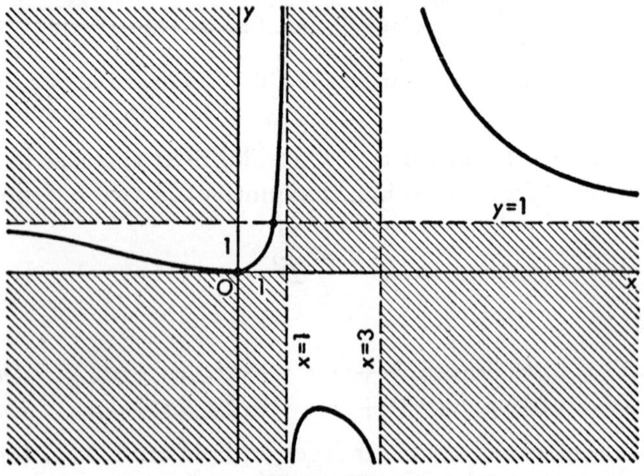

Figure 118

Exercises

In Exs. 1–28, trace the curve.

1. $y = \dfrac{1}{x+2}$.

2. $y = \dfrac{4}{1-x}$.

3. $y = \dfrac{x^2-9}{x}$.

4. $y = \dfrac{1+x}{1-x}$.

5. $y = \dfrac{1}{1-x^2}$.

6. $y = \dfrac{2x}{1-x^2}$.

7. $y = \dfrac{2x}{1+x^2}$.

8. $y = \dfrac{1}{1+x^2}$.

9. $x^3 - xy - 4x - y = 0$.

10. $x^3 + 4x^2 + xy - 2y = 0$.

11. $x^2y + x^2 - 4y - 1 = 0$.

12. $x^2y - 4x^2 - y + 9 = 0$.

13. $y = \dfrac{3-2x}{(x+2)^2}$.

14. $y = \dfrac{x+4}{(x-1)^2}$.

15. $y = \dfrac{x^2+4x+3}{x^2}$.

16. $y = \dfrac{4x-4-x^2}{x^2}$.

17. $y = \dfrac{x(x-1)(x+2)}{(x+1)^2}$.

18. $y = \dfrac{x^2(x-3)}{(x-1)^2}$.

19. $y = \dfrac{x^2-x-2}{x^2-4x}$.

20. $y = \dfrac{x^2-2x-3}{x^2-2x}$.

21. $y = \dfrac{x^2}{(x-2)(x-1)}$.

22. $y = \dfrac{(x-1)^2}{x(x-4)}$.

23. $y = \dfrac{x^3-2x^2-x+2}{(x-3)^2}$.

24. $y = \dfrac{(x-3)^2}{x^3-2x^2-x+2}$.

25. $y = \dfrac{(x-1)(x-2)}{x^2(x-4)}$.

26. $y = \dfrac{(x^2-4)^2}{x^2(x+1)}$.

27. $y = \dfrac{4x^3-32}{2x^3-x^2}$.

28. $y = \dfrac{x(x-2)^2(x+1)^2}{(x-1)^6}$.

29. Draw the curve from which $\tan 2\theta$ may be read off when $\tan\theta$ is given. (Put $x = \tan\theta$, $y = \tan 2\theta$. Cf. Ex. 6.)

30. Represent $\tan 3\theta$ as a function of $\tan\theta$: $\tan 3\theta = \dfrac{3\tan\theta - \tan^3\theta}{1 - 3\tan^2\theta}$. (Put $x = \tan\theta$, $y = \tan 3\theta$.)

31. Represent $\sin 2\theta$ as a function of $\tan\theta$: $\sin 2\theta = \dfrac{2\tan\theta}{1+\tan^2\theta}$. Put $x = \tan\theta$, $y = \sin 2\theta$.

32. Draw a curve from which $\cos 2\theta$ may be read off if $\tan\theta$ is given.

33. A circular cone is circumscribed about a sphere of radius a. Graph the altitude of the cone as a function of the radius. *Ans.* $h = \dfrac{2ar^2}{r^2 - a^2}$.

34. In Ex. 33, graph the volume of the cone as a function of the radius.

Ans. $V = \frac{2}{3}\pi a \cdot \dfrac{r^4}{r^2 - a^2}$.

100. Two-valued functions. If to each value of x there corresponds one and only one value of y, then y is said to be a *one-valued* function. The polynomial, the rational fraction, and many other important functions possess this property.

If to each value of x there correspond two values of y, the function is *two-valued.* Two important classes of functions of this type are the following.

I. Replace y by y^2 in § 98:

$$y^2 = a_0x^n + a_1x^{n-1} + \cdots + a_{n-1}x + a_n \quad (n \geqq 1). \tag{1}$$

The cases $n = 1$, $n = 2$ have been studied at some length: for this reason, we shall now consider chiefly the case $n \geqq 3$.

II. Replace y by y^2 in § 99:

$$y^2 = \frac{N}{D}, \tag{2}$$

where N and D are polynomials in x.

Evidently all curves of either class are symmetric with respect to the x-axis.

The analysis for these curves is not strikingly different from that of §§ 98–99, except in the following vitally important respect. In the former cases, a change of sign of the right member meant a change of sign of y and, for the curve, a crossing of the x-axis — by intersection or by jumping. Here, a change of sign of the right member means, for y^2, a change from positive to negative, and for y, a change from real (positive and negative) to imaginary. Thus, in the successive intervals marked off by the changes of sign, the curve will be alternately *present above and below Ox*, and *missing entirely.*

Example (*a*): Trace the curve $y^2 = (x - 2)(x^2 - 1)$.

1. Symmetric with respect to the x-axis.
2. $(2, 0)$, $(1, 0)$, $(-1, 0)$; $(0, \pm\sqrt{2})$.
3. When x is large positive, y^2 is large positive and y is large positive and negative; when x is large negative, y^2 is negative and y imaginary.

No curve of the type (1) can have either a horizontal or a vertical asymptote.

4. y^2 changes sign at $x = -1, 1, 2$. Since y^2 is negative at the extreme left there will be no curve until the first change of sign is reached, at $x = -1$: the curve appears in the interval from $x = -1$ to $x = 1$, disappears in the interval from $x = 1$ to $x = 2$, is present in the interval $x > 2$.
5. The oval is not symmetric with respect to Oy, the highest point being slightly to the left. (Fig. 119.)

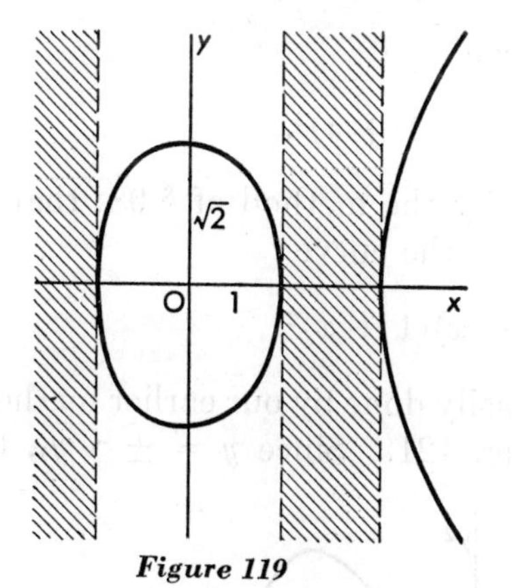

Figure 119

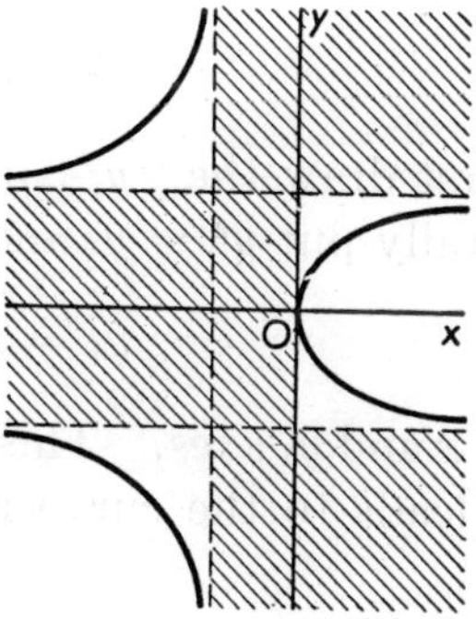

Figure 120

Example (*b*): Trace the curve

$$y^2 = \frac{2x}{x + 1}. \tag{3}$$

1. Symmetric with respect to the x-axis.
2. The only intercept is $(0, 0)$.

3. When x becomes large positive or negative, $y^2 \to 2$: the lines $y = \pm \sqrt{2}$ are asymptotes. (To see this, divide numerator and denominator by x, then let x increase.) For positive x, $y^2 < 2$ and the curve lies between the asymptotes; for negative x, $y^2 > 2$ and the curve lies outside of the strip bounded by those lines. Putting $y = \pm \sqrt{2}$ in (3), we see that there is no intersection with the asymptotes.

The line $x = -1$ is a vertical asymptote.

4. y^2 changes sign as x goes through the values -1, 0. At the extreme left $y^2 > 0$: the curve is present in the interval $x < -1$, absent from $x = -1$ to $x = 0$, present for $x > 0$. This combined with step 3 limits the curve as shown by the shading. (Fig. 120.)

A third example is introduced to illustrate another method for sketching curves of the types (1) and (2).

Example (*c*): Trace the curve

$$y^2 = x(1 - x)(1 + x)^2. \tag{4}$$

First draw the "y^2-curve" by the method of § 98: that is, mentally put $y^2 = y_1$ and draw the curve

$$y_1 = x(1 - x)(1 + x)^2$$

with x and y_1 axes. This is easily done by our earlier methods and leads to the curve in Fig. 121. Since $y = \pm \sqrt{y_1}$, the

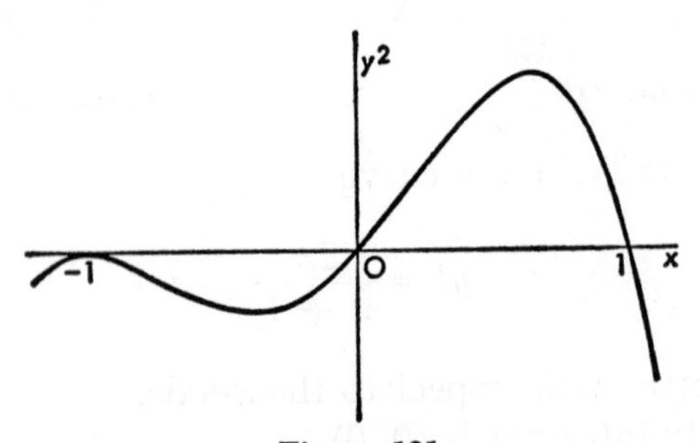

Figure 121

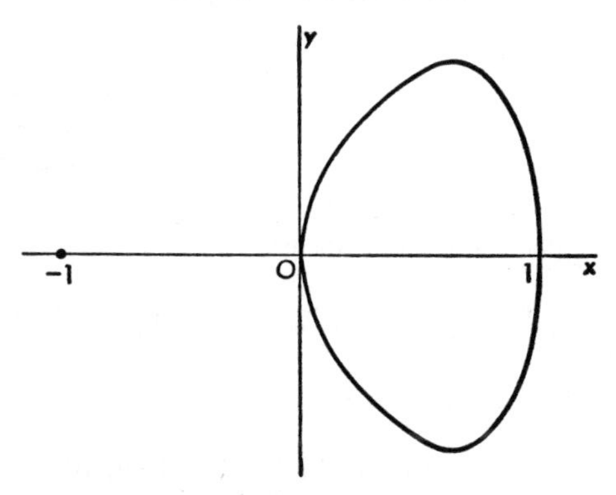

Figure 122

curve (4) in Fig. 122 follows at once from Fig. 121 by replacing each ordinate by its two square roots. In a range such as $x > 1$, where the ordinate of the y^2-curve is negative, the ordinate of the y-curve is imaginary. The point $(-1, 0)$ standing by itself on the curve in Fig. 122 is, quite naturally, called an *isolated point.*

Exercises

In Exs. 1–31, draw the curve.

1. $y^2 = x^3 - 4x.$

2. $y^2 = 4x - x^3.$

3. $y^2 = x(x + 1)^2.$

4. $y^2 = x^2(x + 1).$

5. $y^2 = x^3.$

6. $y^2 = -x^3.$

7. $y^2 = x^3 + 9x.$

8. $y^2 = x^3 - 5x^2 + 2x + 8.$

9. $y^2 = x^2(1 - x^2).$

10. $y^2 = x^2(1 - x)(2 + x).$

11. $y^2 = x(x + 1)(x - 1)^2.$

12. $y^2 = x(x - 2)(x - 1)^2.$

13. $y^2 = x^4 - 2x^3 - 5x^2 + 6x.$

14. $y^2 = 6x - 11x^2 + 6x^3 - x^4.$

15. $y^2 = x^3 - x^4.$

16. $y^2 = x^4 - x^3.$

17. $y^2 = \dfrac{2x}{1 + x^2}.$

18. $y^2 = \dfrac{1}{1 + x^2}.$

19. $y^2 = \dfrac{x^2 - 2x}{x - 1}.$

20. $y^2 = \dfrac{x^2 + 2x}{x - 1}.$

21. $y^2 = \dfrac{(x - 2)^3}{x^4}.$

22. $y^2 = \dfrac{(x - 1)^2(x - 3)}{x^4}.$

23. $y^2 = \dfrac{x^2 - 2x}{(x + 1)^3}.$

24. $y^2 = \dfrac{(x + 2)^2}{x^3}.$

25. $y^2 = \dfrac{x^2 - 8x + 16}{8 - x^3}$. **26.** $y^2 = \dfrac{x^2 - 2x}{x^2 + 4}$.

27. $y^2 = \dfrac{x^2 - x}{x^2 - 4}$. **28.** $y^2 = \dfrac{x^3 - 2x^2 - 9x + 18}{x^3 + 7x^2}$.

29. $y^2 = \dfrac{x^3}{a - x}$, the cissoid of Diocles.

30. $y^2 = \dfrac{x^2(3a - x)}{a + x}$, the trisectrix of Maclaurin.

31. $y^2 = \dfrac{x^2(a + x)}{a - x}$, the strophoid.

32. If the total surface area (including the base) of a circular cone is given as 2π square units, express the altitude as a function of the radius, and draw the graph.

33. A circular cone is circumscribed about a sphere of radius a. Express the radius of the cone as a function of its altitude, and draw the graph.

34. Show that if $\tan \theta$ is given, the values of $\cos \theta$ may be read off from the curve of Ex. 18. Put $x = \tan \theta$, $y = \cos \theta$.

35. Express $\sec \theta$ as a function of $\cos 2\theta$, and draw the graph.

101. Curve tracing in polar coordinates. If r and θ are connected by an equation, we may assign values to θ and compute the corresponding value or values of r. The points thus determined all lie on a definite curve, the *locus* of the equation.

To each equation corresponds a single curve, but due to the fact that a given point may be represented by different pairs of coordinates, a curve may be represented in the polar system by more than one equation. Thus the equations $r = 2$, $r = -2$ both represent a circle of radius 2 with center at O. Since (r, θ) and $(-r, \pi + \theta)$ represent the same point, the equations

$$r = \cos \theta - 1$$

and

$$r = \cos \theta + 1$$

have the same locus. In obtaining intersections of curves given by polar equations, it is often vital that this phenomenon be kept in mind. See Example (*b*) of § 103.

In the great majority of important polar equations, r is expressed in terms of *trigonometric functions of* θ. In such cases an analysis similar to that of § 97 would be comparatively ineffective, except as regards tests for symmetry — see § 102. Instead, we employ our knowledge of trigonometry to *observe the variation of* r *as* θ *varies.* As a rule, not much point-plotting is necessary: in the example of § 21, we plotted points in the first quadrant and from this deduced the entire remainder of the curve.

In this connection it is important to recall that the functions $\sin\theta$, $\cos\theta$, $\csc\theta$, and $\sec\theta$ are *periodic* with period* 2π; the functions $\tan\theta$ and $\cot\theta$ are periodic with period π.

102. Polar equations: tests for symmetry. Although in polar coordinates there is of course no x or y, for brevity we will refer to the polar axis and the perpendicular to this line through O as x-axis and y-axis respectively.

The following theorems are nearly obvious, and may be verified by the reader.

If the equation is unchanged when we change

(*a*) θ *to* $-\theta$, *or* (*b*) θ *to* $\pi - \theta$ *and* r *to* $-r$,	*the curve is symmetric with respect to the*	*x-axis;*
(*c*) θ *to* $\pi - \theta$, *or* (*d*) θ *to* $-\theta$ *and* r *to* $-r$,	*the curve is symmetric with respect to the*	*y-axis;*
(*e*) r *to* $-r$, *or* (*f*) θ *to* $\pi + \theta$,	*the curve is symmetric with respect to the*	*origin.*

Instead of memorizing the above tests, the student should become thoroughly familiar with the geometric meaning of each one, whereupon they will come readily to mind as needed. For instance, test (*b*) operates as follows. Changing θ to $\pi - \theta$ reflects a point P to the position P'; then, changing r to $-r$ reflects P' to P'', the image of P with respect to Ox.

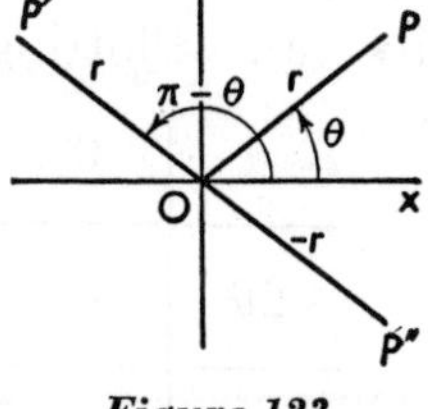

Figure 123

* That is, $\sin(2\pi + \theta) = \sin\theta$ for all values of θ; etc.

As a result of the fact that a point may be represented by more than one pair of coordinates, the converse theorems are not true. For instance, if the equation is changed when r is replaced by $-r$, the curve still may be symmetric with respect to O, by test (f): as an illustration, see Example (a), § 103.

The tests for symmetry afford a check on the remainder of our analysis, and also act as time-savers, often enabling us to obtain the remainder of the curve by reflection after a limited portion has been plotted. Thus, if a curve is symmetric with respect to both axes, then in most cases we may draw the curve for values of θ between 0° and 90°, and deduce the rest of the curve by reflection in both axes.

However, the qualification "in most cases" must not be overlooked. The above procedure does not always give the entire curve, for various reasons. For instance, values of θ in the first quadrant may give points in the third quadrant not corresponding to third-quadrant values of θ: see, for example, Exs. 1–2, p. 170. If due care is exercised, such cases are easily recognized.

103. Polar equations: one-valued functions. We consider first the case in which there is one value of r for each value of θ.

Example (a): Trace the *four-leaved rose* $r = a \cos 2\theta$.

Since $\cos(-2\theta) = \cos 2\theta$, the curve is symmetric with respect to Ox by (a), § 102. Since $\cos 2(\pi - \theta) = \cos 2\theta$, it follows that the curve is symmetric with respect to Oy by (c). Plotting the points found below,* we obtain the

θ	0°	15°	$22\frac{1}{2}$°	30°	45°
2θ	0°	30°	45°	60°	90°
r	a	$\frac{1}{2}\sqrt{3}\,a$	$\frac{1}{2}\sqrt{2}\,a$	$\frac{1}{2}a$	0

* It is of course the values of r and θ, not r and 2θ, that are plotted. Thus the second point is ($\frac{1}{2}\sqrt{3}a$, 15°), etc.

half-loop numbered 1. As θ ranges from 45° to 90°, 2θ ranges from 90° to 180°: thus the values of $\cos 2\theta$ are numerically the same as those found above, but in reverse order and negative. Since r is negative, the corresponding portion of the curve lies not in the first, but in the third quadrant: the half-loop 2. Reflecting in Oy, we obtain the arcs 3, 4; reflecting in Ox, the balance of the curve. Since this covers a complete period of the function $\cos 2\theta$, we have the entire curve.

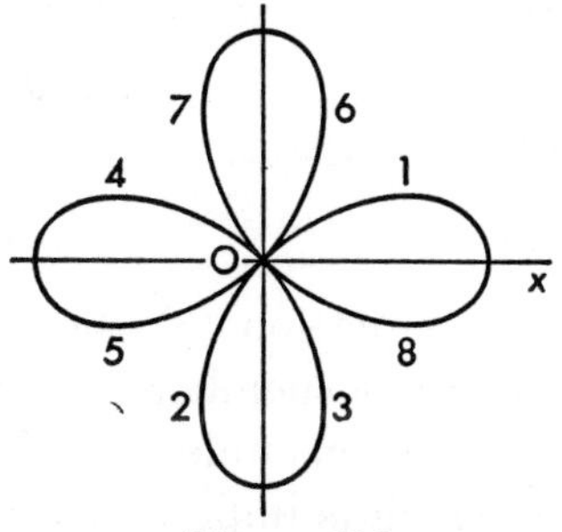

Figure 124

Example (*b*): Find the intersections of the curves

$$r = \sin \theta + 1 \tag{1}$$

and

$$r = \cos \theta - 1. \tag{2}$$

First we sketch the two curves carefully, employing the technique of Example (*a*). Each turns out to be a heart-shaped figure, called a cardioid. The curves are shown in Fig. 125.

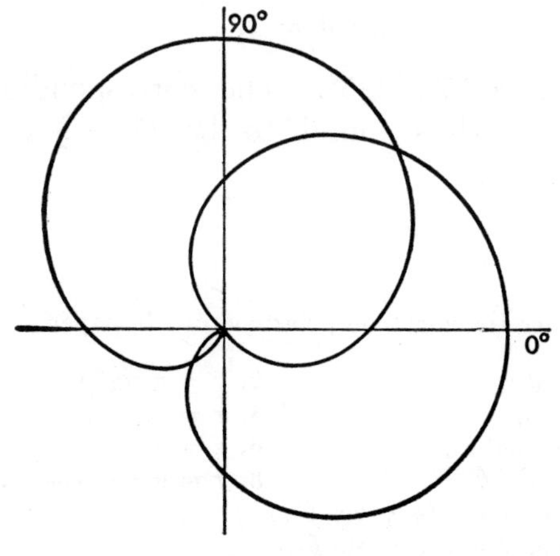

Figure 125

One intersection is at the origin, $r = 0$. To find the others it is natural to solve (1) and (2) as simultaneous equations. Elimination of r yields

$$\sin \theta + 1 = \cos \theta - 1,$$

or

$$2 + \sin \theta = \cos \theta. \tag{3}$$

In equation (3) the left member is never less than unity, the right member never greater than unity. Hence (3) has no solutions unless $\sin \theta = -1$ and $\cos \theta = +1$ for the same θ, which is impossible. Equations (1) and (2) have no common solutions. The curves (1) and (2) do, however, have intersections.

Let us, then, in equation (2) replace θ by $(\pi + \theta)$ and r by $(-r)$, thus obtaining

$$r = \cos \theta + 1. \tag{4}$$

Equation (4) represents the same curve as does (2). Now solve (1) and (4) together, obtaining

$$\sin \theta + 1 = \cos \theta + 1,$$

or

$$\tan \theta = 1,$$

from which $\theta = 45°, 225°$. The corresponding points of intersection are thus found to be $(1 + \frac{1}{2}\sqrt{2}, 45°)$ and $(1 - \frac{1}{2}\sqrt{2}, 225°)$.

Exercises

In Exs. 1–30, trace the curve on polar coordinate paper.

1. $r = a \sin^2 \theta$.
2. $r = a \cos^2 \theta$.
3. $r = a \sin 2\theta$.
4. $r = a(\cos \theta - \sin \theta)$.
5. $r = a(1 + \cos^2 \theta)$.
6. $r = a(2 - \cos^2 \theta)$.
7. $r = a(1 - \sin \theta)$.
8. $r = a(1 - \cos \theta)$.
9. The limaçon $r = a(4 + \cos \theta)$.
10. The limaçon $r = a(3 - \sin \theta)$.

11. The limaçon $r = a(2 \sin \theta - 1)$.
12. The limaçon $r = a(1 + 2 \cos \theta)$.

13. $r(2 - \cos \theta) = a$.
14. $r(2 + \sin \theta) = a$.
15. $r = a \sin 3\theta$.
16. $r = a \cos 3\theta$.
17. $r = a \cos 4\theta$.
18. $r = a \sin 4\theta$.
19. $r = a \cos^2 2\theta$.
20. $r = a \sin^2 2\theta$.
21. $r = a(4 \cos^2 \theta - 3)$.
22. $r = a(1 - 4 \sin^2 \theta)$.
23. $r = a \sec^2 \theta$.
24. $r = a \csc^2 \theta$.
25. $r = a \sin \theta(1 - \sin \theta)$.
26. $r = a \sin \theta(2 - \cos \theta)$.
27. The trisectrix $r = a(4 \cos \theta - \sec \theta)$.

In Exs. 28–30, express θ in radians.

28. The *spiral of Archimedes* $r = a\theta$. Take $a = \dfrac{1}{\pi}$.

29. The spiral $r = a\theta^2$. Take $a = \dfrac{1}{\pi^2}$.

30. The *hyperbolic spiral* $r\theta = a$. Take $a = \pi$.

In Exs. 31–34, show that the two equations represent the same curve.

31. $r = a \sec \theta + b$, $r = a \sec \theta - b$. (*Conchoid.*)
32. $r = a(\sec \theta - \tan \theta)$, $r = a(\sec \theta + \tan \theta)$. (*Strophoid.*)
33. $r = a \cos \theta - b$, $r = a \cos \theta + b$. (*Limaçon.*)
34. $r = a(\sec \theta - \cos \theta)$, $r = a \sin \theta \tan \theta$. (*Cissoid.*)

104. Polar equations; two-valued functions. Consider the case in which r^2 is expressed as a function of θ.

Since r occurs only in an even power, all curves of this class are symmetric with respect to the pole, by (*e*), § 102. Hence, if a curve of this class is symmetric with respect to one axis, it is symmetric with respect to the other also (Ex. 32, p. 34).

Example: Trace the *lemniscate* $r^2 = a^2 \cos 2\theta$.

θ	$0°$	$15°$	$22\frac{1}{2}°$	$30°$	$45°$
2θ	$0°$	$30°$	$45°$	$60°$	$90°$
r^2	a^2	$0.87a^2$	$0.71a^2$	$0.5a^2$	0
r	$\pm a$	$\pm 0.93a$	$\pm 0.84a$	$\pm 0.71a$	0

Since $\cos(-2\theta) = \cos 2\theta$, the curve is symmetric with respect to Ox; hence also to Oy, by the preceding remark. Plotting

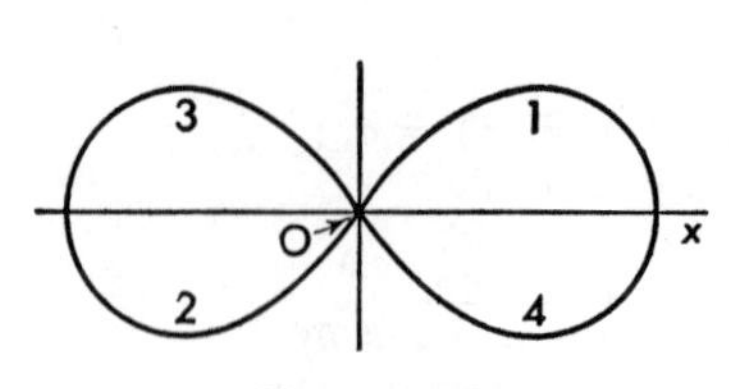

Figure 126

the above points, we get the arcs 1, 2. As θ varies from 45° to 90°, 2θ ranges from 90° to 180°, $\cos 2\theta$ is negative, and r is imaginary. Having examined r through the first quadrant, we obtain the balance of the curve by reflection.

Exercises

In Exs. 1–24, trace the curve.

1. $r^2 = a^2 \cos \theta$.
2. $r^2 = a^2 \sin \theta$.
3. $r^2 = 1 + \cos^2 \theta$.
4. $r^2(1 + \sin^2 \theta) = 2$.
5. $r^2(1 + 4 \sin^2 \theta) = 5$.
6. $r^2 = a^2 \sin 2\theta$.
7. $r^2 \cos 2\theta = a^2$.
8. $r^2 \sin 2\theta = a^2$.
9. $r^2 = 3 - 4 \cos^2 \theta$.
10. $r^2 = 1 - 4 \sin^2 \theta$.
11. $r^2 = a^2(1 + \cos \theta)^3$.
12. $r^2 = a^2(1 - 2 \sin \theta)^3$.
13. $r^2 = a^2(2 - \cos \theta)$.
14. $r^2 = a^2(\sin \theta + 4)$.
15. $r^2 = a^2(1 + \sin \theta)$.
16. $r^2 = a^2(2 \cos \theta - 1)$.
17. $r^2 = a^2 \sin \theta(1 + \sin \theta)$.
18. $r^2 = a^2 \cos \theta(1 + \cos \theta)$.
19. $r^2 = a^2(\sin \theta + \cos \theta)$.
20. $r^2 = a^2 \sin \theta(1 - 2 \sin \theta)$.
21. $r^2 = a^2 \sin \theta \sin 2\theta$.
22. $r^2 = a^2 \cos \theta \cos 2\theta$.
23. $r^2 = a^2 \cos 3\theta$.
24. $r^2 = a^2 \sin 3\theta$.

In Exs. 25–28, change the equation to polar form (§ 43) and sketch the curve.

25. $(x^2 + y^2)^3 = a^4x^2$. (Ex. 1.)
26. $(x^2 + y^2)^3 = a^4y^2$. (Ex. 2.)
27. $(x^2 + y^2)^2 = 2a^2xy$. (Ex. 6.)
28. $(x^2 + y^2 - 1)(x^2 + y^2) = x^2$. (Ex. 3.)

In Exs. 29–37, transform the equation to rectangular form.

29. Ex. 4.
30. Ex. 5.
31. Ex. 7.
32. Ex. 8.
33. Ex. 9.
34. Ex. 10.
35. Ex. 15.
36. Ex. 21.
37. Ex. 22.

105. Applications. In so elementary a course as this, we are heavily handicapped in trying to apply our results to other sciences, for two reasons: if the student is to understand fully the nature of the problem, considerable knowledge of the science in question — engineering, physics, chemistry, etc. —

is apt to be needed; and as a rule, more advanced mathematics is required in deriving the formulas that occur. Nevertheless it seems worth while to mention an occasional problem in order to give some idea of ways in which our science is useful to others besides mathematicians. The reader must take the formulas on faith; as a partial offset to this, we shall try to select problems in which at least the meaning of the variables is a matter of common knowledge, and in which the reader can appreciate intuitively the reasonableness of the results.

106. Bending of beams. Curves of the type (1), § 98, occur quite often in physical problems. An important instance is in connection with the bending of beams.

Consider a beam of length l (Fig. 127), rigidly fixed in a horizontal position (nailed down, or embedded in concrete) at the left end O, supported freely — i.e., not fastened — at the right end A, the two supports being at the same level. Then, if there is no load, the beam lies in the horizontal line OA. (This assumes that the beam does not bend perceptibly under its own weight.) The distance by which any point of the beam sags below the line OA, under a load, is called the *deflection* at that point.

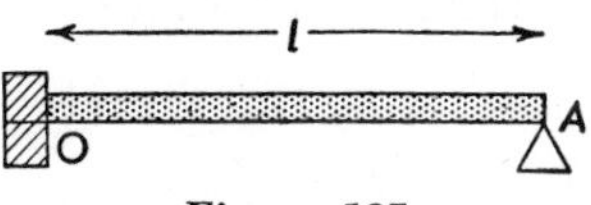

Figure 127

Suppose that the beam bears a load uniformly distributed along its whole length.* With the origin at the fixed end and the y-axis *positive downward*, the deflection at distance x from the fixed end is given by the formula

$$y = \frac{k}{48l^3}(3l^2x^2 - 5lx^3 + 2x^4),$$

where k is a constant depending on the magnitude of the load and on the physical characteristics of the beam. Analysis of this equation will be left to the reader. In Fig. 128 the

* The weight of the beam itself, if it causes appreciable deflection, would evidently come under this heading.

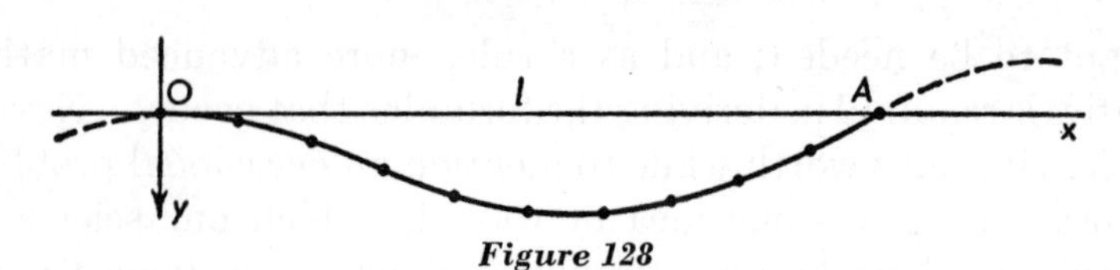

Figure 128

interval occupied by the beam, from $x = 0$ to $x = l$, is shown in detail; points are plotted for $x = 0.1l$, $x = 0.2l$, etc., with the y-scale greatly exaggerated.

107. Gravitational attraction. It is a physical law that two material particles attract each other with a force proportional to the product of the masses and inversely proportional to the square of the distance between them:

$$F = \frac{km_1m_2}{d^2}. \tag{1}$$

For simplicity in writing, we shall put

$$km_1m_2 = c,$$

so that (1) becomes

$$F = \frac{c}{d^2}.$$

From this law, using more advanced mathematics, we can in simple cases determine the attraction between two bodies of various shapes — a particle and a straight rod, two spheres, etc.

The attraction (or repulsion) between two electric charges or two magnetic poles likewise follows the "law of inverse squares."

Exercises

In Exs. 1–7, draw the curve of the beam, with the origin at the left end and the y-axis positive downward. In Exs. 3–6, note that for physical reasons the curve must be symmetric with respect to the line $x = \frac{1}{2}l$. Find the maximum deflection in Exs. 1–6.

1. Cantilever beam (fixed at one end, unsupported at the other — for example, a springboard) bearing a load at the free end:

$$y = \frac{k}{6l^2}(3lx^2 - x^3), \qquad 0 \leqq x \leqq l. \qquad \textit{Ans. } y_{\max} = \tfrac{1}{3}kl.$$

2. Cantilever beam bearing a uniformly distributed load:

$$y = \frac{k}{24l^3}(6l^2x^2 - 4lx^3 + x^4), \qquad 0 \leqq x \leqq l. \qquad \textit{Ans. } y_{\max} = \tfrac{1}{8}kl.$$

3. Beam fixed at both ends, bearing a load at the midpoint:

$$y = \frac{k}{48l^2}(3lx^2 - 4x^3), \qquad 0 \leqq x \leqq \tfrac{1}{2}l. \qquad \textit{Ans. } y_{\max} = \tfrac{1}{192}kl.$$

4. Beam fixed at both ends, bearing a uniformly distributed load:

$$y = \frac{k}{24l^3}(l^2x^2 - 2lx^3 + x^4), \qquad 0 \leqq x \leqq l. \qquad \textit{Ans. } y_{\max} = \tfrac{1}{384}kl.$$

5. Beam freely supported at both ends, bearing a load at the midpoint:

$$y = \frac{k}{48l^2}(3l^2x - 4x^3), \qquad 0 \leqq x \leqq \tfrac{1}{2}l. \qquad \textit{Ans. } y_{\max} = \tfrac{1}{48}kl.$$

6. Beam freely supported at both ends, bearing a uniformly distributed load:

$$y = \frac{k}{24l^3}(l^3x - 2lx^3 + x^4), \qquad 0 \leqq x \leqq l. \qquad \textit{Ans. } y_{\max} = \tfrac{5}{384}kl.$$

7. Beam fixed at both ends, bearing a pile of bricks increasing from height zero at the left end to height l at right end:

$$y = \frac{k}{60l^4}(2l^3x^2 - 3l^2x^3 + x^5), \qquad 0 \leqq x \leqq l.$$

In Exs. 8–12, draw the curve.

8. The attraction between a sphere of radius a and a particle at distance x from the center:

$$F = \frac{cx}{a^3}, \qquad 0 \leqq x \leqq a; \qquad F = \frac{c}{x^2}, \qquad x \geqq a.$$

9. A body falls to earth from a height of 1 mile. Considering the earth as a homogeneous sphere of radius 4000 mi., compute the amount by which the force on the body changes during the fall.* (Ex. 8.) *Ans.* $\frac{1}{20}$ of 1%.

* The result shows that for moderate heights, and for most purposes, the earth's attraction may be considered *constant*. This assumption is nearly always satisfactory in engineering practice, and results in tremendous simplification.

10. The attraction between two particles at distance x apart:

$$F = \frac{c}{x^2}, \qquad x > 0.$$

11. The attraction between a thin spherical shell of radius a and a particle at distance x from the center:

$$F = 0, \quad 0 \leqq x < a; \qquad F = \frac{c}{2a^2}, \quad x = a; \qquad F = \frac{c}{x^2}, \quad x > a.$$

12. The attraction between a hollow sphere of inner radius b, outer radius a, and a particle at distance x from the center:

$$F = 0, \quad 0 \leqq x \leqq b; \quad F = \frac{c(x^3 - b^3)}{x^2(a^3 - b^3)}, \quad b \leqq x \leqq a; \quad F = \frac{c}{x^2}, \quad x \geqq a.$$

In drawing the curve, take $a = 3b$.

108. On higher plane curves. One interesting method for gaining some insight as to shapes which may be taken by curves of degree greater than two will now be illustrated.

Consider the cubic

$$(1) \qquad (x + y - 1)(x^2 + y^2 - 1) = 0,$$

which, since its left member is factorable, degenerates into the line

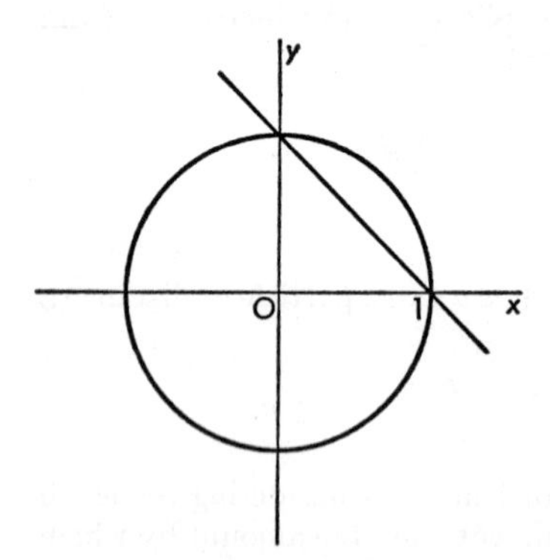

Figure 129

$$(2) \qquad x + y = 1$$

and the circle

$$(3) \qquad x^2 + y^2 = 1.$$

Associate with (1) the cubic

$$(4) \qquad (x + y - 1)(x^2 + y^2 - 1) = \epsilon,$$

where ϵ is to be taken constant, not zero, but as small as we wish. It is intuitively evident, and can be proved with the aid of calculus, that by taking ϵ sufficiently small numerically the curve (4) can be made to approximate the curve (1) as closely as desired. But the cubics (1) and (4) have no intersections, since ϵ is being chosen different from zero. It follows that the cubic (4) has one of the two shapes exhibited in Figs. 130–131.

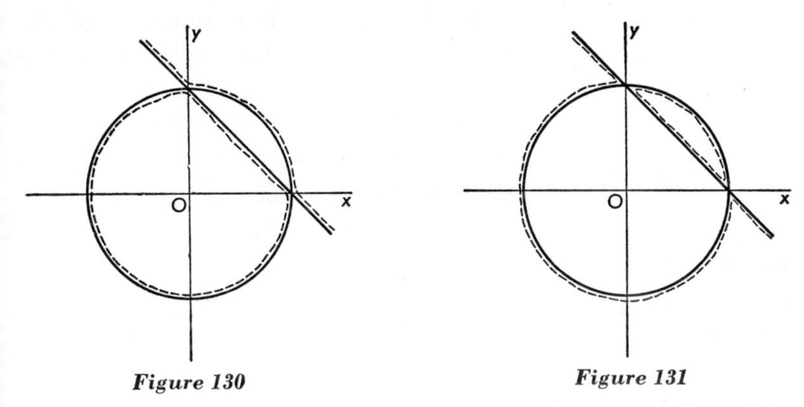

Figure 130　　*Figure 131*

For our purpose it is unessential, though easily determined, that Fig. 130 corresponds to small positive ϵ, Fig. 131 to small negative ϵ. To see this, use $x = 0$, $y = 1.1$ in the left member of (4) and thus determine that with $\epsilon = 0.021$ the cubic (4) passes through (0, 1.1).

In drawing the approximate curves, it is helpful to note that any algebraic curve

$$f(x, y) = 0 \tag{5}$$

divides the plane into two regions, one in which f is negative, one in which f is positive. Each time a moving point (x, y) crosses the curve (5), $f(x, y)$ changes sign. Thus for any negative ϵ, small or not, the locus of equation (4) is restricted to the same "side" of the cubic (1) as is the dotted curve in Fig. 131. If ϵ is not small, however, the present argument gives no hint as to the shape of the cubic (4).

In Ex. 11 below, for instance, a logical first step in treating

$$xy(x + y - 1) = \epsilon \tag{6}$$

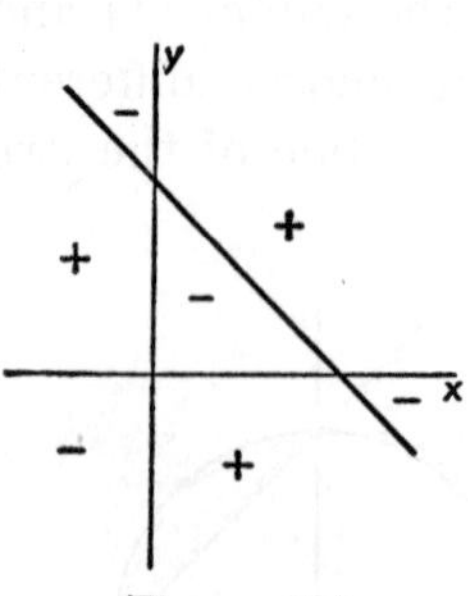

Figure 132

is first to determine the regions corresponding to $\epsilon > 0$ and to $\epsilon < 0$. Choose a random point, say $(-1, -1)$. For this point ϵ is negative. At once we conclude that the various portions of the plane have associated with them the signs (of ϵ) indicated in Fig. 132.

The following exercises are meant to be instructive and amusing but by no means exhaustive (or exhausting).

Exercises

In each exercise, draw the degenerate curve for $\epsilon = 0$, then draw approximate curves yielded by sufficiently small nonzero ϵ.

1. $xy = \epsilon$.
2. $(x - y)(x + y) = \epsilon$.
3. $(x - 1)(y^2 - 4x) = \epsilon$.
4. $(y - 1)(4x^2 + y^2 - 4) = \epsilon$.
5. $(x - 2)(x^2 + y^2 - 1) = \epsilon$.
6. $(x - 2)(x^2 - y^2 - 1) = \epsilon$.
7. $(x - 1)(x - 2)(y - 2) = \epsilon$.
8. $(x^2 + y^2 - 1)(x^2 + y^2 - 2x) = \epsilon$.
9. $(x^2 - 4y)(x^2 + y^2 - 5) = \epsilon$.
10. $(y^2 - x)(x^2 - y) = \epsilon$.
11. $xy(x + y - 1) = \epsilon$.
12. $(x^2 + y^2 - 1)(y^2 - 4x + 8) = \epsilon$.
13. $(x^2 + 4y^2 - 4)(x^2 - y^2 - 1) = \epsilon$.
14. $(4x^2 + y^2 - 1)(x^2 - y^2 - 1) = \epsilon$.
15. $(x^2 + y^2 - 4)[(x - 2)^2 + y^2 - 1][(x + 2)^2 + y^2 - 1] = \epsilon$.
16. $xy(x^2 + y^2 - 1) = \epsilon$.
17. $(x - 1)(x - 2)(x^2 + y^2 - 16) = \epsilon$.
18. $(x - 1)(y - 2)(y^2 - 4x) = \epsilon$.
19. $(y^2 - 4x)(x^2 - y^2 - 4) = \epsilon$.
20. $(x - 1)(x^2 + y^2 - 1) = \epsilon$.
21. $(r - a)(r - a + a \cos \theta) = \epsilon$.
22. $(r - a)(r^2 - a^2 \cos 2\theta) = \epsilon$.

CHAPTER 12 *Parametric Equations*

109. Parametric equations. In both pure and applied mathematics, a curve often arises most naturally as the locus of points whose coordinates are determined by two equations

$$x = f(t), \qquad y = g(t), \tag{1}$$

giving x and y in terms of a third variable t. The variable t is then called a parameter; the equations (1) are parametric equations of the curve. To obtain the rectangular equation of the curve, we need to eliminate the parameter. As we shall see, such elimination may not be feasible; it may not be wise even when feasible.

A curve may be drawn by plotting points directly from its parametric equations, assigning suitable values to the parameter and computing corresponding values of x and y. The curve may be drawn by eliminating the parameter and then tracing the locus of the rectangular equation by our usual methods. Sometimes it is desirable to combine these two techniques.

Parametric equations for a curve are not unique. The number of different parametric representations for any specified curve is unlimited.

Example (*a*): Draw the curve whose parametric equations are

$$x = -2 - 3t, \qquad y = 1 + t. \tag{2}$$

Elimination of the parameter at once yields

$$x + 3y = 1, \tag{3}$$

so that the locus is the straight line shown in Fig. 133, assuming that there are no restrictions on the parameter, except that it be real.

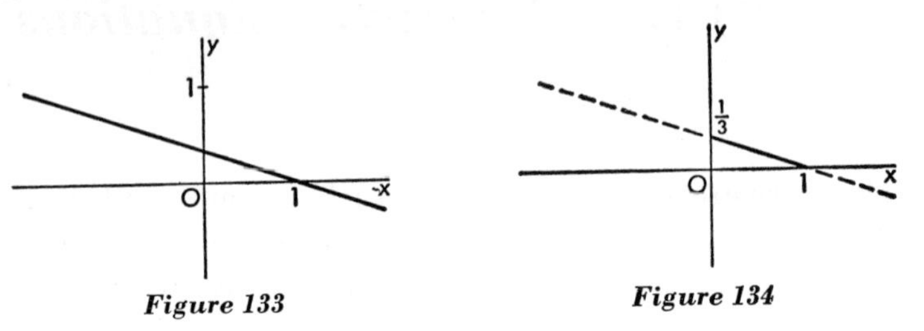

Figure 133

Figure 134

Example (*b*): Draw the curve

$$x = \cos^2 t, \qquad y = \tfrac{1}{3} \sin^2 t. \tag{4}$$

Since $\cos^2 t + \sin^2 t = 1$, we are again, as in Example (a), led to the equation

$$x + 3y = 1. \tag{3}$$

This time, however, the properties of the trigonometric functions automatically restrict x and y. For real t, $\cos^2 t$ lies between zero and unity. Hence the locus of (4) is that portion of the line (3) for which

$$0 \leqq x \leqq 1,$$

the portion shown solid in Fig. 134. Note the restriction on y, $0 \leqq y \leqq \frac{1}{3}$, obtained in a similar manner.

110. Point-plotting. To plot, by points, a curve represented by parametric equations, we merely assign suitable values to the parameter and compute the corresponding values of x and y.

Example: Plot the curve

$$x = a \cos^3 \theta, \qquad y = a \sin^3 \theta. \tag{1}$$

θ	0	$\frac{1}{6}\pi$	$\frac{1}{4}\pi$	$\frac{1}{3}\pi$	$\frac{1}{2}\pi$
x	a	$\frac{3}{8}\sqrt{3}\,a$	$\frac{1}{4}\sqrt{2}\,a$	$\frac{1}{8}a$	0
y	0	$\frac{1}{8}a$	$\frac{1}{4}\sqrt{2}\,a$	$\frac{3}{8}\sqrt{3}\,a$	a

Plotting these points, we get the portion of Fig. 135 lying in the first quadrant. By taking the cube root of the square of each member, the given equations may be written in the form

(2) $x^{\frac{2}{3}} = a^{\frac{2}{3}} \cos^2 \theta, \qquad y^{\frac{2}{3}} = a^{\frac{2}{3}} \sin^2 \theta;$

adding, we get the Cartesian equation

$$x^{\frac{2}{3}} + y^{\frac{2}{3}} = a^{\frac{2}{3}}.$$

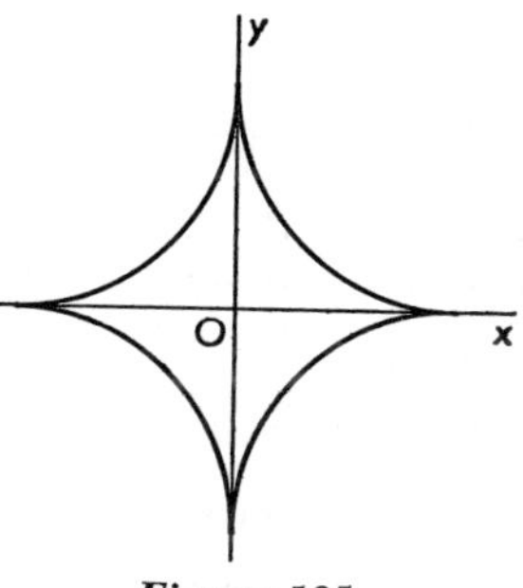

Figure 135

From this it appears that the curve is symmetric with respect to both axes, whence the balance of the curve may be obtained by reflection. This curve is the *hypocycloid of four cusps.*

The student is warned not to confuse parametric representation with polar coordinates. The parameter θ here occurring is quite different from the polar angle θ.

Exercises

In Exs. 1–14, find the rectangular equation and draw the curve. Assuming the parameter restricted to real values, determine whether the parametric equations represent all or part of the curve.

1. $x = 1 - t,\ y = 1 + t.$
2. $x = 2t,\ y = 3 - 4t.$
3. $x = 1 - t^2,\ y = 1 + t^2.$
4. $x = 2t^2,\ y = 3 - 4t^2.$
5. $x = 1 + t,\ y = t^2.$
6. $x = 4t^2,\ y = t - 2.$
7. $x = 2 \sin^2 \varphi,\ y = \cos 2\varphi.$
8. $x = \tan^2 \varphi,\ y = \sec^2 \varphi.$
9. $x = \cos \theta,\ y = \sec \theta.$
10. $x = (1 + t)^{-2},\ y = (1 + t)^2.$
11. $x = \log_{10} n^2,\ y = \log_{10} 100n.$
12. $x = 1 + 2^{-t},\ y = 2^t.$
13. $x = \tan \beta,\ y = \sin 2\beta.$
14. $x = \frac{1}{2}(1 + t^2),\ y = \sqrt{1 - t^4}.$

In Exs. 15–20, plot the curve by points, using the parametric equations with real parameter. Obtain the rectangular equation.

15. $x = t^2 - 1,\ y = t^3 - 5t.$ *Ans.* $y^2 = (x + 1)(x - 4)^2.$

16. $x = 2 - t^2,\ y = t + t^3.$ *Ans.* $y^2 = (2 - x)(3 - x)^2.$

17. $x = \dfrac{1}{1 + t},\ y = \dfrac{2}{1 + t^2}.$ *Ans.* $y = \dfrac{2x^2}{2x^2 - 2x + 1}.$

18. $x = \dfrac{1}{1 - t},\ y = \dfrac{1}{1 - t^2}.$ *Ans.* $x^2 - 2xy + y = 0.$

19. $x = \cos\varphi,\ y = \cos\varphi + \sin\varphi.$ *Ans.* $2x^2 - 2xy + y^2 = 1.$

20. $x = \cos\varphi,\ y = \cos\varphi - \sin\varphi.$ *Ans.* $2x^2 - 2xy + y^2 = 1.$

In Exs. 21–26, plot by points that portion of the curve corresponding to $t \geqq 0$. Obtain the rectangular equation.

21. Ex. 5. **22.** Ex. 6. **23.** Ex. 10.
24. Ex. 12. **25.** Ex. 17. **26.** Ex. 18.

111. Parametric equations of a straight line. In terms of a real parameter t, one set of parametric equations for a straight line are the equations

$$(1) \qquad x = x_1 + at, \qquad y = y_1 + bt,$$

in which x_1, y_1, a, b are fixed quantities. The point (x_1, y_1) lies on the line (1) because it corresponds to the value $t = 0$. If $a \neq 0$ in (1), then $\dfrac{b}{a}$ is the slope of the line. If $a = 0$ in (1), then the line is $x = x_1$ and the slope does not exist.

For some purposes parametric equations for a line are convenient. For instance, given

$$(2) \qquad x = 2 - 3t, \qquad y = 1 + 4t,$$

we can rapidly obtain several points on the line by using specific values of t. The values $t = 0, -\frac{1}{2}, 2$, for example, yield the points $(2, 1)$, $(\frac{7}{2}, -1)$, $(-4, 9)$ with little work.

On the other hand, parametric equations are not convenient for such tasks as obtaining the intersection of two lines, essentially because there is no reason to expect the parameters in two instances to be the same. Given line (2) and another line

(3) $$x = 1 + \beta, \qquad y = 3 - 6\beta,$$

the intersection can be obtained by eliminating both t and β, but the parametric form has contributed nothing to the solution (because our first step is to eliminate those parameters).

112. Parametric equations of a circle. The equation of any circle can be put in the form

(1) $$(x - h)^2 + (y - k)^2 = r^2,$$

or

(2) $$\left(\frac{x - h}{r}\right)^2 + \left(\frac{y - k}{r}\right)^2 = 1.$$

We know from trigonometry two functions, sine and cosine, which have the property that the sum of their squares is unity. Therefore we can at once write equations

(3) $$\frac{x - h}{r} = \cos \varphi, \qquad \frac{y - k}{r} = \sin \varphi,$$

which will lead to (2), or (1), when the parameter φ is eliminated. Thus, from (3), we obtain a set of parametric equations

(4) $$\boldsymbol{x = h + r \cos \varphi, \qquad y = k + r \sin \varphi,}$$

with φ as parameter, for the circle (1).

113. Parametric equations of conics. Following the technique developed in the previous section, a set of parametric equations for the ellipse

(1) $$\frac{(x - h)^2}{a^2} + \frac{(y - k)^2}{b^2} = 1$$

can be obtained mentally. The desired equations are

(2) $$\boldsymbol{x = h + a \cos \beta, \qquad y = k + b \sin \beta,}$$

with parameter β.

In seeking parametric equations for the hyperbola

$$\frac{(x-h)^2}{a^2} - \frac{(y-k)^2}{b^2} = 1, \tag{3}$$

we turn again to trigonometry. Recall that

$$\sec^2 \alpha - \tan^2 \alpha = 1,$$

which suggests the corresponding parametric representation

$$x = h + a \sec \alpha, \qquad y = k + b \tan \alpha, \tag{4}$$

with parameter α, for the hyperbola (3).

For the parabola

$$(y-k)^2 = 4a(x-h), \tag{5}$$

a simple set of parametric equations is

$$x = h + at^2, \qquad y = k + 2at, \tag{6}$$

with parameter t. See also the type exhibited in the next section.

The student should keep in mind that any number of different parametric equations can be written down for any curve whose equation is given. The examples in this chapter are merely widely used specific instances.

114. Motion in a plane curve. As an important application of parametric representation we cite the following problem of mechanics.

When a point moves in a plane curve under the action of a given force* or system of forces, an especially convenient way of studying the motion is to express the rectangular coordinates of the point as functions of the time t. The equations giving x and y in terms of t are *parametric equations of the path.*

* The "point" is supposed to be endowed with mass — a "material particle." Further, the argument applies to a body of any size or shape, provided that for present purposes the motion of the entire body is completely characterized by the motion of one of its points. This would be the case, for instance, in computing the range of a projectile, or determining the orbit of a planet.

Example: Determine the path of a point moving according to the laws

$$x = \cos t,$$
$$y = 4 \cos t + \cos 2t.$$

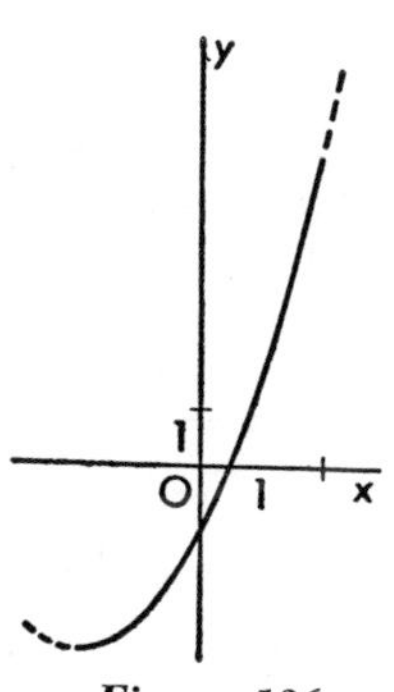

Figure 136

Since $\cos 2t = 2 \cos^2 t - 1$, we have

$$y = 4 \cos t + 2 \cos^2 t - 1$$
$$= 4x + 2x^2 - 1.$$

In standard form, this equation is

$$(x + 1)^2 = \tfrac{1}{2}(y + 3):$$

a parabola with vertex at $(-1, -3)$, opening upward. When $t = 0$, $x = 1$, $y = 5$. As $\cos t$ ranges from 1 to -1, the point moves along the parabola to $(-1, -3)$; it then returns to the starting point, and subsequently repeats the same cycle indefinitely.

Exercises

In Exs. 1–4, locate three points on the line whose parametric equations are given. Find the slope and draw the line.

1. $x = 2 - t$, $y = 1 + 2t$.
2. $x = 4 + 3t$, $y = 7 + t$.
3. $x = 4t$, $y = 1 - 2t$.
4. $x = -1 + t$, $y = 2t$.
5. Find the parametric equations of the straight line in terms of the parameter $k = \dfrac{P_1P}{P_1P_2}$, where $P:(x, y)$ is any point on the line, and $P_1:(x_1, y_1)$, $P_2:(x_2, y_2)$ are two given points on the line. (See Ex. 19, p. 12.)
Ans. $\boldsymbol{x = x_1 + k(x_2 - x_1),\ y = y_1 + k(y_2 - y_1)}$.

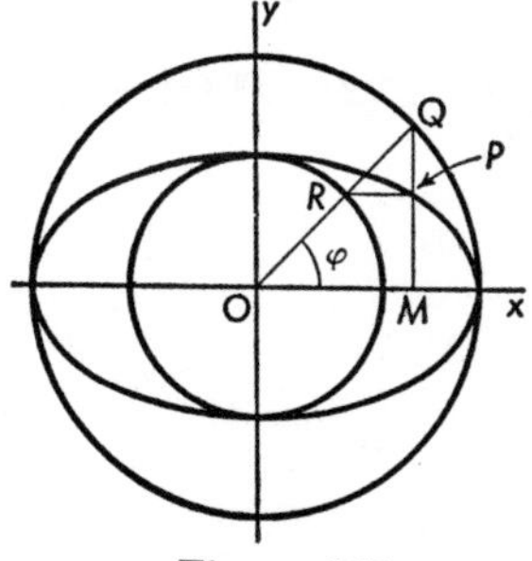

Figure 137

6. A circle is drawn on the major axis of the ellipse $\dfrac{x^2}{a^2} + \dfrac{y^2}{b^2} = 1$ as diameter; the ordinate MP of any point P is produced to Q on the circle; the line OQ is drawn, making with Ox an angle φ, called the *eccentric angle corresponding to the point P*. Find the parametric equations of the ellipse in terms of the eccentric angle. *Ans.* $\boldsymbol{x = a \cos \varphi,\ y = b \sin \varphi}$.

In Exs. 7–14, identify the curve; eliminate the parameter and draw the curve.

7. $x = 1 + 2 \cos \varphi,\ y = -2 + 3 \sin \varphi.$
8. $x = -3 + \cos \varphi,\ y = 4 - 2 \sin \varphi.$
9. $x = 3 \sin \varphi,\ y = 1 + \cos \varphi.$
10. $x = -3 + \cos \varphi,\ y = 4 \sin \varphi.$
11. $x = 2 + 3 \sec t,\ y = -4 + \tan t.$
12. $x = \tan 3t,\ y = 2 + \sec 3t.$
13. $x = 4 - \lambda^2,\ y = -3 - 2\lambda.$
14. $x = -1 + 2\lambda^2,\ y = 1 + 4\lambda.$

In Exs. 15–24, find at least two different parametric representations for the given curve.

15. $2x - y = 7.$
16. $x + 4y = 9.$
17. $x^2 + (y - 1)^2 = 9.$
18. $(x + 2)^2 + y^2 = 16.$
19. $4x^2 + y^2 = 25.$
20. $(x - 1)^2 + 9y^2 = 16.$
21. $y^2 = -8(x - 1).$
22. $(x + 2)^2 = 4 + (y - 3)^2.$
23. $y^2 - x^2 = a^2.$
24. $(x + 4)^2 = 12(y + 1).$

In Exs. 25–37, a point moves in a plane according to the given laws. Trace the motion from time $t = 0$. In Exs. 25–36, find the rectangular equation of the path.

25. $x = t + 2,\ y = 4t + 8.$
26. $x = -1 + 3t,\ y = 2t + 1.$
27. $x = \sin^2 t,\ y = -\cos^2 t.$
28. $x = 1 + 2t^2,\ y = -t^2.$
29. $x = t^2,\ y = t - t^2.$
30. $x = t^2,\ y = t - t^3.$
31. $x = \sin t,\ y = \cos 2t.$
32. $x = \sin t,\ y = \sin 2t.$
33. $x = 1 - \cos t,\ y = \cos 2t.$
34. $x = \sin t + \cos t,\ y = \sin t.$
35. $x = a \cos^4 t,\ y = a \sin^4 t.$ (Ex. 22, p. 129.)
36. $x = \sin t + \cos t,\ y = \frac{1}{2} \sin 4t.$ *Ans.* $y^2 = x^2(2 - x^2)(x^2 - 1)^2.$
37. $x = t - \sin t,\ y = 1 - \cos t.$ (The *cycloid.*)

CHAPTER 13 *Trigonometric Functions*

115. Introduction. The elementary transcendental functions are of prime importance in the mathematical sciences: in fact, the physicist, chemist, or engineer is apt to be more familiar with certain types of transcendental curves than with algebraic curves (beyond the straight line and conic sections).

In this chapter we shall see how trigonometric functions may be represented graphically. But before we proceed, one possible source of confusion should be cleared up. In the work on polar coordinates, trigonometric functions were very prominent. Perhaps for this reason, the student sometimes tends to think that we are dealing now with polar coordinates. On the contrary, all the work of this chapter is in rectangular coordinates. As abscissa x, we lay off the *radian measure* of the angle; as ordinate y the corresponding value of the function. Note that the curve $y = \cos x$ (Fig. 139) is an infinite succession of waves along the x-axis, while the curve $r = \cos \theta$ (Fig. 36) is a circle.

116. Graphs of trigonometric functions. For the curve

$$y = \sin x$$

the analysis of § 97 works out as follows.

1. Since $\sin(-\alpha) = -\sin \alpha$, the curve is symmetric with respect to the origin.

2. x-intercepts, $x = 0, \pm\pi, \pm 2\pi, \cdots$; y-intercept, $y = 0$.

3. The sine function is periodic, $\sin(x + 2\pi) = \sin x$, so the curve will consist of endless repetitions of the arc from $x = 0$ to $x = 2\pi$.

4. Since the extreme values of the sine are -1 and 1, the curve lies between the lines $y = -1$, $y = 1$.

5.

x	0	$\frac{1}{6}\pi$	$\frac{1}{4}\pi$	$\frac{1}{3}\pi$	$\frac{1}{2}\pi$
y	0	$\frac{1}{2}$	$\frac{1}{2}\sqrt{2}$	$\frac{1}{2}\sqrt{3}$	1

The formula

$$\sin(\pi - \alpha) = \sin \alpha$$

shows that the above values repeat, in reverse order, in the interval $\frac{1}{2}\pi \leqq x \leqq \pi$. The formula

$$\sin(\pi + \alpha) = -\sin \alpha$$

shows that the arc OAB already drawn is repeated in the interval $\pi \leqq x \leqq 2\pi$, except that the sign of y is changed: the arc BCD. Having covered a full period of the function, we extend the curve each way by repeating the part already drawn. (Fig. 138.)

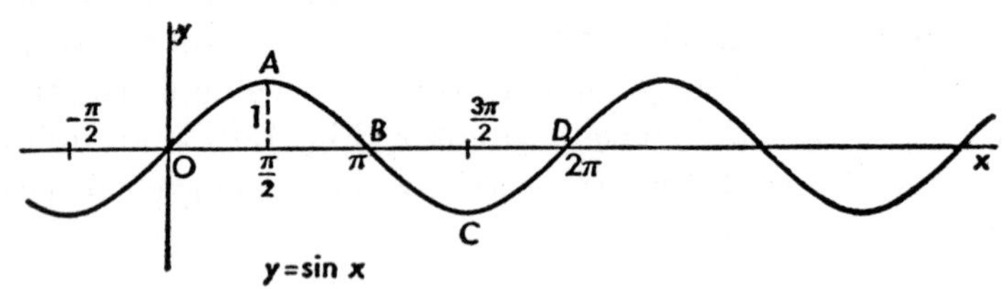

Figure 138

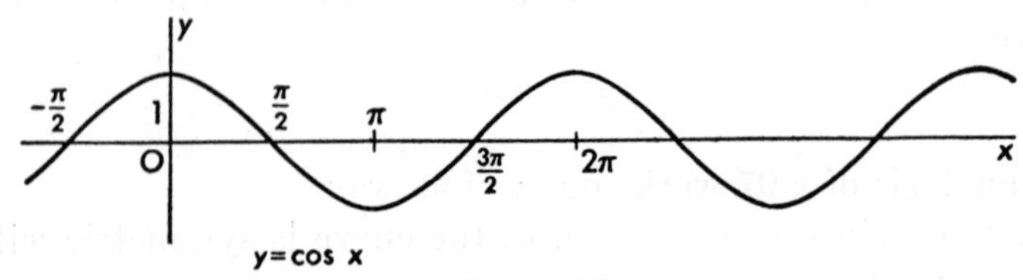

Figure 139

The curves (Figs. 139–140)

$$y = \cos x,$$
$$y = \tan x$$

may be obtained by similar analysis.

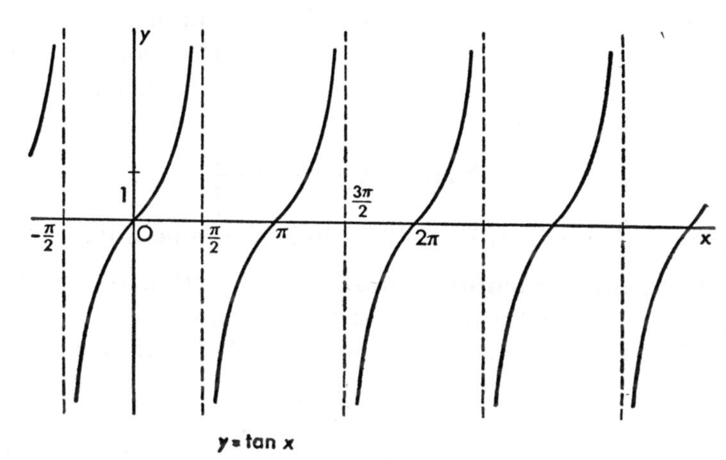

Figure 140

117. Change of scale. To trace the curve

$$y = a \sin \frac{x}{a} \qquad (a > 0), \tag{1}$$

let us introduce two new variables x_1, y_1 by the formulas

$$x_1 = \frac{x}{a}, \qquad y_1 = \frac{y}{a}, \qquad \text{or} \qquad x = ax_1, \qquad y = ay_1.$$

This evidently reduces (1) to the form

$$y_1 = \sin x_1.$$

It follows that the graph of (1) is exactly the curve of Fig. 138, except that *all coordinates are multiplied by* a — a mere change of scale.

More generally, the curve

(2) $$y = b \sin \frac{x}{a} \qquad (b > 0)$$

may be obtained at once from (1) by merely noting that in (2) each value of y is $\frac{b}{a}$ times the corresponding value in (1). (Compare Ex. 28, p. 111.) That is, the curve (2) has the same general appearance as Fig. 138, but the coordinates of the first wave-crest A are $(\frac{1}{2}\pi a, b)$ instead of $(\frac{1}{2}\pi, 1)$, etc.

Of course the curves

$$y = b \cos \frac{x}{a},\ y = b \tan \frac{x}{a}$$

may be deduced from Figs. 139–140 in the same way.

118. Simple harmonic motion. A spring of natural length L, offering the same resistance to compression as to extension, lies on a smooth horizontal table,* with one end S fixed and a weight P attached to the free end O. Let the spring be stretched a distance a and the bob then released. By Hooke's Law, the subsequent motion of the bob is due to a force (exerted by the spring) always directed toward the point of equilibrium O and proportional to the distance from O. The distance OP at time t is given by an equation of the form

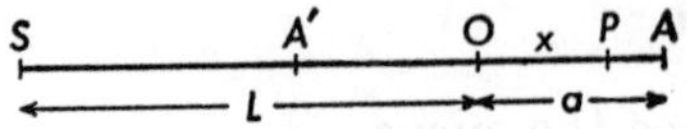

Figure 141

(1) $$x = a \cos kt.$$

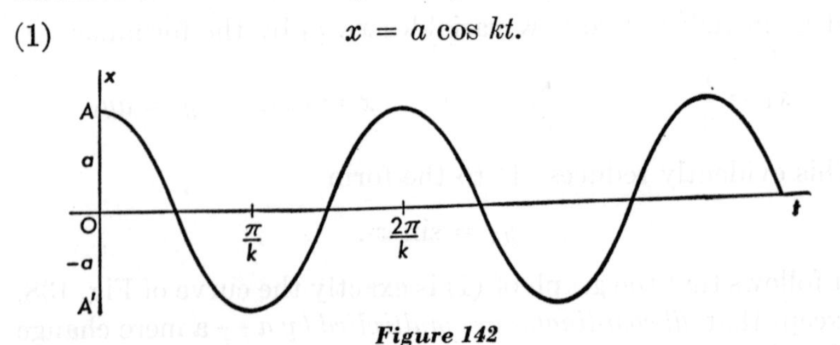

Figure 142

* Smooth, to eliminate friction; horizontal, to eliminate the effect of gravity. But see Ex. 27, p. 190.

From the known properties of the cosine, or from Fig. 142, we see that the weight P oscillates between the points $x = a$ and $x = -a$, the period (time for a complete oscillation) being $\frac{2\pi}{k}$.

When the formula for distance as a function of time is of form (1), P is said to perform simple harmonic motion with O as center. Problems involving this type of motion, or leading to the same mathematics, arise frequently in nature.

Each of the equations

$$x = h + a \cos (kt + \alpha), \tag{2}$$

$$x = h + a \sin (kt + \alpha) \tag{3}$$

can be reduced to form (1) by a translation of axes (Exs. 21–22 below). A translation of the x-axis means, physically, that we are measuring distance from a different point in the line of motion; a translation of the t-axis means that we are measuring time from a different initial instant. Since neither of these changes affects the motion, it follows that the motions (2) and (3) are simple harmonic.

Exercises

In Exs. 1–20, trace the curve.

1. $y = a \tan \frac{x}{a}$. **2.** $y = a \cos \frac{x}{a}$.

3. $y = a \csc \frac{x}{a}$. **4.** $y = a \cot \frac{x}{a}$.

5. $y = 3 \sin 2x$. **6.** $y = -\frac{1}{2} \sin 3x$.

7. $y = - \sin 4x$. **8.** $y = \sin \frac{1}{2}x$.

9. $y = 6 \cos \frac{1}{2}x$. **10.** $y = \frac{1}{2} \cos \frac{1}{4}x$.

11. $y = -2 \cos 3x$. **12.** $y = - \cos \frac{1}{3}x$.

13. $y = \cos^2 x - \sin^2 x$. **14.** $y = \cos x \sin x$.

15. $y = \sin x - 2$. **16.** $y = 1 - \cos x$.

17. $y = \cos^2 x$. **18.** $y = \sin^2 x$.

19. $y = (\cos x - \sin x)^2$. **20.** $y = (\cos x + \sin x)^2$.

21. Reduce the equation $y = k + b \cos \left(\frac{x}{a} + \alpha\right)$ to the form $y_1 = b \cos \frac{x_1}{a}$ by translation of axes.

22. Reduce the equation $y = k + b \sin\left(\frac{x}{a} + \alpha\right)$ to the form $y_1 = b \cos \frac{x_1}{a}$ by translation of axes.

23. In Fig. 138, what curve is obtained by translating the y-axis a distance $\frac{1}{2}\pi$?

24. In Fig. 140, what curve is obtained by translating the y-axis a distance $\frac{1}{2}\pi$? Check by Ex. 4.

25. Draw the curve $y = a \sec \frac{x}{a}\cdot$ (In Ex. 3, translate the y-axis.)

26. In Fig. 141, if P is initially at O and is given an initial velocity $\boldsymbol{v_0}$ **in** the line of the spring, the formula is

$$x = \frac{v_0}{k} \sin kt.$$

Discuss the motion.

27. When a spring is suspended vertically (Fig. 143), the effect of the weight is to stretch the spring a distance $OO_1 = h$. If the weight is displaced to A and then released, its distance from O at time t is

$$x = h + a \cos\left(t\sqrt{\frac{g}{h}}\right).$$

Show that the motion is simple harmonic with O_1 as center. (See Ex. 21.)

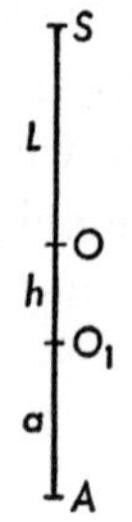

Figure 143

28. A particle moves in a plane subject to a force directed toward a fixed point and proportional to the distance from that point. With the fixed point as origin, and the particle initially at $(a, 0)$ with velocity v_0 parallel to the y-axis, its coordinates at time t are

$$x = a \cos kt, \qquad y = \frac{v_0}{k} \sin kt.$$

Show that the path is an ellipse. When does the ellipse become a circle?

119. Sums of sines and cosines. To graph a function expressed as a sum of sines or cosines, or both, we rely heavily on the method of composition of ordinates (§ 81).

Example: Trace the curve

$$y = \sin x - \tfrac{1}{2} \sin 2x. \tag{1}$$

The curve

$$y = \sin x \tag{2}$$

is merely Fig. 138; the curve

(3) $$y = -\tfrac{1}{2}\sin 2x$$

is (1), § 117, with $a = \frac{1}{2}$, and reflected in the x-axis. In (2) the period is 2π, in (3) it is π: thus (1) has a period 2π. In Fig. 144 the curve is traced through the first period, and of course can be extended at will to right or left. A glance at the figure shows that our usual analysis (§ 97) would have been inadequate. Note, however, that the curve is symmetric with respect to the origin, also (Ex. 1 below) with respect to the point $(\pi, 0)$; thus it would have been sufficient to compound ordinates in the interval $0 \leqq x \leqq \pi$ and then obtain the graph for the other half-period by reflection in $(0, 0)$ or $(\pi, 0)$.

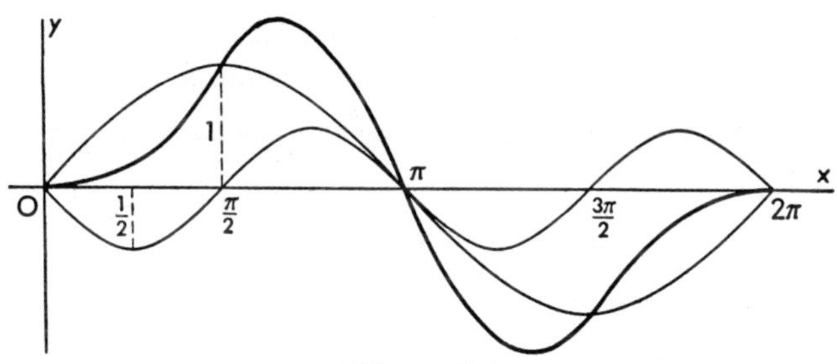

Figure 144

In § 82 only two component curves were involved. In practice, it is often necessary to compound several ordinates. Evidently the method may be used to combine any number of terms — all at once if feasible, or piecemeal if necessary in order to keep the drawing clear. (Exs. 11–12 below.)

120. Composition of simple harmonic motions. It is frequently necessary to combine two or more simple harmonic motions in the same straight line. No elaborate discussion is needed. Clearly, the problem of compounding the two simple harmonic motions

$$x = \sin t, \qquad x = -\tfrac{1}{2}\sin 2t$$

is mathematically identical with the example of § 119. Likewise, if we merely take time t as abscissa in place of x, the formulas of Exs. 3–12 below represent compound harmonic motions.

It can be seen that Fig. 136, p. 183, is the result of combining a simple harmonic motion on the x-axis with a compound harmonic on the y-axis. In that case the result was merely a parabolic arc, but the path of a point subject to two harmonic motions at right angles may turn out to be an extremely complicated curve.

Exercises

1. Show that the curve obtained by compounding any number of ordinates of the form $y = b \sin \dfrac{nx}{a}$ (n an integer) is symmetric with respect to the point $(\pi a, 0)$. (Translate the origin to that point.)

2. Show that the curve obtained by compounding any number of ordinates of the form $y = b \cos \dfrac{nx}{a}$ (n an integer) is symmetric with respect to the line $x = \pi a$.

In Exs. 3–16, trace the curve. Note Exs. 1–2.

3. $y = 2 \cos x + \cos 2x$.

4. $y = \cos 2x - \cos x$.

5. $y = 4 \sin x - \sin 2x$.

6. $y = \sin x - \sin 2x$.

7. $y = \cos x - \sin x$.

8. $y = \cos x + 4 \sin x$.

9. $y = \sin 3x + 2 \cos x$.

10. $y = \sin x - \frac{1}{2} \cos 2x$.

11. $y = \sin x - \frac{1}{2} \sin 2x + \frac{1}{3} \sin 3x$.

12. $y = \sin x - \frac{1}{2} \sin 2x + \frac{1}{3} \sin 3x - \frac{1}{4} \sin 4x$.

13. $y = \frac{1}{2}x + \sin 2x$.

14. $y = \frac{1}{2}x - \cos x$.

15. $y^2 = \cos \frac{1}{2}\pi x$. On the same axes, draw the circle $x^2 + y^2 = 1$.

16. $y^2 = \sin \pi x$. Are the ovals ellipses?

CHAPTER 14 *Exponentials and Logarithms*

121. The exponential function. The number a^n $(a > 0)$ is defined in elementary algebra for all *rational* values of the exponent. In more advanced mathematics it becomes necessary to assign a meaning to the function

$$y = a^x \qquad (a > 0) \tag{1}$$

as x varies continuously.

Let k be any given *irrational* number, positive or negative. When x approaches k assuming rational values only, the function a^x approaches a definite limiting value. (This statement, highly reasonable on its face, can be strictly proved.) This limit we denote by a^k. The function (1), called the *exponential function*, thus becomes defined for all (real) values of x. It obeys the laws of exponents:

$$a^x \cdot a^t = a^{x+t}, \tag{2}$$
$$(a^x)^t = a^{xt}. \tag{3}$$

In calculus and the mathematical sciences we have frequent occasion to deal with a certain irrational number, denoted by e, defined by the formula*

$$\boldsymbol{e = \lim_{z \to \infty} \left(1 + \frac{1}{z}\right)^z,}$$

*** Proof that this limit exists will be omitted; but see Ex. 28 below.**

and having the value

$$e = 2.71828 \cdots .$$

In fact, on account of its outstanding importance the function

$$y = e^x \tag{4}$$

will be studied here, rather than the general function (1).

For the graph of (4), the analysis follows.

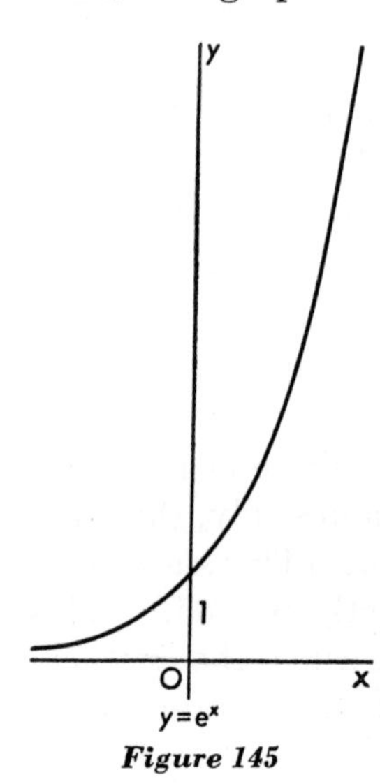

Figure 145

1. There is no symmetry.

2. Since any power of a positive number is positive, there is no x-intercept; when $x = 0$, $y = 1$.

3. As x increases positively, y increases at a tremendous rate (see Ex. 1 below). As x increases negatively, y approaches 0: the negative x-axis is an asymptote.

4. As noted above, y is always positive: the curve lies in the first and second quadrants.

5. The function e^x, its reciprocal e^{-x}, and various other exponential forms (see § 122) have been very extensively tabulated. With such a table* in hand, we may plot any desired number of points.

122. Hyperbolic functions. Certain exponential combinations occur so frequently that they are denoted by special names and symbols, and have been tabulated to a high degree of accuracy. These are the *hyperbolic functions,*† denoted by the symbols $\sinh x$ (read hyperbolic sine of x), $\cosh x$,

* For instance, the *Macmillan Logarithmic and Trigonometric Tables,* Revised (New York, The Macmillan Company).

† An adequate explanation of the reason for this name is rather beyond our horizon at present. We merely remark that these functions are related to the equilateral hyperbola $x^2 - y^2 = a^2$ in much the same way that the trigonometric functions are related to the circle $x^2 + y^2 = a^2$. As one instance, if $x = a \cos t$, $y = a \sin t$, then $x^2 + y^2 = a^2$; if $x = a \cosh t$, $y = a \sinh t$, then (Ex. 13 below) $x^2 - y^2 = a^2$.

tanh x, etc.:

$$\text{(1)} \qquad \sinh x = \frac{e^x - e^{-x}}{2},$$

$$\text{(2)} \qquad \cosh x = \frac{e^x + e^{-x}}{2},$$

$$\text{(3)} \qquad \tanh x = \frac{\sinh x}{\cosh x} = \frac{e^x - e^{-x}}{e^x + e^{-x}};$$

the respective reciprocals are csch x, sech x, coth x.

To discover the elementary properties of these functions, we have direct recourse to the definitions.

Example: Show that

$$\sinh 2x = 2 \sinh x \cosh x.$$

In (1), replace x by $2x$:

$$\sinh 2x = \frac{e^{2x} - e^{-2x}}{2} = \frac{(e^x - e^{-x})(e^x + e^{-x})}{2}$$
$$= 2\left(\frac{e^x - e^{-x}}{2}\right)\left(\frac{e^x + e^{-x}}{2}\right) = 2 \sinh x \cosh x.$$

The graphs of the six functions are simple curves, all well within our reach. To trace the curve

$$\text{(4)} \qquad y = \cosh x = \tfrac{1}{2}(e^x + e^{-x}),$$

we might employ the customary analysis, supplemented by point-plotting from a table. A faster method: Reflect the curve (Fig. 145)

$$\text{(5)} \qquad y = e^x$$

in the y-axis to obtain the curve

$$\text{(6)} \qquad y = e^{-x}.$$

By the midpoint formulas (§ 7) the curve (4), for every value of x, is midway between the curves (5) and (6).

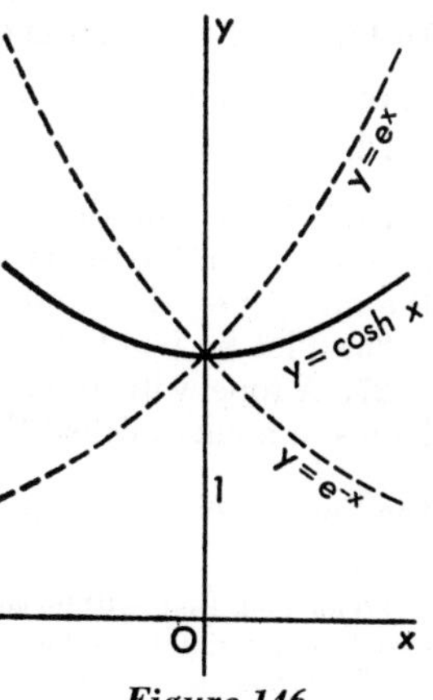

Figure 146

Exercises

1. Given $\log_{10} e = 0.434$, compute roughly, as to order of magnitude, the numbers e^{10}, e^{20}, e^{30}. *Ans.* $e^{30} > 10$ trillion.

2. Show that Fig. 145 represents the curve $y = ae^{\frac{x}{a}}$ if the y-intercept is marked as a instead of 1. How may the curve $y = be^{\frac{x}{a}}$ then be obtained? (§ 117.)

In Exs. 3–11, trace the curve. Note Ex. 2.

3. $y = 2e^{-3x}$.
4. $y = 3e^{\frac{1}{2}x}$.
5. $y = e^{-\frac{1}{x}}$.
6. $y = e^{-x^2}$.
7. $y = e^x + 2e^{-x}$.
8. $y = 2e^{-x} - \frac{1}{2}e^{2x}$.
9. $y = e^{\sin x}$.
10. $y = e^{\tan x}$.
11. $y = e^{\frac{1}{x^2-1}}$.

12. Plot by points on a large scale, in the interval $-1.5 \leqq x \leqq 1.5$, the curves $y = e^x$, $y = 1 + x + \frac{1}{2}x^2$.

13. Show that $\cosh^2 x - \sinh^2 x = 1$. [Square (2), § 122; square (1); subtract.]

14. Solve Ex. 13 another way. (Factor the left member.)

15. Show that $\text{sech}^2 x = 1 - \tanh^2 x$. (Ex. 13.)

16. Show that $\cosh 2x = \cosh^2 x + \sinh^2 x$.

17. Show that $\tanh 2x = \dfrac{2 \tanh x}{1 + \tanh^2 x}$. (Ex. 16, and Example, § 122.)

18. Trace the curve $y = \sinh x$. (Average the ordinates of the curves $y = e^x$, $y = -e^{-x}$.)

In Exs. 19–24, trace the curve. (§ 97.)

19. $y = \sinh x$.
20. $y = \cosh x$.
21. $y = \tanh x$.
22. $y = \coth x$.
23. $y = \text{csch}\, x$.
24. $y = \text{sech}\, x$.

25. The arc of the curve $y = \cosh x$ drawn in Fig. 146 looks like a parabola. In the interval $-2 \leqq x \leqq 2$, plot the curve by points; also, on the same axes, the parabola* $y = 1 + \frac{1}{2}x^2$.

26. Plot by points the curve $y = \sinh x - x$, $-1 \leqq x \leqq 1$.

27. A rope, wire, or chain suspended from two of its points under gravity hangs in a curve called the *catenary*. With the origin at distance a below the lowest point, the equation is $y = a \cosh \dfrac{x}{a}$. Draw the curve.†

* The fact that within sufficiently narrow limits the hyperbolic cosino can be closely approximated by a quadratic function is often useful in science.

† Thus the quickest way to visualize the graph of the hyperbolic cosine is to take a piece of string between the thumb and forefinger of each hand and let it hang freely.

28. Using logarithms, compute $\left(1 + \frac{1}{z}\right)^z$ for $z = 10, 20, 50, 100$.

123. The natural logarithm. The reader is already acquainted with the common logarithm, or logarithm to the base 10, denoted by $\log_{10} x$, and defined by the statement that

$$y = \log_{10} x \qquad \text{if} \qquad x = 10^y.$$

Common logarithms are useful in many phases of numerical computation.

In the more advanced applications of mathematics, we have much more often to deal with the so-called *natural logarithm*, or logarithm to the base e, denoted by $\log_e x$, or simply by $\ln x$, and defined by the statement that

$$(1) \qquad \boldsymbol{y = \ln x} \qquad \textit{if} \qquad \boldsymbol{x = e^y}.$$

Theoretically we are at liberty to choose any positive number a as the base of a system of logarithms, but in practice only the bases 10 and e have been found useful.*

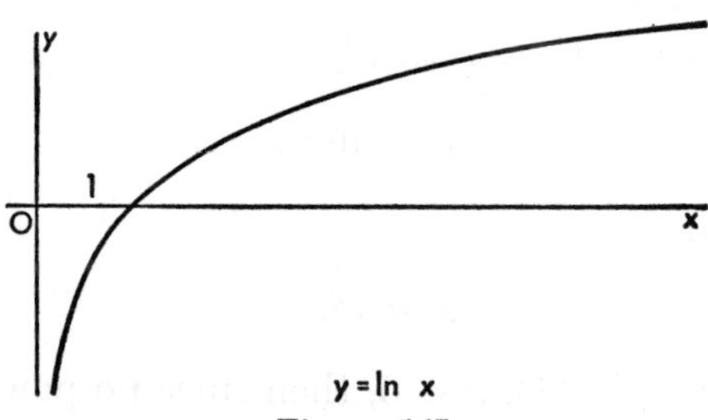

Figure 147

The curve

$$(2) \qquad y = \ln x$$

is very easily traced. By the definition (1), equation (2) is equivalent to

$$x = e^y.$$

But this is equation (4), § 121, *with x and y interchanged.* It

* At first sight it seems strange that such an awkward number as e should be chosen as a base. But in calculus (see any calculus text), the choice of e as base is practically forced upon us by an intrinsic property of the exponential function, whence the name "natural logarithm."

follows by the theorem on interchange of variables (Ex. 33, p. 34) that the curve (2) is the reflection of Fig. 145 in the 45°-line.

124. Properties of logarithms. The following properties are valid for any system of logarithms with base $a > 1$.

(*a*) Negative numbers have no (real) logarithms.
(*b*) Numbers between 0 and 1 have negative logarithms.
(*c*) Numbers greater than 1 have positive logarithms.
(*d*) As x approaches 0 through positive values, the logarithm increases negatively without limit.
(*e*) The logarithm of 1 is 0.
(*f*) As x increases indefinitely, the logarithm does likewise.

For any base (we use e), these properties may be read off at once from the graph — for instance, (*a*) follows from the fact that there is no curve to the left of the y-axis — and are easily remembered by recalling the graph. The properties may be established analytically if we examine, instead of

$$y = \ln x, \tag{1}$$

the equivalent

$$x = e^y. \tag{2}$$

To prove (*a*): If, in (1), $x < 0$, then since no power of a positive number can be negative, (2) becomes impossible.

Exercises

Verify the properties of logarithms stated in Ex. 1–4.

1. $\ln bc = \ln b + \ln c$ $(b > 0, c > 0)$. Use $e^p \cdot e^q = e^{p+q}$ to show that $\ln (e^p \cdot e^q) = \ln e^{p+q} = p + q$; put $e^p = b$, $e^q = c$.

2. $\ln \frac{b}{c} = \ln b - \ln c$ $(b > 0, c > 0)$.

3. $\ln c^n = n \ln c$ $(c > 0)$.

4. $\ln \frac{1}{c} = -\ln c$ $(c > 0)$.

In Exs. 5–16, trace the curve.

5. $y = 2 \ln \frac{1}{2}x$.

6. $y = \frac{1}{3} \ln 3x$.

7. $y = \ln \sqrt{x}$.

8. $y = \ln x^3$.

9. $y = \ln (1 - x)$.

10. $y = \ln (1 + x)$.

11. $y = \ln (1 - x)^2$.

12. $y = \ln kx^2$.

13. $y = \ln (1 - x^2)$. (§ 81.)

14. $y = \ln (x^2 + x)$. (§ 81.)

15. $y = \ln \dfrac{1 - x}{1 + x}$. (§ 81.)

16. $y = \ln \dfrac{1 - x}{x^2}$. (§ 81.)

CHAPTER 15 *Families of Curves*

125. Family of lines. Up to this point, in considering such an equation as

$$(1) \qquad y = \tfrac{1}{2}x + c,$$

we have thought of c as having a definite, fixed (though as yet unspecified) value. From this point of view, (1) represents a single straight line — the line of slope $\frac{1}{2}$ and y-intercept c, whatever c may be.

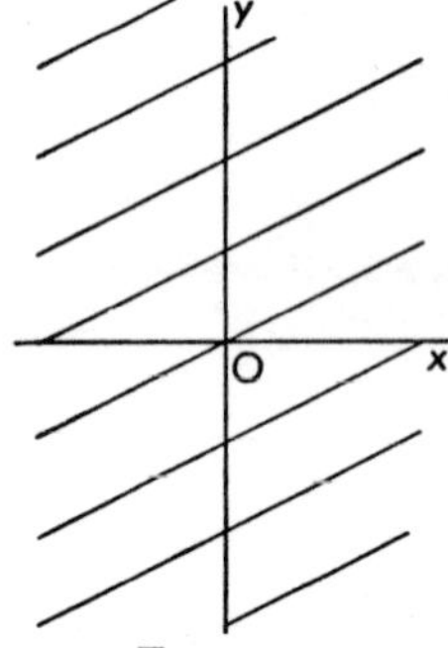

Figure 148

Instead, in some problems we may wish to think of c as an *arbitrary constant* — i.e., a constant to which any value may be assigned at will. Then, as c ranges through all real values, (1) may be considered as standing for an infinity of equations, representing an infinity of straight lines, all straight lines of slope $\frac{1}{2}$. The constant c is called a *parameter;* the aggregate of lines is a *one-parameter family*, or *system*, of lines.

126. Families of curves. Instead of the linear equation above, consider an equation in x and y of any form, involving a parameter c: in our usual functional notation,

$$(1) \qquad f(x, y, c) = 0.$$

For any one value of c, equation (1) represents some one definite curve; as c assumes all real values, we obtain a *one-parameter family* of curves.

Example: Discuss the family

$$(x - c)^2 = 4ay.$$

For $c = 0$, this is the standard parabola

$$x^2 = 4ay \tag{2}$$

with vertex at O, opening upward. For any other value of c the same parabola is obtained, except that the vertex is at $(c, 0)$. Thus to produce the entire family, it is only necessary to draw the parabola (2) and then translate it right and left.

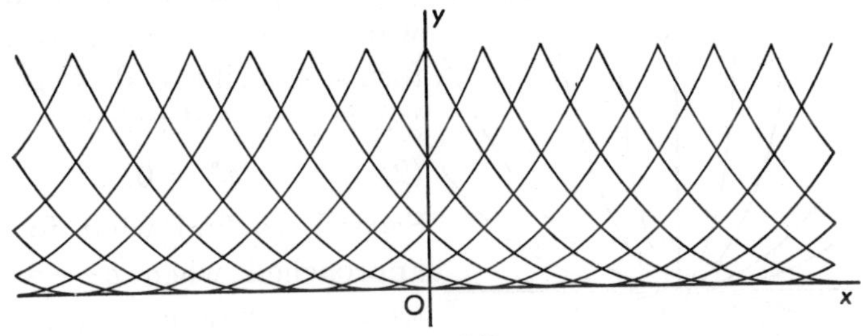

Figure 149

In this chapter the letter c will always denote an arbitrary constant. Such letters as a, b, p, etc., will as usual denote given (i.c., fixed) constants. For simplicity we confine our attention to cases in which the equation of the family is algebraic in x and y, and is algebraic of first or second degree in c.

It may happen that the family-equation has no locus for values of c in a certain range: for instance, in Ex. 3 below, there will be no curves for $c < 0$. Such values of c will be excluded, since equations having no locus are of no interest in our present study. With this exception, c will be considered as free to assume any real value.

Of course there is no limit to the number of arbitrary constants that may be present. For instance, the equation

$$(x - c_1)^2 + (y - c_2)^2 = a^2$$

represents the two-parameter system consisting of all circles of radius a.

127. Exceptional forms. Certain particular values of c may produce curves quite different in kind from the other curves of the family. Also, it may happen that as c increases without limit, the successive curves approach a limiting form different from the general form.

Example: Discuss the family

$$x^2 = cy. \tag{1}$$

This family evidently includes all parabolas with vertical axis and vertex at the origin, opening upward if $c > 0$, downward if $c < 0$. As exceptional forms, first: if $c = 0$, we get the y-axis (counted twice)

$$x^2 = 0. \tag{2}$$

Second, to find what form is approached when $c \to \infty$, introduce a new constant c_1 by the formula

$$c = \frac{1}{c_1}, \qquad \text{or} \qquad c_1 = \frac{1}{c}.$$

Then (1) becomes

$$y = c_1x^2. \tag{3}$$

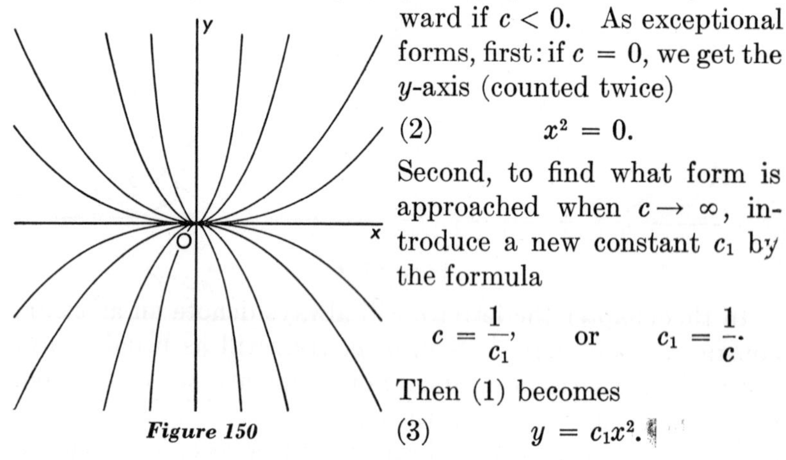

Figure 150

As $c \to \infty$, $c_1 \to 0$: thus the limiting form is the x-axis

$$y = 0. \tag{4}$$

From the above presentation it might seem that the x-axis does not "belong," in quite the same sense as does the y-axis, since (2) corresponds to the value $c = 0$, while there is no value of c that will produce (4). The fact is that both exceptional forms enter by the same door, so that logically we should include both or neither. To see this, note that we could just as well have written the equation in the first place in the form (3) instead of (1). Then, we get (4) corresponding to $c_1 = 0$, while no value of c_1 gives (2).

Exercises

1. Show that if the equation of a family of curves can be so written that x and c occur only in the combination $x - c$, the entire family can be obtained by horizontal translation of any one member. (Example, § 126.)

2. Show that if the equation of a family of curves can be so written that y and c occur only in the combination $y - c$, the entire family can be obtained by vertical translation of any one member.

In Exs. 3–20, discuss the family of curves, and draw an adequate number of them. Note Exs. 1–2. Look for exceptional forms.

3. $x^2 + y^2 = c.$

4. $y = cx + b.$

5. $y = cx - c^2.$

6. $x \cos c + y \sin c = p.$

7. $xy = c.$

8. $(x - c)^2 - y^2 = a^2.$

9. $x^2 + y^2 + 2cy = 0.$

10. $4(x - c)^2 + y^2 = 4a^2.$

11. $(x + c)^2 + (y - c)^2 = a^2.$

12. $(x - c)^2 + (y - c)^2 = 2c^2.$

13. $(x - c)^2 + 4y^2 = 4c^2.$

14. $(x + c)^2 = 4cy.$

15. $xy = cx + (1 - c)y.$

16. $cx^2 + (1 - c)y^2 = 1.$

17. $y - c = x(1 - x^2).$

18. $(y - c)^2 = x(1 - x^2).$

19. $y = cx + \frac{a}{c}\cdot$ (§ 89.)

20. $(y - cx)^2 = a^2c^2 + b^2.$ (§ 89.)

128. Equations linear in the parameter. If in the equation

$$f(x, y, c) = 0 \tag{1}$$

we substitute for x and y the coordinates of some given point (x_1, y_1), the result, of course, is an equation in c only. The roots of this equation, substituted for c in (1), evidently give the equations of those curves of the family that *pass through* (x_1, y_1). As an example, the reader may find that curve of the family (Fig. 150)

$$x^2 = cy$$

that passes through $(2, -\frac{1}{2})$. Note that this problem is identical, except in point of view, with Ex. 18, p. 100.

Suppose now that equation (1) is of *first degree in c*. Then, corresponding to a random (x, y)-pair we get one and only one value of c, which means that, in general, through every point of the plane there passes *one and only one curve* of the family. See, for instance, Fig. 150, where the plane is filled once, except that all the curves pass through $(0, 0)$.

129. Lines through the intersection of two lines. Given two lines

$$A_1x + B_1y + C_1 = 0, \tag{1}$$
$$A_2x + B_2y + C_2 = 0, \tag{2}$$

intersecting in a point $P_1:(x_1, y_1)$, consider the equation formed by multiplying one of the equations by a parameter c and adding: that is,

$$A_1x + B_1y + C_1 + c(A_2x + B_2y + C_2) = 0. \tag{3}$$

Since (3) is of first degree, it represents a family of straight lines. All these lines have one property in common: they *pass through the intersection* P_1 *of the* original lines. To see this, note that since P_1 lies on both of the lines (1) and (2), we have the relations

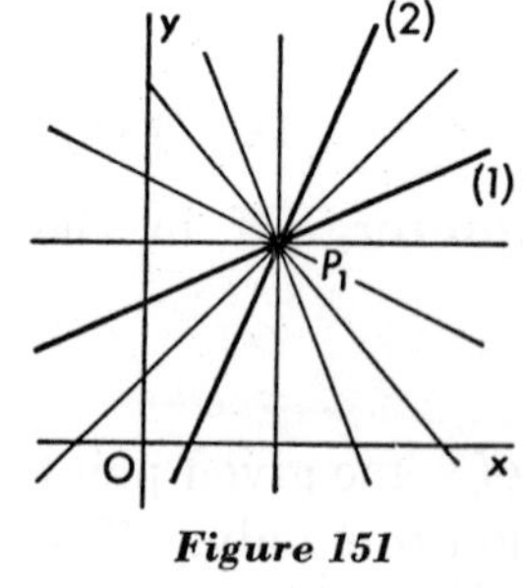

Figure 151

$$A_1x_1 + B_1y_1 + C_1 = 0, \tag{4}$$
$$A_2x_1 + B_2y_1 + C_2 = 0. \tag{5}$$

Now, substituting the coordinates (x_1, y_1) in (3), we find that P_1 will lie on the lines if

$$A_1x_1 + B_1y_1 + C_1 + c(A_2x_1 + B_2y_1 + C_2) = 0.$$

But this is true, by virtue of (4) and (5).

THEOREM: *If* $A_1x + B_1y + C_1 = 0$ *and* $A_2x + B_2y + C_2 = 0$ *are two intersecting lines, the equation*

$$A_1x + B_1y + C_1 + c(A_2x + B_2y + C_2) = 0 \tag{6}$$

represents a family of lines through the point of intersection of the given lines.

Although there is no value of c that will yield the line (2), we shall consider that line as a member of the family by special agreement (§ 127, last paragraph).

Given any line through P_1, excepting (2), let Q be any other point of that line. Substituting the coordinates of Q in (6), we unfailingly determine a value of c, which means that our line P_1Q belongs to the family. Since (2) is already in the family by agreement, we have proved that the family (6) *includes all lines through* P_1.

Our theorem enables us to solve many problems having to do with the point of intersection of two lines, *without finding that point.*

Example: Find the equation of the line through the intersection of the lines

$$(7) \qquad x + y = 4,$$

$$(8) \qquad y = 2x - 3,$$

and parallel to the line

$$(9) \qquad x - 2y = 4.$$

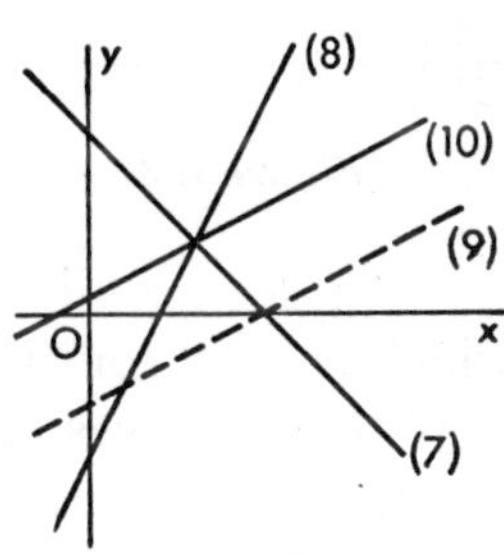

Figure 152

Our required line is one of the family

$$x + y - 4 + c(y - 2x + 3) = 0,$$

or

$$(1 - 2c)x + (1 + c)y - 4 + 3c = 0.$$

The slope of any member of the family is

$$m = -\frac{1 - 2c}{1 + c};$$

the slope of the line (9) is $\frac{1}{2}$. Equate these values and solve for c:

$$-\frac{1 - 2c}{1 + c} = \frac{1}{2}, \qquad c = 1.$$

This substituted in the family-equation gives the answer

$$(10) \qquad x - 2y + 1 = 0.$$

130. Curves through the intersections of two curves. Let there be given any two curves, intersecting in certain points $P_1:(x_1, y_1)$, $P_2:(x_2, y_2)$, etc. The equations of these curves will be denoted by*

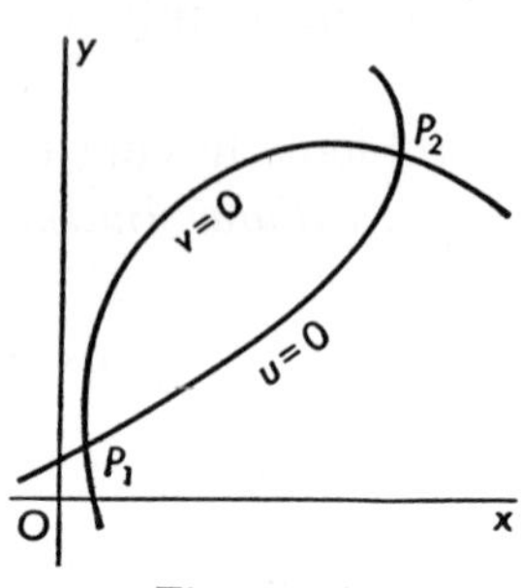

Figure 153

(1) $u(x, y) = 0, \qquad v(x, y) = 0.$

Now consider the family of curves

(2) $u(x, y) + cv(x, y) = 0.$

Since P_1 lies on each of the curves (1), we have the relations

(3) $u(x_1, y_1) = 0, \qquad v(x_1, y_1) = 0.$

Substituting the coordinates of P_1 in (2), we get

$$u(x_1, y_1) + cv(x_1, y_1) = 0,$$

which is true, for all values of c, by virtue of (3). It follows that all the curves (2) pass through P_1, and by the same argument, through all the other intersections of the base-curves (1).

From here on, for brevity we shall denote $u(x, y)$ and $v(x, y)$ simply by u and v.

THEOREM: *If $u = 0$ and $v = 0$ are two intersecting curves, the equation*

(4) $$u + cv = 0$$

represents a family of curves passing through all the intersections of the original curves.

Although no value of c produces the curve $v = 0$, we include that curve by agreement.

Since (4) is of first degree in c, it follows by § 128 that the curves fill the plane once and only once, so that *no two of the curves can intersect*, except, of course, that they all pass through the common points P_1, P_2, etc.

* Step-by-step comparison with § 129 should be helpful in reading the following argument. There, of course, $u(x, y) = A_1x + B_1y + C_1$, $v(x, y) = A_2x + B_2y + C_2$. In fact, to make such comparison possible was the only reason for introducing § 129, since the theorem of that article is a mere special case of the one forthcoming.

Example: Find the equation of a family of parabolas through the intersections of the curves (Fig. 154)

$$y^2 = x, \qquad x = 1.$$

One such family of parabolas is

$$y^2 - x + c(x - 1) = 0.$$

It is a simple exercise for the reader to verify the drawing. Note the exceptional forms $x = 1$, included by agreement, and $y = \pm 1$, when $c = 1$.

Even if the base-curves do not intersect, it is clear that (4) will still represent a system of curves, but the relationship of those to the original curves is not always so obvious geometrically. As an extreme case, see Ex. 21 below.

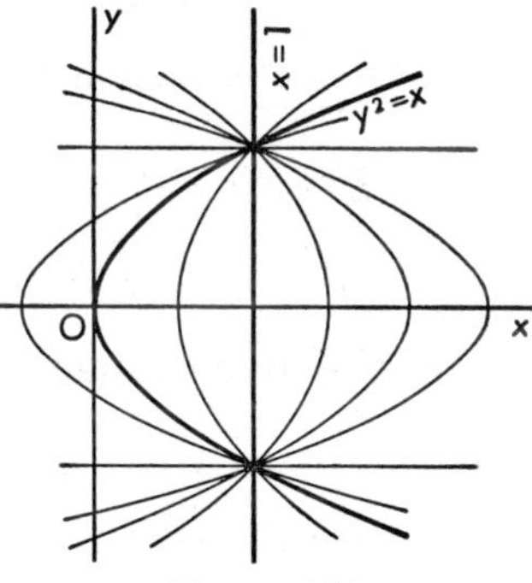

Figure 154

When the base-curves are straight lines, we have seen (§ 129) that the family $u + cv = 0$ includes all lines through the common point P_1. For other curves the analogous theorem is not true, in general. In the example above, consider a parabola through the intersections P_1, P_2, and a third point Q. Since three points do not determine a parabola (Ex. 31, p. 130), there are an infinite number of parabolas through P_1, P_2, and Q, of which only one belongs to the family (4).

In § 129, since the family was *complete* — i.e., since it included *all* lines through P_1 — we were assured in advance, in the example of that section, that our family would include a line parallel to the line (9). To save tiresome repetition, we merely remark that in all cases where we shall need this property of completeness (§131, and many of the exercises), its presence can be shown very easily by argument similar to that of § 129.

131. Applications. Although space is not available for a full development of the subject, our theorem can be used to solve a variety of problems both old and new. Methods based on the theorem usually have the disadvantage (if it is one) that they are not straightforward, but depend on a special device of some kind. On the other hand, such methods are in many cases remarkable for their speed; as a rule the proofs as well are extremely simple and compact.

Example: Find the equation of the circle tangent to the line $2x + y = 8$ at P_1: (4, 0) and passing through P_2: (7, 3).

Write the equation $u = 0$ of the point-circle (4, 0):

$$(x - 4)^2 + y^2 = 0,$$

or

$$x^2 + y^2 - 8x + 16 = 0.$$

Using the given tangent as $v = 0$, write the family-equation

$$x^2 + y^2 - 8x + 16 + c(2x + y - 8) = 0. \tag{1}$$

Substitute the coordinates (7, 3) to determine c:

$$18 + 9c = 0, \qquad c = -2,$$

and the answer is

$$x^2 + y^2 - 12x - 2y + 32 = 0.$$

Proof: All the circles (1) pass through the intersection (4, 0) of the point-circle and the given line; since these circles cannot intersect the line elsewhere, they must be tangent to it.

Exercises

In Exs. 1–5, find the equation of the line.

1. Through the intersection of the lines $7x + 12y = 4$, $8x - y = 4$, and through (2, 1). *Ans.* $9x - 14y = 4$.

2. Through the intersection of the lines $13x + 17y = -2$, $4x + 3y = 8$, and through (0, 2). *Ans.* $85x + 71y = 142$.

3. Through the intersection of the lines of Ex. 1, and perpendicular to the line $x - y = 8$. *Ans.* $103x + 103y = 56$.

4. Through the intersection of the lines of Ex. 2, and perpendicular to the line $x - 2y + 21 = 0$. *Ans.* $58x + 29y = 172$.

5. Through the intersection of the lines of Ex. 1 and parallel (*a*) to the x-axis; (*b*) to the y-axis; (*c*) to the line through $(-3, 4)$ and $(22, -2)$.

6. What is represented by equation (6), § 129, when the given lines are parallel?

In Exs. 7–18, using the given curves as $u = 0$ and $v = 0$, draw a number of curves of the family $u + cv = 0$. Look for exceptional forms.

7. $x^2 + y^2 = 1$, $x = 0$.
8. $y^2 = x$, $y = 1$.
9. $y = x^2$, $y = x$.
10. $y = x^3$, $y = x$.
11. $y = x^3$, $x = 0$.
12. $y = x^2$, $x = 0$.
13. $y^2 = x$, $x = -1$.
14. $xy = 1$, $x = 0$.
15. $y = x^3$, $x^2 = 0$.
16. $y = x^4$, $x^2 = 0$.
17. $x^2 + y^2 = 1$, $x^2 - y^2 = 1$.
18. $2x^2 + y^2 = 3$, $x^2 + y^2 = 2$.

19. If $u = 0$ is a circle (perhaps a point-circle) and $v = 0$ a line, show that $u + cv = 0$ represents a family of circles.

20. If $u = 0$ and $v = 0$ are two circles (including the possibility of point-circles), show that $u + cv = 0$ is a family of circles. What can be said of the family if the given circles are (*a*) concentric? (*b*) Tangent to each other?

21. If $u = 0$ and $v = 0$ are two equations of second degree having no locus, show that $u + cv = 0$ represents a family of central conics (exceptionally, a family of parallel lines).

In Exs. 22–33, find the equation of the circle.

22. Through the intersections of $x^2 + y^2 + x - y = 2$ and $x^2 + y^2 = 5$, and passing through $(-2, 2)$. *Ans.* $x^2 + y^2 + 3x - 3y + 4 = 0$.

23. Through the intersections of $x^2 + y^2 = 6x - 2y - 1$ and $x^2 + y^2 = 4y$, and passing through $(3, 1)$. *Ans.* $11x^2 + 11y^2 - 36x - 8y + 6 = 0$.

24. Through the intersections of $x^2 + y^2 = 16$ and $3x + 2y = 8$, and through $(4, 2)$.

25. Through the intersections of $x^2 + y^2 = 2x + 4y$ and $y = x - 1$, and through $(-1, 2)$.

26. Through the intersections of $x^2 + y^2 = x$ and $x^2 + y^2 = y$, and with radius $\frac{1}{2}\sqrt{5}$. *Ans.* $x^2 + y^2 = 2x - y$; $x^2 + y^2 = 2y - x$.

27. Through $(1, 1)$, $(2, -1)$, $(2, 3)$. (Example, § 52. As $u = 0$, take the circle with P_1P_2 as diameter; as $v = 0$, the line P_1P_2.)

28. Ex. 2, p. 86.
29. Ex. 9, p. 86.
30. Ex. 10, p. 86.
31. Ex. 11, p. 86.
32. Ex. 12, p. 87.
33. Ex. 21, p. 87.

CHAPTER 16 *Curve-Fitting*

132. Empirical equations. In many applications we are concerned with a set of pairs of values of two variables as determined by observation or experiment. A few examples follow.

(1) Ballistics: The range of a projectile, as a function of the initial velocity.

(2) Chemistry: The amount of a given element present during a chemical reaction, as a function of the time.

(3) Mechanical Engineering: The gasoline consumption of an automobile, as a function of the speed.

(4) Biology: The rate of growth of a culture of bacteria as a function of time.

By plotting the observed pairs of values on rectangular coordinate paper and drawing a smooth curve through the points, we obtain a more or less accurate graphic representation. The problem of this chapter is to obtain an analytic representation, an equation which corresponds to the empirically determined curve. Such an equation is called an *empirical equation*, or *empirical formula*, and the process of finding that equation is called *curve-fitting*.

We shall find that different methods produce different equations for the same data. An empirical formula is merely an approximation to the true relation between the variables.

Frequently a close correspondence can be found; but if important extraneous factors enter, only a rough approximation is obtainable. Even then, however, the method may yield useful results.

133. The method of selected points.

Example: A set of eight cylindrical bars, all of the same radius and material but of different lengths, are weighed with results as tabulated* (L in feet, W in pounds). Express W as a function of L.

L	1	2	3	4	5	6	7	8
W	0.38	0.68	1.13	1.47	1.78	2.18	2.60	2.84

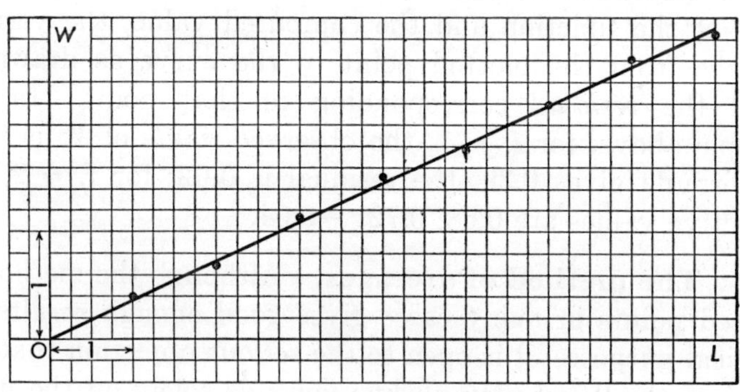

Figure 155

Plotting the points (1, 0.38), (2, 0.68), etc., with 4 spaces as the horizontal and 5 as the vertical unit, we find that the points follow more or less closely a straight line through the origin. In this example, we know beforehand that the weight is proportional to the length:

* By intention, the data given are quite inaccurate; for if not, in the small-scaled drawing to which we are limited in the text, all the points would appear to lie on the line. In this chapter, the student should make all his drawings on a very generous scale.

$$W = bL.$$

Now, laying a transparent ruler on the paper, we draw the straight line (through the origin) which, on the average, according to our best judgment, most closely follows the set of points. The ordinate of the last point on the line segment drawn is about 14.4 spaces, or 2.88 units; the equation of the line through (0, 0), (8, 2.88) is, by § 34,

$$W = 0.360L.$$

This is known as the *method of selected points*. That is, we select two points on our estimated "best-fitting line" — in the example, the points (0, 0), (8, 2.88) — and find the equation of the line through those points by § 34.

134. Residuals. The difference between the observed value of the function and the computed value, the vertical distance from the plotted point to the curve as drawn, is called the *residual* corresponding to that point. In the example above, for $L = 1$, the observed value of W is 0.38, the formula gives 0.36, the residual is 0.02; for $L = 2$, the respective values are 0.68, 0.72, -0.04.

135. The method of averages. Although the method of selected points in many cases gives good enough results for practical purposes, it is open to the serious objection that the choice of the "best-fitting" line depends solely on the judgment of the operator, and hence is to some extent a matter of guesswork. We shall, therefore, study two other methods, each an explicit mathematical procedure for fitting a curve of specified form to a given set of data. The first of these procedures is called the *method of averages*.

To fit an empirical equation containing one undetermined constant to given data by the method of averages, we determine the constant so that *the algebraic sum of the residuals is zero*.

136. Linear equation with one constant. In the equation

$$y = a + bx, \tag{1}$$

assume that a is known, and b is to be determined from a set of observations. The discussion applies without essential change to problems where b is known and a to be found.

Let the given pairs of values be (x_1, y_1), (x_2, y_2), (x_3, y_3), $\cdots$, (x_n, y_n). At the first point the observed value is y_1, the value computed by (1) is $a + bx_1$, the residual r_1 is

$$r_1 = y_1 - a - bx_1.$$

The succeeding residuals are

$$\begin{aligned} r_2 &= y_2 - a - bx_2, \\ r_3 &= y_3 - a - bx_3, \\ &\cdots \\ r_n &= y_n - a - bx_n. \end{aligned}$$

By the method of averages the sum of the residuals must be zero:

$$r_1 + r_2 + r_3 + \cdots + r_n = 0. \tag{2}$$

For compactness, we shall use the summation sign to indicate a sum over all the terms of which a representative one is written; for instance,

$$\begin{aligned} \Sigma r_i &= r_1 + r_2 + \cdots + r_n, \\ \Sigma x_i &= x_1 + x_2 + \cdots + x_n, \end{aligned}$$

etc. Then (2) may be written

$$\Sigma r_i = 0. \tag{3}$$

Since $r_i = y_i - a - bx_i$ for each of $i = 1, 2, 3, \cdots, n$, equation (3) yields

$$\Sigma y_i - na - b\Sigma x_i = 0, \tag{4}$$

an equation from which b is to be found.

RULE I: To make the line $y = a + bx$ fit the n points (x_1, y_1), (x_2, y_2), $\cdots$, (x_n, y_n) by the method of averages, determine the unknown constant b from the equation

$$\Sigma y_i = na + b\Sigma x_i. \tag{5}$$

This rule is easily modified to apply to an equation of any form containing a single unknown constant.

Example: Solve the example of § 133 by the method of averages.

We are to fit the line $W = bL$ to the data in the following table.

L	1	2	3	4	5	6	7	8
W	0.38	0.68	1.13	1.47	1.78	2.18	2.60	2.84

By Rule I, b is to be determined so that

$$\Sigma W_i = b\Sigma L_i.$$

From the given data,

$$\Sigma L_i = 36, \qquad \Sigma W_i = 13.06.$$

Hence

$$13.06 = 36b,$$

so that $b = 0.363$, and the fitted formula is

$$W = 0.363L.$$

Exercises

In Exs. 1–6, by plotting, verify the existence of an approximate linear relation. Assuming that the first point is exactly on the line, determine the equation (*a*) by selected points; (*b*) by the method of averages.

1.

x	0	1	2	3	4	5
y	0	3	5	8	10	13

Ans. (*b*) $y = 2.6x$.

2.

x	0	1	2	3	4	5
y	0	0.5	1.1	1.7	2.3	2.9

Ans. (*b*) $y = 0.57x$.

3.

x	0	1	2	3	4	5	6
y	5	4.4	4.3	3.8	3.6	3.4	2.6

Ans. (*b*) $y = 5 - 0.376x$.

4.

x	0	1	2	3	4	5	6
y	−3	−1.0	0.0	1.8	3.6	5.4	6.6

Ans. (*b*) $y = 1.64x - 3$.

5.

x	0	2	5	10	20	30	50
y	10	9.6	9.3	8.9	7.6	7.1	4.8

Ans. (*b*) $y = 10 - 0.109x$.

6.

x	0	5	10	20	30	50	100
y	100	95.7	92.1	81.6	74.0	56.3	10.2

Ans. (*b*) $y = 100 - 0.884x$.

Solve Exs. 7–10 by the method of averages.

7. An airplane flies at practically constant speed. By landmarks, the pilot estimates his distance (in miles) from the starting point at the end of each hour as shown. Find (*a*) the speed; (*b*) the distance after $6\frac{1}{2}$ hrs.

t	1	2	3	4	5	6	7
d	100	205	300	450	550	640	720

Ans. (*a*) 106 mi. per hr.

8. As a means of computing π, a class of high school students were required to measure the circumferences of a set of circular disks of known radii. The measurements, averaged for the whole class for each disk, were as shown. Find the error in the computed value of π.

r	3	4	5	10
c	18.829	25.092	31.390	62.853

Ans. -0.0015.

9. When a body slides from rest down a smooth inclined plane, the velocity acquired in t seconds is approximately

$$v = gt \sin \alpha,$$

where g is the acceleration of gravity (32.16 ft. per sec. per sec.) and α is the angle of inclination of the plane. For the plane on which the given observations were made, compute b in the formula $v = bt$, and find α.

t	1	2	3	4	5	6	7	8
v	1.31	2.52	3.52	4.72	5.87	7.02	8.17	9.42

Ans. $\alpha = 2°6'$.

10. A steel spring 8 in. long is suspended vertically and a weight is attached. The length of the spring for various weights is measured. By Hooke's Law, L is a linear function of W; determine the function.

W	0.5	2	5	10	15	17.5
L	8.37	9.46	11.68	14.90	18.52	20.02

Ans. $L = 0.70W + 8$.

137. Nonlinear equations. When the points do not follow a straight line, no linear relation exists and we must try to represent the data by an equation of some other form.

As an example, suppose we have some reason to believe that the variables satisfy the relation

$$\log_{10} y = a + bx^2. \tag{1}$$

Let us introduce two new variables u, v such that

$$u = x^2, \qquad v = \log_{10} y.$$

Equation (1) then becomes

$$v = a + bu,$$

which is linear in u and v. For each pair of values of x and y we compute x^2 and $\log_{10} y$, and plot these as abscissa and ordinate — briefly, we plot $(x^2, \log_{10} y)$ instead of (x, y). If the points plotted follow a straight line, the data given may be represented by an empirical equation of the form (1). If they do not follow a straight line, no equation of the form (1) can be made to fit the data.

RULE II: To test a formula

$$v = a + bu, \tag{2}$$

where u and v are functions of x and y:

1. For each pair of values x and y given, compute the corresponding values of u and v.
2. Plot the values thus obtained as abscissa and ordinate respectively; that is, plot the points (u, v).
3. Determine whether the points (u, v) follow a straight line. If they do, the data can be represented by an equation of the form (2).

By changing the form of the equation, we may be able to discover other pairs of coordinates by which (2) can be tested. For examples, see the last paragraph of § 138. We try, naturally, to choose a pair of coordinates whose use involves the least numerical work.

The straight line is of no importance except as a means to an end. The only thing of final interest is the equation of the curve fitting the data given.

138. Nonlinear equation with one constant. First, let us consider equations of the form

$$v = a + bu$$

where either a or b (usually a) is known.

Example: When a body slides from rest down a smooth inclined plane (cf. Ex. 9, p. 216), it travels in time t a distance

$$x = \tfrac{1}{2}gt^2 \sin \alpha. \tag{1}$$

Represent x as a function of t, and find α, for the plane yielding the results given in the first two lines of the table.

t	1	2	3	4	5
x	1.10	4.35	10.05	17.15	26.25
t^2	1	4	9	16	25

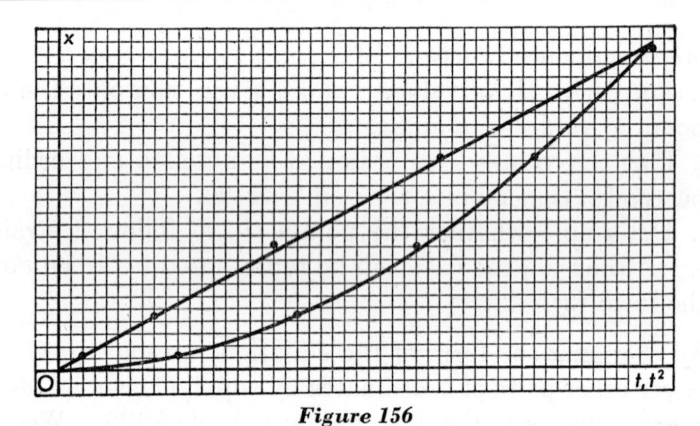

Figure 156

According to (1), we must expect the data to satisfy an equation of the form

$$x = bt^2. \tag{2}$$

This formula can evidently be tested by plotting (t^2, x). Thus the first step is to compute the values of t^2 as shown in the third line of the table. Plotting t^2 as abscissa with 2 spaces as the unit, and x as ordinate with 1 space as the unit, we find that the points follow a straight line quite closely. By the method of averages we find

$$58.90 = 55b, \qquad b = 1.071, \qquad x = 1.071t^2.$$

Computing values of x from this equation, we draw the curve, with 10 spaces as the t-unit and 1 as the x-unit. The points given in the table are shown by dots. Finally,

$$\tfrac{1}{2}g \sin \alpha = b = \frac{58.90}{55}, \qquad \sin \alpha = \frac{2(58.90)}{55(32.16)},$$
$$\log_{10} \sin \alpha = 8.82347 - 10, \qquad \alpha = 3°49'.$$

Other coordinates by which (2) may be tested are easily found: divide through by t, plot $\left(t, \frac{x}{t}\right)$; take the square root of both members, plot $(t, \sqrt{x})$; take the logarithm of both members, plot $(\log_{10} t, \log_{10} x)$.

Exercises

1. Solve the example of § 138 by plotting $\left(t, \frac{x}{t}\right)$.

2. Discover by plotting that the following data suggest a relation of the form $y = bx^3$, verify this by plotting (x^3, y), determine the constant, and plot the curve.

x	0.6	0.8	1.0	1.2	1.4
y	0.161	0.374	0.806	1.302	2.150

Ans. $y = 0.773x^3$.

3. Discover by plotting that the data suggest an equilateral hyperbola, verify this assumption, and determine the constant.

x	0.5	0.8	1.0	1.5	2.0
y	7.11	4.23	3.69	2.20	1.64

4. The kinetic energy of a body of mass m moving with velocity v is

$$E = \tfrac{1}{2}mv^2.$$

Find a value of m to fit the following data, and plot the curve.

v	0.8	1.2	1.5	1.7	2.2
E	2.17	4.92	7.95	10.22	17.80

Ans. 7.14.

5. In Ex. 8, p. 215, the disks were made of sheet metal weighing 11.03 oz. per sq. ft. The disks were weighed and the results averaged over the whole class, as follows (r in inches, W in ounces). What was the error in the computed value of π?

r	3	4	5	10
W	2.167	3.853	6.022	24.080

Ans. +0.0023.

6. In the inclined-plane experiment (Ex. 9, p. 216), the body acquires at a distance x from the starting point a velocity given by the formula

$$v^2 = 2g\,x \sin \alpha.$$

For the following experiment, verify the formula by plotting (x, v^2), find v in terms of x, plot the curve, and determine α.

x	2	4	6	8	10
v	3.4	4.8	5.9	6.9	7.8

Ans. $v^2 = 5.93x$; $\alpha = 5°17'$.

7. Solve Ex. 6 by using the coordinates $(\sqrt{x}, v)$.

Ans. $v^2 = 5.90x$; $\alpha = 5°16'$.

8. Discover by plotting that the data suggest an equilateral hyperbola, verify this assumption, and find the equation.

x	0.20	0.40	0.81	1.15	1.45
y	0.63	0.73	1.01	1.31	1.57

9. It is known from physics that the attraction between two magnetic poles is inversely proportional to the square of the distance between them: $A = \frac{k}{d^2}$. Find k for the following experiment, and plot the curve.

d	0.5	0.6	0.75	1.0	1.5	2.0
A	0.0352	0.0243	0.0162	0.0087	0.0041	0.0022

10. Solve Ex. 9 by plotting $\left(\frac{1}{d}, dA\right)$.

11. Plot the curve $y^2 = x$ accurately in the interval $0 < x < 4$. By counting squares, determine the area bounded by the curve, the x-axis, and the line $x = h$ for $h = 1, 2, 3, 4$. Show that the data fit a formula $A = kh^{\frac{3}{2}}$, and find k to four decimal places.

12. Discover by plotting that the data suggest an equilateral hyperbola with asymptotes $x = 0$, $y = 1$. Verify the assumption $y = \frac{x + b}{x}$ by plotting $\left(\frac{1}{x}, y\right)$ if a table of reciprocals is at hand, otherwise by plotting (x, xy); determine b; plot the curve.

x	0.23	0.51	1.04	2.52	4.10	6.20	8.02
y	8.91	5.20	2.97	1.80	1.48	1.30	1.24

In Exs. 13–19, find a pair of coordinates to test the given formula.

13. $y = ax^b$. *Ans.** $(\log x, \log y)$.

14. $y = ae^{bx}$. *Ans.*† $(x, \log y)$.

15. A projectile thrown with initial velocity v_0 against a wall h ft. distant:

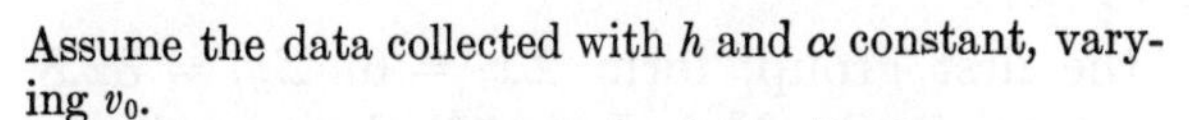

$$y = h \tan \alpha - \frac{gh^2}{2v_0^2} \sec^2 \alpha.$$

Assume the data collected with h and α constant, varying v_0.

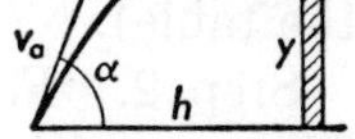

Figure 157

16. Ex. 15 by another pair of coordinates.

17. Ex. 15 with v_0 and α constant, h varying.

18. Ex. 15 with h and v_0 constant, α varying. *Ans.* $(\sin 2\alpha, y \cos^2 \alpha)$.

19. Cantilever beam bearing a load at the free end (Ex. 1, p. 173):

$$y = ax^2 + bx^3.$$

* These curves appear so often that specially ruled paper, called logarithmic paper, has been constructed and is obtainable on the market. When this paper is used, the data are plotted directly. If the formula is applicable, the points will follow a straight line, no matter what logarithmic base is used.

† These data may be plotted directly on the kind of specially ruled paper called semi-logarithmic paper.

139. Equations containing two constants. Consider such equations as

$$y = a + bx, \qquad x = a + bt^2, \qquad S = ar^2 + br,$$

where both a and b are unknown. If we apply the method of averages, Rule I, § 136, we obtain only one equation. We need two equations to solve for a and b. Hence we divide the data into two groups and apply the method of averages to each group separately. Usually the best result is obtained if the two groups are of equal size or nearly so.

When the assumed formula involves three constants we divide the data into three groups, and so on.

Example: Fit the equation $y = a + bx$ to the given data.

x	1	2	3	4	5	6
y	11.6	11.1	10.1	9.6	9.0	8.1

The student should plot the points on a large scale, verifying the existence of an approximate linear relation, and determine the equation of a well-fitting line by the method of selected points. By the method of averages:

Step 1. Divide the data into two groups (as indicated in the table).

Step 2. For the first group, form $\Sigma x_i = 6$, $\Sigma y_i = 32.8$. Then a and b must satisfy (by Rule I, § 136) the equation

$$(1) \qquad 3a + 6b = 32.8.$$

Step 3. For the second group, form $\Sigma x_i = 15$, $\Sigma y_i = 26.7$. Then a and b must satisfy the equation

$$(2) \qquad 3a + 15b = 26.7.$$

Step 4. Solve for a and b: $b = -0.678$, $a = 12.29$. The desired line is $y = 12.29 - 0.678x$.

140. The method of least squares: derivation. An important method for curve-fitting is the *method of least squares*, in which the constants are determined so that the *sum of the squares of the residuals is a minimum.*

Consider the problem of fitting the equation

$$y = a + bx \tag{1}$$

to the set of n points $(x_1, y_1), (x_2, y_2), \cdots, (x_n, y_n)$ by the method of least squares. For any point (x_i, y_i), the residual is

$$r_i = y_i - a - bx_i$$

and its square is

$$r_i^2 = y_i^2 + a^2 + b^2x_i^2 - 2ay_i - 2bx_iy_i + 2abx_i, \tag{2}$$

for each of $i = 1, 2, \cdots, n$.

Let R be the sum of the squares of the residuals,

$$R = \Sigma r_i^2 = r_1^2 + r_2^2 + \cdots + r_n^2, \tag{3}$$

the expression to be minimized. Using the n equations (2) to form R, we obtain

$$R = \Sigma y_i^2 + na^2 + b^2\Sigma x_i^2 - 2a\Sigma y_i - 2b\Sigma x_iy_i + 2ab\Sigma x_i. \tag{4}$$

Equation (4) gives R as a quadratic in a,

$$R = na^2 + 2a(b\Sigma x_i - \Sigma y_i) + \Sigma(y_i - bx_i)^2. \tag{5}$$

By Ex. 22, p. 105, R takes on its minimum value when

$$a = -\frac{2(b\Sigma x_i - \Sigma y_i)}{2n};$$

that is, when

$$na + b\Sigma x_i = \Sigma y_i. \tag{6}$$

Equation (4) also gives R as a quadratic in b,

$$R = b^2\Sigma x_i^2 + 2b(a\Sigma x_i - \Sigma x_iy_i) + \Sigma(y_i - a)^2. \tag{7}$$

By Ex. 22, p. 105, R takes on its minimum value when

$$b = -\frac{2(a\Sigma x_i - \Sigma x_i y_i)}{2\Sigma x_i^2};$$

that is, when

$$(8) \qquad a\Sigma x_i + b\Sigma x_i^2 = \Sigma x_i y_i.$$

Equations (6) and (8) are two linear equations for the determination of a and b. For the a and b which satisfy equations (6) and (8), the sum of the squares of the residuals will be less than for any other values of a and b.

141. The method of least squares: application. The derivation of § 140 yields

RULE III: To make the line $y = a + bx$ fit the n points (x_i, y_i); $i = 1, 2, \cdots, n$, in the sense of least squares, solve for a and b in the equations

$$(1) \qquad \boldsymbol{na + b\Sigma x_i = \Sigma y_i},$$

$$(2) \qquad \boldsymbol{a\Sigma x_i + b\Sigma x_i^2 = \Sigma x_i y_i}.$$

Example: Use the method of least squares to fit the line $y = a + bx$ to the given data (Example, § 139).

x	1	2	3	4	5	6
y	11.6	11.1	10.1	9.6	9.0	8.1

Here $n = 6$. We first form the sums involved:

$$\Sigma x_i = 21, \qquad \Sigma y_i = 59.5, \qquad \Sigma x_i^2 = 91, \qquad \Sigma x_i y_i = 196.1,$$

and then solve the equations

$$6a + 21b = 59.5,$$
$$21a + 91b = 196.1,$$

for a and b. The solution is $a = 12.35$, $b = -0.694$. The required line is $y = 12.35 - 0.694x$.

Exercises

In Exs. 1–8, fit a linear equation to the given data (*a*) by selected points; (*b*) by the method of averages. In (*b*) use groups of equal size when the number of points given is even.

1.

x	1	2	3	4	5	6
y	8	6	5	3	2	0

Ans. (*b*) $y = 9.44 - 1.56x$.

2.

x	1	2	3	4	5	6
y	1.0	2.0	3.0	3.5	4.5	5.5

Ans. (*b*) $y = 0.33 + 0.83x$.

3.

x	2	3	4	6	7	8
y	0.5	1.5	2.5	4.1	5.0	6.0

Ans. (*b*) $y = -1.15 + 0.88x$.

4.

x	1	2	3	4	5	6
y	2.4	3.1	3.4	4.5	5.2	5.5

Ans. (*b*) $y = 1.57 + 0.70x$.

5.

x	0	1	2	3	4	5
y	2.6	4.2	5.7	7.4	8.9	10.4

Ans. (*b*) $y = 2.59 + 1.58x$.

6.

x	1	3	4	6	7	9
y	8	7	6	5	4	3

Ans. (*b*) $y = 8.71 - 0.64x$.

7. Group the first four and the last three pairs.

x	1	2	3	4	5	6	7
y	1.82	4.19	6.90	9.21	11.65	14.36	16.72

Ans. (*b*) $y = 2.49x - 0.69$.

8. Solve Ex. 7, grouping the first three and the last four pairs.

Ans. (*b*) $y = 2.48x - 0.66$.

9. The force necessary to lift a weight by means of a pulley was measured as shown. Find F in terms of W, and tabulate the values of F at intervals of 5 lbs. from $W = 5$ to $W = 30$.

W	5	8	12	20	25	30
F	1.65	2.15	2.65	4.20	5.10	5.90

10. In the example of § 49, the following data were obtained by placing a weight of 10 lbs. at various distances x from A. Verify that F is a linear function of x; find w and L.

x	0	2	4	8	12	15
F	41.1	42.6	43.7	46.2	48.8	50.5

Ans. $w = 5$ lbs. 2 oz. per ft.; $L = 16.0$ ft.

11. In Ex. 17, p. 79, the following data were obtained with a lever weighing 4 lbs. per ft. Verify that F is a linear function of W; find L_1, L_2.

W	15	20	30	50	75	100
F	1.6	3.9	9.0	18.8	31.3	44.2

12. The work done in changing the velocity of a moving body is

$$W = \tfrac{1}{2}mv^2 - \tfrac{1}{2}mv_0^2,$$

where m is the mass of the body, v_0 the initial velocity, and v the final velocity. For the data given, verify that there exists a relation $W = a + bv^2$; determine the constants and plot the curve; find m and v_0.

v	8	10	12	14	16	18	20
W	26	115	227	358	515	684	880

Ans. $m = 5.10$; $v_0 = 7.41$.

In Exs. 13–18, use the method of least squares.

13. Ex. 1. *Ans.* $y = 9.40 - 1.54x$.
14. Ex. 2. *Ans.* $y = 0.20 + 0.87x$.
15. Ex. 3. *Ans.* $y = -1.22 + 0.90x$.
16. Ex. 4. *Ans.* $y = 1.73 + 0.65x$.
17. Ex. 5. *Ans.* $y = 2.62 + 1.57x$.
18. Ex. 6. *Ans.* $y = 8.71 - 0.64x$.

Solid Analytic Geometry

CHAPTER 17 *Coordinates in Space*

142. Rectangular coordinates. To determine the position of a point in three-dimensional space, three magnitudes must be given. Thus to locate a point in the interior of a room, we may give its height above the floor and its distances from two adjacent walls; the position of a point in the interior of the earth may be fixed by its depth below the surface together with the latitude and longitude of the point on the surface above it.

Usually the most convenient way of fixing the position of a point in space is by means of its *distances from three mutually perpendicular planes*, as in the first example cited. These distances are the *rectangular coordinates* of the point: the three planes are the *coordinate planes*, their three lines of intersection are the *coordinate axes*, and their point of intersection is the *origin*. The coordinates are denoted by the letters x, y, z, and are written $P:(x, y, z)$, or merely (x, y, z). The three axes are called the x-axis, the y-axis, and the z-axis; the three planes are the xy-plane (containing the x- and y-axes), the yz-plane, and the zx-plane. Of course a definite positive sense must be chosen for each coordinate: that is, the coordinates are *directed segments*, as in plane geometry.

Space is divided by the coordinate planes into eight compartments, or *octants*. The region in which all three coordinates are positive is called the *first octant;* there will be no occasion to refer to the others by number.

143. Figures. In the system of drawing adopted in this and most books, *parallel lines are represented by parallel lines:* i.e., if two lines in space are parallel, they are shown in the figure by lines that are actually parallel.

Two of the axes are represented by perpendicular lines, while the third, which of course is supposed to be perpendicular to the other two, is shown by a line drawn in any suitable direction. The axes will be placed as in Fig. 158 (except when some other arrangement is more convenient), the positive half of each axis being the part drawn in full. Figures in the yz-plane, or in a plane parallel to that plane, are drawn in their true form and proportions; all others are distorted, due to foreshortening.

144. Distance between two points. Given any two points P_1: (x_1, y_1, z_1), P_2: (x_2, y_2, z_2), we note that, in Fig. 159,

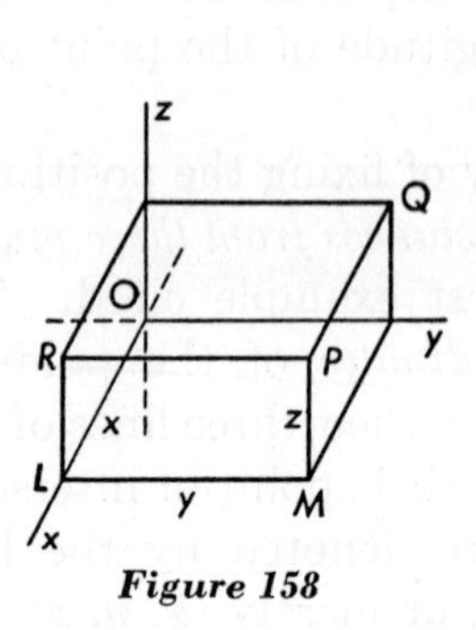

Figure 158

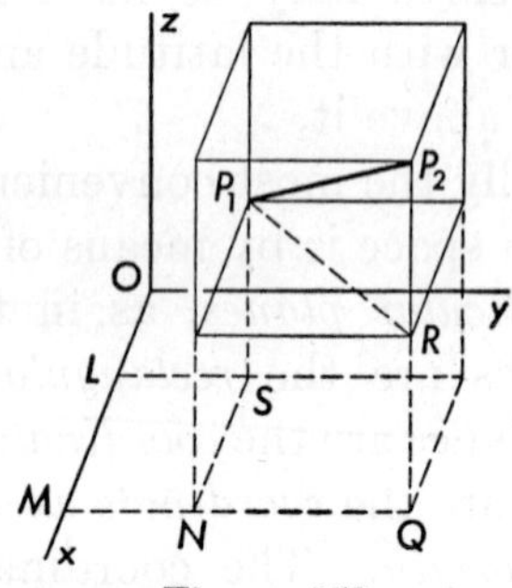

Figure 159

$$LM = x_2 - x_1,$$
$$NQ = y_2 - y_1,$$
$$RP_2 = z_2 - z_1.$$

Since

$$\overline{P_1P_2} = \sqrt{\overline{P_1R}^2 + \overline{RP_2}^2} = \sqrt{\overline{SN}^2 + \overline{NQ}^2 + \overline{RP_2}^2},$$

the length of the segment P_1P_2 is

$$(1) \qquad d = \sqrt{(x_2 - x_1)^2 + (y_2 - y_1)^2 + (z_2 - z_1)^2}.$$

145. Midpoint of a line segment. To find the point $P:(x, y, z)$ bisecting the line segment joining $P_1:(x_1, y_1, z_1)$ and $P_2:(x_2, y_2, z_2)$, let us drop perpendiculars PL, P_1L_1, P_2L_2 from these points to Ox. Then

$$OL = OL_1 + \tfrac{1}{2}L_1L_2:$$

that is,

$$x = x_1 + \tfrac{1}{2}(x_2 - x_1) = \tfrac{1}{2}(x_1 + x_2).$$

In this way we obtain the formulas

$$(1) \qquad \begin{cases} x = \tfrac{1}{2}(x_1 + x_2), \\ y = \tfrac{1}{2}(y_1 + y_2), \\ z = \tfrac{1}{2}(z_1 + z_2). \end{cases}$$

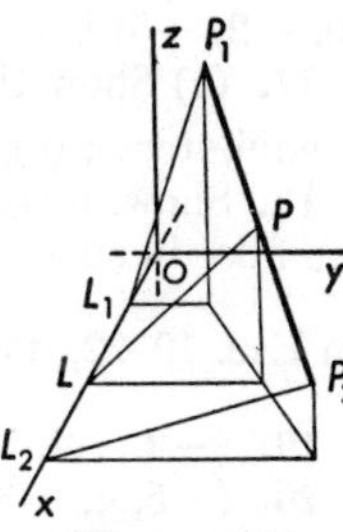

Figure 160

Exercises

1. Plot the points (3, 4, 1), (−3, 4, 1), (−3, −4, 1).

2. Plot the points (2, 3, 4), (2, 3, −4), (−2, 3, −4).

3. From the point (2, 3, 4) draw perpendiculars to each coordinate axis; find the length of each perpendicular segment.

4. From the point (3, 2, 3) draw perpendiculars to each coordinate axis; find the length of each perpendicular segment.

5. In each coordinate plane, draw a line through O making an angle of 45° with each of the axes, and (*a*) a parallel, (*b*) a perpendicular to this line through an arbitrary point in that plane.

6. Draw a box with its edges parallel to the axes, having the points (0, 0, 0), (3, 4, 2) as ends of a diagonal.

7. In each coordinate plane, plot by points a circular quadrant of radius 5 with its center at the origin.

8. In the xy-plane, sketch the parabola $y^2 = 4x$.

9. In the yz-plane, sketch the parabola $y^2 = 16z$.

10. In the xz-plane, sketch the ellipse $x^2 + 4z^2 = 4$.

11. What is the distance of the point (x, y, z) from Ox? From Oy? From Oz? From O?

12. Where is a point situated if

(*a*) $x = 0$?	(*b*) $z = 0$?	(*c*) $x = y = 0$?
(*d*) $y = z = 0$?	(*e*) $x = 2$?	(*f*) $x = 2, y = 1$?
(*g*) $x = z$?	(*h*) $y = z, x = 0$?	(*i*) $x = y = z$?

13. (*a*) Show that the triangle with vertices (2, 4, 1), (1, 2, −2), (5, 0, −2) is right-angled; (*b*) find its area. *Ans.* (*b*) $\sqrt{70}$.

14. (*a*) Show that the triangle with vertices (5, 9, 11), (0, −1, −4), (5, −11, 1) is right-angled; (*b*) find its area. *Ans.* (*b*) $25\sqrt{21}$.

15. (*a*) Show that the triangle with vertices (2, 0, 8), (8, −4, 6), (−4, −2, 4) is isosceles; (*b*) find its area. *Ans.* (*b*) $6\sqrt{19}$.

16. (*a*) Show that the triangle with vertices (6, 2, 3), (1, −3, 2), (0, −2, −5) is isosceles; (*b*) find its area. *Ans.* (*b*) $\sqrt{638}$.

17. (*a*) Show that the triangle with vertices (1, 3, 3), (2, 2, 1), (3, 4, 2) is equilateral; (*b*) find its area. Note Ex. 22, p. 21. *Ans.* (*b*) $\frac{3}{2}\sqrt{3}$.

18. Show, by counterexamples, that the theorems of Exs. 21–22, p. 20–21, are false in space of three dimensions.

In Exs. 19–22, do the points lie on a straight line?

19. (−2, −4, 7), (2, 2, 3), (4, 5, 1).
20. (−5, 4, −3), (1, 1, 3), (−9, 6, −7).
21. (−1, −1, 1), (7, 11, 9), (−6, −9, −4).
22. (−5, −10, 9), (−1, −5, 5), (11, 10, −9).
23. Find the midpoint between (2, −1, 6), (−4, 9, 1).
24. Find the midpoint between (4, 0, −3), (−1, 8, 13).

In Exs. 25–28, solve by another method (§ 145).

25. Ex. 13 (*a*). **26.** Ex. 14 (*a*).
27. Ex. 15 (*a*). **28.** Ex. 16 (*a*).

29. Show that the quadrilateral with vertices (5, 1, 1), (3, 1, 0), (4, 3, −2), (6, 3, −1) is a rectangle.

30. Show that the quadrilateral with vertices (3, 2, 5), (1, 1, 1), (4, 0, 3), (6, 1, 7) is a parallelogram.

31. Show that the quadrilateral with vertices (3, −3, 5), (1, 3, 4), (−5, 4, 6), (−1, 1, 2) is not a parallelogram.

32. Show that the quadrilateral with vertices (3, 5, 1), (2, 4, 6), (3, −1, 7), (0, 2, 4) is not a parallelogram.

33. Find the fourth vertex of the parallelogram having as consecutive vertices the first three points of Ex. 31. *Ans.* (−3, −2, 7).

34. Find the fourth vertex of the parallelogram having as consecutive vertices the first three points of Ex. 32. *Ans.* (4, 0, 2).

35. Prove that the straight lines joining the midpoints of adjacent sides of any quadrilateral (not necessarily plane) form a parallelogram.

146. Direction cosines; radius vector. Given a directed line L passing through the origin, the angles α, β, γ formed by this line with the positive x-, y-, and z-axes are called the *direction angles* of the line, and the cosines of these angles are the *direction cosines* of the line. Direction cosines are denoted by l, m, n: that is,

$$l = \cos\alpha, \qquad m = \cos\beta, \qquad n = \cos\gamma.$$

More generally, if the given line does not pass through the origin, its direction angles and direction cosines are defined as *equal to those of the parallel line through the origin.*

Let $P:(x, y, z)$ be any point on the line L. The segment OP, denoted by ρ, is called the *radius vector* of P. By the distance formula it follows that

$$(1) \qquad \rho = \sqrt{x^2 + y^2 + z^2}.$$

From the triangles OAP, OBP, OCP we read off

$$(2) \qquad \cos\alpha = l = \frac{x}{\rho}, \qquad m = \frac{y}{\rho}, \qquad n = \frac{z}{\rho}.$$

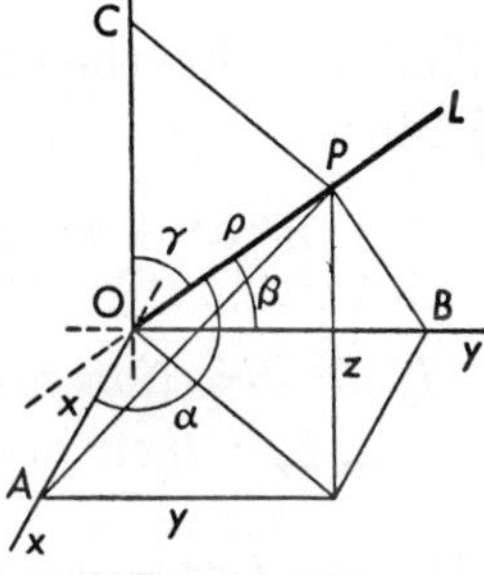

Figure 161

Hence:

The direction cosines of the radius vector of a point are proportional to the coordinates of the point.

Clearing of fractions in (2), we see that

$$(3) \qquad x = l\rho, \qquad y = m\rho, \qquad z = n\rho.$$

Square both members of (1) and substitute the values (3):

$$l^2\rho^2 + m^2\rho^2 + n^2\rho^2 = \rho^2.$$

The direction cosines of any line satisfy the relation

$$(4) \qquad l^2 + m^2 + n^2 = 1.$$

By (1) and (4), the radius vector of the point (l, m, n) is

$$\rho = \sqrt{l^2 + m^2 + n^2} = 1.$$

Hence, given any set of numbers l, m, n satisfying (4), there will always be a line having those numbers as direction cosines, viz., the line through $(0, 0, 0)$ and (l, m, n).

If the positive sense on the line be reversed, its direction angles are replaced by their supplements, and the signs of the direction cosines are changed.

147. Direction components. Consider any line (not necessarily through the origin) whose direction cosines are *proportional to three numbers a, b, c:*

$$l = ka, \qquad m = kb, \qquad n = kc.$$

To determine the proportionality constant k, substitute in (4), § 146:

$$k^2(a^2 + b^2 + c^2) = 1, \qquad k = \frac{1}{\sqrt{a^2 + b^2 + c^2}}.$$

If the direction cosines of a line are proportional to three numbers a, b, c, their actual values are

$$(1) \qquad l = \frac{a}{\sqrt{a^2 + b^2 + c^2}}, \qquad m = \frac{b}{\sqrt{a^2 + b^2 + c^2}},$$

$$n = \frac{c}{\sqrt{a^2 + b^2 + c^2}}.$$

The ambiguity of sign mentioned in § 146 appears here in the fact that we might equally well choose the negative sign before each radical.

Given a line with direction cosines l, m, n, any set of numbers a, b, c proportional to l, m, n are called *direction components* for that line. Although the direction cosines of a given line are definite, fixed numbers (except that all the signs may be changed), it is clear that every line has infinitely many sets of direction components: viz., any set of numbers proportional to l, m, n.

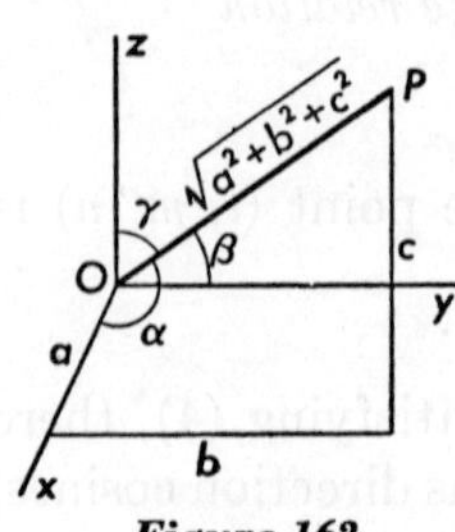

Figure 162

Given any three numbers a, b, c (not all zero), the formulas (1) produce values of l, m, n, no matter what a, b, c may be: thus there will always be a line through the origin having a, b, c as direction components. In fact, this of course is the line through (0, 0, 0) and P: (a, b, c), as shown in Fig. 162.

148. Direction components of the line through two points. Let d be the distance between the two points P_1: (x_1, y_1, z_1), P_2: (x_2, y_2, z_2). Then the direction cosines of the line P_1P_2 are

$$l = \frac{P_1L}{d} = \frac{x_2 - x_1}{d},$$

$$m = \frac{P_1M}{d} = \frac{y_2 - y_1}{d},$$

$$n = \frac{P_1N}{d} = \frac{z_2 - z_1}{d}.$$

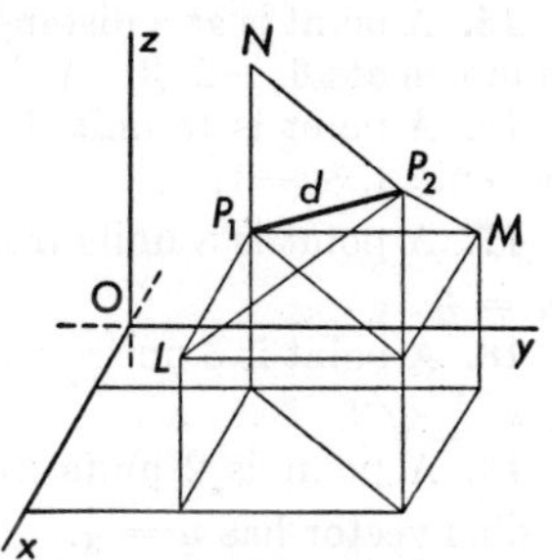

Figure 163

Hence:

A set of direction components of the line joining the points (x_1, y_1, z_1) *and* (x_2, y_2, z_2) *is* $x_2 - x_1$, $y_2 - y_1$, $z_2 - z_1$.

Exercises

1. A line makes an angle of 45° with Oy and 60° with Oz. What angle does it make with Ox?

2. A line has the direction cosines $l = \frac{3}{10}$, $m = \frac{2}{5}$. What angle does it make with Oz? *Ans.* 30°.

3. A line has direction cosines $l = \frac{7}{10}$, $n = \frac{1}{10}$. What angle does it make with Oy? *Ans.* 45°.

4. In Fig. 161, if the point P lies in the xy-plane, show that the relation $l^2 + m^2 + n^2 = 1$ reduces to $\cos^2 \alpha + \sin^2 \alpha = 1$.

5. For each of the following points, find the length and direction cosines of the radius vector: (*a*) (2, 1, 3); (*b*) (−3, 2, 4); (*c*) (−1, 4, 0).

6. Where must a point lie if its radius vector has (*a*) $m = 0$? (*b*) $m = \frac{1}{2}$? (*c*) $l = n = 0$? (*d*) $m = 1$? (*e*) $m = n = \frac{1}{2}\sqrt{2}$?

7. Can a line be drawn at an angle of 30° with Ox and 45° with Oz?

8. Show that it is impossible to draw a line two of whose direction angles shall be less than 45°.

In Exs. 9–12, draw the line.

9. Through O, with direction components 2, 3, 1.

10. Through O, with direction components 4, 2, 5.

11. Through (1, 3, 2), with direction components 4, 2, 1.

12. Through (5, 4, 6), with direction components 2, 2, 3.

In Exs. 13–26, find the coordinates of the points satisfying the given conditions.

13. A point is at a distance 6 from O, and its radius vector makes an angle of 45° with Ox and 60° with Oz.

14. A point is at a distance 12 from O, and its radius vector has $m = \frac{1}{3}$, $n = \frac{5}{6}$. *Ans.* $(\pm 2\sqrt{7}, 4, 10)$.

15. A point is at a distance 21 from O, and its radius vector has direction components 6, -2, 3. *Ans.* $(18, -6, 9)$; $(-18, 6, -9)$.

16. A point is 18 units from O, and its radius vector has direction components 4, 8, -1. *Ans.* $(8, 16, -2)$; $(-8, -16, 2)$.

17. A point is 8 units from O, $2\sqrt{5}$ from Ox, and its radius vector has $m = \frac{1}{2}$. *Ans.* $(\pm 2\sqrt{11}, 4, 2)$; $(\pm 2\sqrt{11}, 4, -2)$.

18. A point is 3 units from the xz-plane, and its radius vector has $l = \frac{1}{4}$, $n = \frac{1}{2}\sqrt{3}$. *Ans.* $(\sqrt{3}, 3, 6)$.

19. A point is 2 units from the xz-plane, 6 from the yz-plane, and its radius vector has $n = \frac{1}{3}$. *Ans.* $(6, 2, \sqrt{5})$.

20. A point is 6 units from the xy-plane, 3 from the xz-plane, and its radius vector has $l = -\frac{1}{2}$. *Ans.* $(-\sqrt{15}, 3, 6)$.

21. A point is 2 units from O, and its radius vector has $\beta = 60°$, $\gamma = 45°$. *Ans.* $(\pm 1, 1, \sqrt{2})$.

22. A point is at distance 5 from O, 4 from the yz-plane, and its radius vector has $n = \frac{1}{5}$. *Ans.* $(4, \pm 2\sqrt{2}, 1)$.

23. A point is distant $4\sqrt{10}$ from Ox, and its radius vector has direction components 3, 2, 6.

24. A point is at the distance $\sqrt{5}$ from the z-axis, 1 from the xy-plane, and its radius vector makes an angle $\frac{1}{4}\pi$ with Ox.

25. A point is at the distance $\sqrt{2}$ from Oz, $\sqrt{3}$ from Ox, and its radius vector makes an angle of 45° with Oz.

26. A point is at the distance $2\sqrt{10}$ from Oz, $2\sqrt{7}$ from Oy, and its radius vector has $l = \frac{1}{4}$.

In Exs. 27–32, solve by a new method.

27. Ex. 19, p. 234.
28. Ex. 20, p. 234.
29. Ex. 21, p. 234.
30. Ex. 22, p. 234.
31. Ex. 31, p. 234.
32. Ex. 32, p. 234.

33. Show that it is impossible to draw a line so that the sum of any two (positive) direction angles is less than 90°. (Show that if $0° < \alpha + \beta < 90°$, then $\cos\alpha > \sin\beta$, whence $\cos^2\alpha + \cos^2\beta > 1$.)

149. Projections. The *projection* of a point P upon any line is defined as the *foot of the perpendicular* from P to that line. The projection of a line segment P_1P_2 upon any line is the segment joining the projections of the endpoints P_1, P_2 upon that line.

The projection of a broken line upon any line is the sum of the projections of the segments forming the broken line. *The projection of a broken line $P_1P_2 \cdots P_n$ upon any line is equal to the projection of the closing line P_1P_n upon that line.* Thus, in Fig. 164,

$$L_1L_2 + L_2L_3 + L_3L_4 = L_1L_4.$$

Note that the segments need not all lie in the same plane.

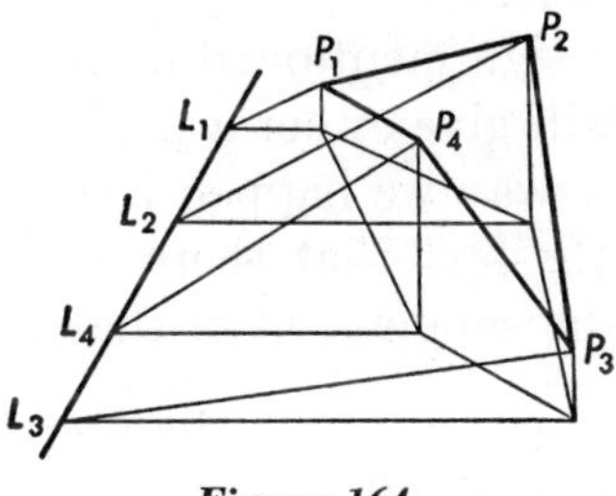

Figure 164

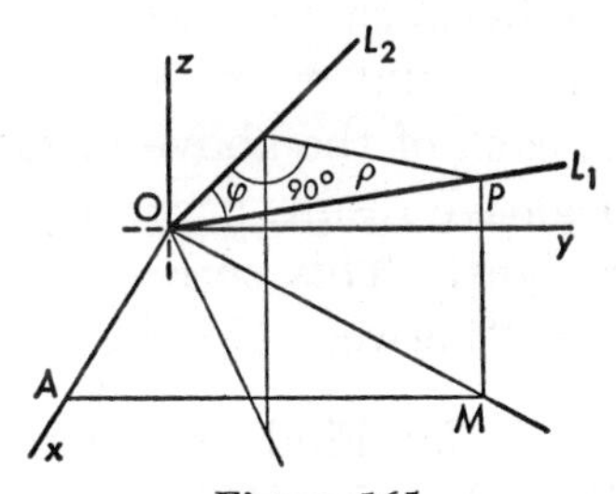

Figure 165

150. Angle between two lines. To find the angle φ between any two lines L_1, L_2 intersecting at the origin, let us denote the direction cosines of L_1 by l_1, m_1, n_1, those of L_2 by l_2, m_2, n_2, and choose on L_1 any point $P:(x, y, z)$ with radius vector ρ. Then, in Fig. 165,

$$OA = x = l_1\rho, \qquad AM = y = m_1\rho, \qquad MP = z = n_1\rho.$$

Now the projection on L_2 of the broken line $OAMP$ equals the projection on L_2 of the closing line OP:

$$OP \cos \varphi = OA \cdot l_2 + AM \cdot m_2 + MP \cdot n_2,$$
$$\rho \cos \varphi = l_1\rho l_2 + m_1\rho m_2 + n_1\rho n_2:$$

that is,

$$(1) \qquad \mathbf{\cos \varphi = l_1l_2 + m_1m_2 + n_1n_2.}$$

If the two lines have direction components a_1, b_1, c_1 and a_2, b_2, c_2, the direction cosines may be found by § 147. Substituting in formula (1), we find

$$\text{(2)} \qquad \cos \varphi = \frac{a_1a_2 + b_1b_2 + c_1c_2}{\sqrt{a_1^2 + b_1^2 + c_1^2} \cdot \sqrt{a_2^2 + b_2^2 + c_2^2}}.$$

Of course these results apply at once (§ 146, second paragraph) to any two intersecting lines. Further, the angle between two non-intersecting lines is defined as *equal to the angle between two intersecting lines that are respectively parallel to the given lines.* With this convention the formulas give the angle between any two lines in space.

On account of the ambiguity of sign mentioned in §§ 146–147, each of the above formulas will give either a positive or a negative result, according to the way we happen to choose the signs. This corresponds to the fact that there are two angles "between two lines," one the supplement of the other.

Example: Find the angle between the lines joining the points (3, 1, 2), (4, 0, 4) and (−2, 4, 4), (0, −1, 3).

By § 148, the direction components of the lines are respectively 1, −1, 2 and 2, −5, −1. By (2), we find

$$\cos \varphi = \frac{2 + 5 - 2}{\sqrt{6} \cdot \sqrt{30}} = \frac{1}{6}\sqrt{5}.$$

151. Perpendicular lines. In formula (1), § 150, if $\cos \varphi = 0$, then $\varphi = 90°$, and vice versa.

THEOREM: *Two lines having the direction cosines l_1, m_1, n_1 and l_2, m_2, n_2 are perpendicular if and only if*

$$\text{(1)} \qquad l_1l_2 + m_1m_2 + n_1n_2 = 0.$$

COROLLARY: *Two lines having direction components a_1, b_1, c_1 and a_2, b_2, c_2 are perpendicular if and only if*

$$\text{(2)} \qquad a_1a_2 + b_1b_2 + c_1c_2 = 0.$$

Exercises

1. Find the angle between two lines whose direction components are 2, 1, -1 and 1, -1, -2. *Ans.* 60°.

2. Find the angle between two lines whose direction components are 1, 4, 8 and 1, -2, 2. *Ans.* $\cos \varphi = \frac{1}{3}$.

3. Find the angle between the radius vectors of the points (1, 1, 0) and (3, 4, -5). *Ans.* $\cos \varphi = \frac{7}{10}$.

4. Find the angle between the radius vectors of the points (1, 1, -2) and (1, 1, 0). *Ans.* $\cos \varphi = \frac{1}{3}\sqrt{3}$.

In Exs. 5–8, find the angle between the line joining the first pair of points and the line joining the second pair of points.

5. (6, 4, 3), (4, 3, 5) and (3, -1, 4), (2, -2, 4). *Ans.* 45°.

6. (-1, 9, 4), (-3, 2, 5) and (6, 3, 2), (1, 4, -1).

7. (6, 4, 2), (2, 3, -1) and (-3, 7, 4), (-1, 2, 3).

8. (4, 4, 4), (2, 5, 2) and (3, 1, 2), (2, 2, 2). *Ans.* 45°.

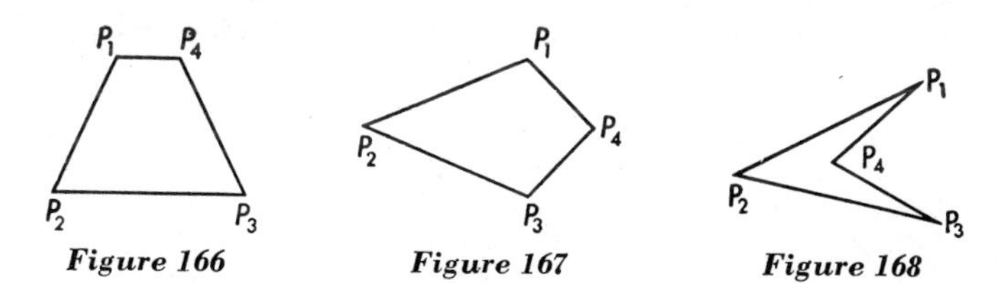

Figure 166 *Figure 167* *Figure 168*

In Exs. 9–14, show that the quadrilateral with vertices P_1, P_2, P_3, P_4 as given has the shape indicated. Find the area. (Figures not drawn to scale.)

9. (4, 5, 4), (7, 8, 4), (-1, 0, -4), (0, 1, 0): isosceles trapezoid. (Fig. 166.) *Ans.* $18\sqrt{2}$.

10. (4, 0, 7), (5, 1, -1), (0, -4, 3), (1, -3, 7): kite. (Fig. 167.) *Ans.* $24\sqrt{2}$.

11. (2, 2, 2), (4, 4, -8), (-2, -2, -2), (3, 3, -6): arrowhead. (Fig. 168.) *Ans.* $6\sqrt{2}$.

12. (0, 0, 0), (4, 2, -4), (6, 6, 0), (2, 4, 4): square. *Ans.* 36.

13. (7, 1, 3), (5, 0, 0), (4, -2, 3), (6, -1, 6): rhombus. *Ans.* $3\sqrt{19}$.

14. (1, 3, 5), (-3, -3, 3), (3, -2, 2), (5, 1, 3): right-angled trapezoid. (Fig. 169.) *Ans.* $6\sqrt{21}$.

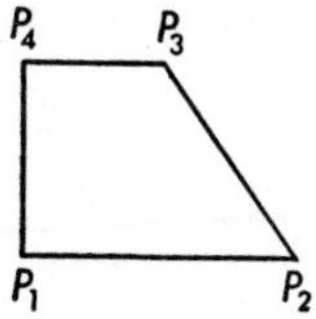

Figure 169

In Exs. 15–17, find the area of the triangle with the given vertices, using the formula $A = \frac{1}{2}S_1S_2 \sin \varphi$. (Fig. 170.)

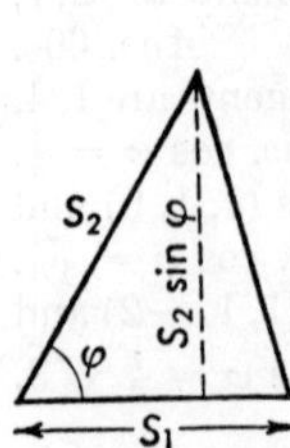

Figure 170

15. (2, 2, −1), (3, 1, 2), (4, 2, −2). *Ans.* $\frac{3}{2}\sqrt{6}$.

16. (3, 1, 0), (6, 4, −1), (5, 2, 2). *Ans.* $\frac{1}{2}\sqrt{122}$.

17. $(0, 0, 0)$, (x_1, y_1, z_1), (x_2, y_2, z_2).

$$Ans.\ A = \frac{1}{2}\left\{\begin{vmatrix} x_1 & x_2 \\ y_1 & y_2 \end{vmatrix}^2 + \begin{vmatrix} z_1 & z_2 \\ x_1 & x_2 \end{vmatrix}^2 + \begin{vmatrix} y_1 & y_2 \\ z_1 & z_2 \end{vmatrix}^2\right\}^{\frac{1}{2}}.$$

18. Find the area of a triangle with vertices (0, 0, 0), $(x_1, y_1, 0)$, $(x_2, y_2, 0)$, (*a*) by Ex. 17; (*b*) by § 12.

19. Show, with the aid of a figure, that when the axes are translated to a point $O_1{:}\,(h, k, l)$ as new origin, the old and new coordinates are connected by the following formulas. Compare with § 61.

$$x = x_1 + h, \qquad y = y_1 + k, \qquad z = z_1 + l.$$

20. In Ex. 15, find the coordinates of the vertices referred to (2, 2, −1) as new origin; find the area by Ex. 17. *Ans.* (0, 0, 0), (1, −1, 3), (2, 0, −1).

21. In Ex. 16, find the coordinates of the vertices referred to (3, 1, 0) as new origin; find the area by Ex. 17. *Ans.* (0, 0, 0), (3, 3, −1), (2, 1, 2).

CHAPTER 18 *Surfaces. Curves*

152. The locus of an equation. If x, y, and z are connected by an equation, we may assign values at pleasure to two of the variables and compute the third, thus determining certain sets of values of x, y, and z satisfying the equation. Each of these sets of values may be considered as the rectangular coordinates of a point. The points whose coordinates satisfy the equation are not scattered at random throughout space; instead, they form a definite *surface*, called the *locus* of the equation:

In space of three dimensions, the locus of an equation is a surface containing those points, and only those points, whose coordinates satisfy the equation.

Example (*a*): If the coordinates satisfy the equation

$$x^2 + y^2 + z^2 = 25,$$

the point must be at the distance 5 from the origin. Hence the locus is a sphere of radius 5 with center at O.

Example (*b*): If the coordinates satisfy the equation

$$x = y,$$

the point must be equidistant from the yz- and zx-planes. Hence the locus is a plane containing the z-axis and bisecting the angle between the yz- and zx-planes.

Exceptionally, the locus may reduce to a line or a point, etc.; or there may be no locus at all. See Exs. 31–34, p. 247.

153. Planes. It will be shown in § 167, and assumed meanwhile, that *every equation of the first degree represents a plane.*

The equation

$$x = k$$

represents a plane parallel to the yz-plane at a distance k from it; for that plane contains all those points, and only those points, whose x-coordinate is k. An analogous result holds for an equation of the same form in y or in z. Hence:

An equation of the first degree in one variable represents a plane parallel to the plane of the other two variables.

154. Intercepts; traces. To find the intercepts of a surface on the axes, we *equate two of the variables to zero and solve for the third:* to find the x-intercepts, set y and z equal to zero; etc.

A plane and a surface intersect in general in a curve, called the *section* of the surface by the plane. Of particular importance are the sections of a surface by the coordinate planes; these sections will be called for brevity the *traces* of the surface.

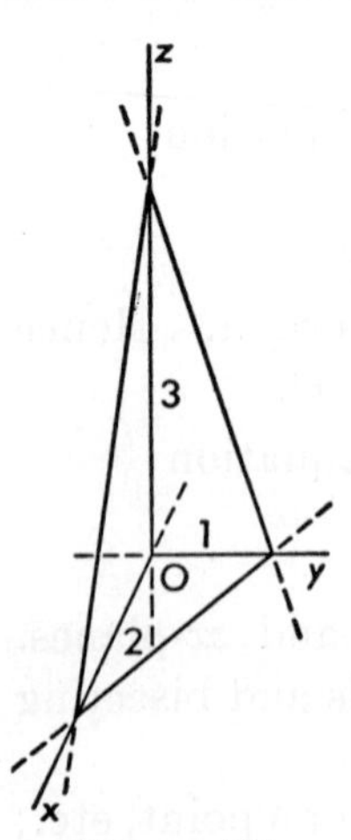

Figure 171

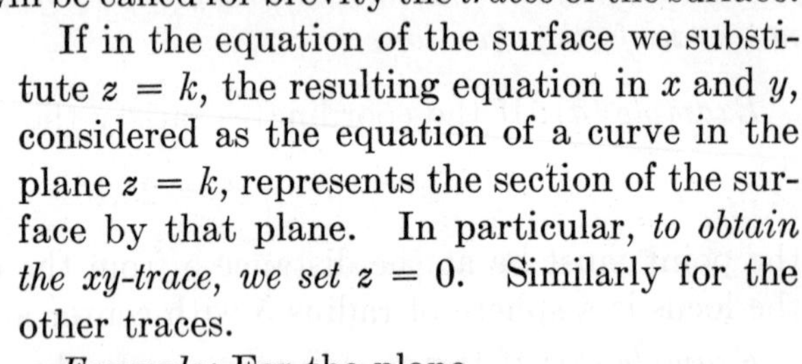

If in the equation of the surface we substitute $z = k$, the resulting equation in x and y, considered as the equation of a curve in the plane $z = k$, represents the section of the surface by that plane. In particular, *to obtain the xy-trace, we set* $z = 0$. Similarly for the other traces.

Example: For the plane

$$3x + 6y + 2z = 6,$$

the x-intercept is 2, y-intercept 1, z-intercept 3. The traces are,

in the xy-plane, the line $x + 2y = 2$;
in the yz-plane, the line $3y + z = 3$;
in the zx-plane, the line $3x + 2z = 6$.

155. Symmetry. To reduce the mechanical difficulties of draftsmanship, we shall confine our sketches chiefly to the first octant. Fortunately, in many applications involving three variables, all the variables are restricted to positive values. If not, considerations of symmetry may be helpful.

Two points P_1, P_2 are said to be *symmetric with respect to a plane* if the plane is perpendicular to the line P_1P_2 at its mid-point — i.e., if P_2 is the image, or reflection, of P_1 in that plane. A geometric figure is symmetric with respect to a plane if corresponding to every point P_1 of the figure the image P_2 also belongs to the figure.

The definitions of line and center of symmetry laid down in § 21 apply without change to figures in space.

The principal tests for symmetry are as follows.

THEOREM I: *A surface is symmetric with respect to the yz-plane if x can be replaced by $-x$ without changing the equation; and conversely.* Similarly for symmetry with respect to the zx- and xy-planes.

THEOREM II: *A surface is symmetric with respect to the x-axis if y and z can be replaced by $-y$ and $-z$ simultaneously without changing the equation: and conversely.* Similarly for symmetry with respect to the y- and z-axes.

156. Sketching by parallel plane sections. One of the simplest methods of sketching a surface is by means of a series of plane sections parallel to a coordinate plane. It often happens that one particular set of sections will form a clearer picture of the surface than any other set, so that care should be taken to make the best choice.

It is well to begin by carrying out the following steps:

1. *Test the surface for symmetry.*
2. *Find the intercepts on the axes.*
3. *Determine the traces on the coordinate planes.*
4. *Examine the sections parallel to each coordinate plane.*

Example (*a*): Discuss the surface

$$x^2 = 1 - yz,$$

and make a first-octant sketch.

1. The surface is symmetric with respect to the yz-plane, the x-axis, and the origin.

2. The x-intercepts are ± 1; no y- or z-intercept.

3. The xy-trace is the pair of lines $x = \pm 1$; yz-trace, the equilateral hyperbola $yz = 1$; zx-trace, the lines $x = \pm 1$.

4. Sections $x = k$ are hyperbolas; $y = k$, parabolas; $z = k$, parabolas.

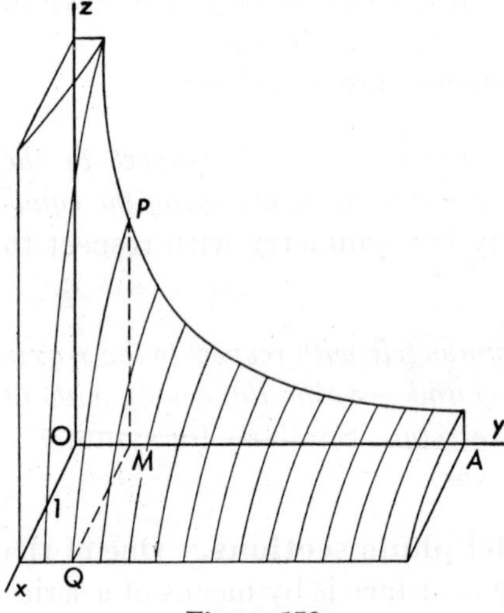

Figure 172

In Fig. 172 the surface is pictured by means of the sections $y = k$. The construction is as follows. On the axis that is *perpendicular to the cutting planes* — here, the y-axis — mark the total interval OA that is to be included, and divide this into a suitable number of (preferably equal) sub-intervals. At each point of division, such as M, draw the traces MP, MQ of the cutting plane, intersecting the traces of the surface at P and Q. The parabolic arc PQ is the required section.

In the drawing of Fig. 172, the xy- and yz-traces play a vital role: in any plane $y = k$, we have merely to draw a parabolic arc terminated by those traces. It may happen that two such useful traces are not present. Then, as guide-curve we may take a section of the surface by any convenient plane parallel to a coordinate plane (or two such sections if necessary).

Example (*b*): Discuss and sketch the surface

$$x^2 = yz.$$

The general discussion will be left to the reader. The only traces are the y- and z-axes. Take a section in a horizontal plane — say the plane $z = 1$ (the curve AQB in Fig. 173). Then, in each plane $y = k$ draw a parabolic arc such as the arc PQ, from the y-axis to the section AQB.

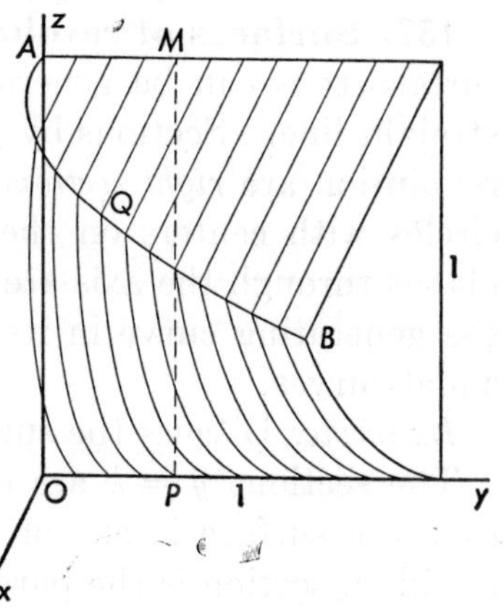

Figure 173

Exercises

In Exs. 1–8, mark the intercepts; draw the traces, and any other lines that may be needed to show the position of the plane.

1. $2x + 3y + z = 12.$
2. $x + y + 3z = 6.$
3. $x + y - 3z = 3.$
4. $x - 2y + z = 4.$
5. $2x + y = 6.$
6. $2y + 9z = 18.$
7. $y - z = 1.$
8. $y = x + z.$

In Exs. 9–30, discuss the surface and make a first octant sketch.

9. $4x^2 + y^2 + z^2 = 4.$
10. $x^2 + 4y^2 + 16z^2 = 16.$
11. $y^2 + z^2 = 2x.$
12. $x^2 + z^2 = 6y.$
13. $x^2 - y^2 + z^2 = 9.$
14. $z^2 - 4x^2 - y^2 = 4.$
15. $z^2 = 9x^2 + y^2.$
16. $x^2 = y^2 + z^2.$
17. $z = x^2.$
18. $y^2 + z^2 = 4y.$
19. $x^2 + z^2 = 8 - y.$
20. $y^2 + z^2 = 4 - x.$
21. $z = 1 - x - y^2.$
22. $z = 1 - xy.$
23. $z = y - xy.$
24. $z = y^2 - xy.$
25. $z^2 = y - xy.$
26. $z^2 = y^2 - xy.$
27. $z = x^2y.$
28. $z^2 = x^2y.$
29. $z = y^2(1 - x).$
30. $z^2 = (1 - x^2)(1 - y^2).$

In Exs. 31–34, discuss the locus.

31. $y^2 + z^2 = 0.$
32. $x^2 + y^2 + z^2 = 0.$
33. $y^2 + z^2 + a^2 = 0.$
34. $x^2y^2 + z^2 = 0.$

157. Surfaces of revolution. A *surface of revolution* is a surface that can be generated by rotating a curve about a straight line. Sections by planes perpendicular to the axis of revolution are *right sections*, or *parallels;* evidently these are circles with centers on the axis of revolution. Sections by planes through the axis are *meridians;* since these are merely the generating curve in its successive positions, they are all equal curves.

Example: Discuss the surface $x^2 + z^2 = ay$.

The sections $y = k$ are circles with centers on the y-axis: thus the surface is one of revolution around the y-axis. A meridian section is the parabola $x^2 = ay$ in the xy-plane: the surface is a "paraboloid of revolution." (Fig. 174.)

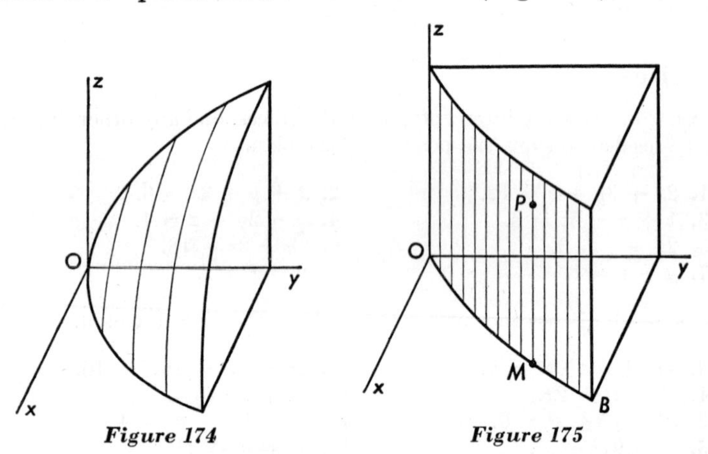

Figure 174 *Figure 175*

The equation of a surface of revolution around the x-axis always has the form (or can be reduced to the form)

$$y^2 + z^2 = f(x).$$

For if not, the sections $x = k$ could not be circles with centers in Ox. A similar result holds, of course, for surfaces of revolution around Oy or Oz. Hence, whenever the surface is one of revolution around a coordinate axis, that fact must be apparent from a glance at the equation.

158. Cylinders. A *cylinder* is the surface described by a moving line which remains parallel to its original position and always intersects a fixed curve, called the *directing curve.* Thus the cylinder is completely covered by straight lines, called *generators,* all of which are parallel.

The section by any plane perpendicular to the generators is a *right section;* all right sections, and in fact all parallel plane sections, are equal curves. If the right section has a center, the line through this center parallel to the generators is the *axis* of the cylinder.

159. Equations in two variables: cylinders perpendicular to a coordinate plane. Consider an equation of the form

$$(1) \qquad f(x, y) = 0:$$

that is, an equation not containing z. In the xy-plane this equation represents, of course, a curve of some kind — say the curve OMB (Fig. 175). At any point M of this curve erect a perpendicular to the xy-plane, and let $P:(x, y, z)$ be any point of the perpendicular. Now the coordinates of M satisfy (1), and those of P must do the same, since its x- and y-coordinates are the same as those of M, and z does not occur in the equation. Thus P is on the surface, and since P is any point of the perpendicular at M, that entire line must lie in the surface. Further, since M is any point of the curve, the surface includes all lines perpendicular to the xy-plane through points of the curve. On the other hand, if a point does not lie in one of these lines its coordinates cannot satisfy the equation, since its x- and y-coordinates cannot be identical with those of any point on the curve. Thus the equation represents a *cylinder with generators perpendicular to the xy-plane,* the directing curve being the curve represented by the given equation in that plane.

Evidently a strictly analogous result holds for equations involving y and z only, or x and z only.

THEOREM: *An equation in two variables represents a cylinder whose generators are perpendicular to the plane of the two variables and whose directing curve is the curve represented by the given equation in that plane.*

COROLLARY: *An equation of the first degree in two variables represents a plane perpendicular to the plane of the two variables.*

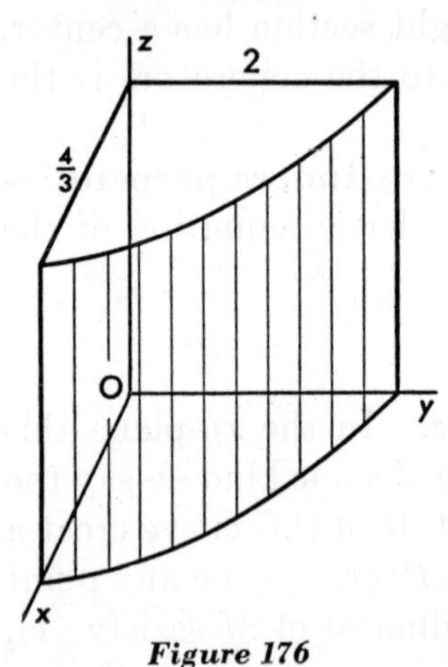

Figure 176

The corollary, of course, is merely that case of the theorem in which the directing curve is a straight line.

Example: Discuss the surface

$$y^2 = 4 - 3x. \tag{2}$$

Since z is missing, we know at once that the surface is a cylinder with generators perpendicular to the xy-plane, whose base is the parabola represented by (2) in that plane. Sections $x = k$ are pairs of straight lines; $y = k$, straight lines; $z = k$, parabolas.

Another easily recognized case is that in which the equation represents a cylinder with generators *parallel* to a coordinate plane. For, in the process of examining the three sets of sections, we are bound to uncover a family of straight lines all parallel to each other and to one of the coordinate planes.

Exercises

In Exs. 1–12, show that the equation represents a surface of revolution. Sketch the surface.

1. $x^2 + y^2 + 4z^2 = 4a^2$. **2.** $4x^2 + y^2 + 4z^2 = 4a^2$.
3. $x^2 + z^2 = y^2$. **4.** $x^2 + y^2 = (1 - z)^2$.
5. $x^2 + y^2 - z^2 = a^2$. **6.** $x^2 - 4y^2 - 4z^2 = 4a^2$.
7. $x^2 + y^2 + z^2 = 2ay$. **8.** $y^4 = x^2 + z^2$.
9. $y^3 + x^2 + z^2 = y$. **10.** $x^2y^2 + y^2z^2 = (1 - y)^2$.
11. $x^2y + yz^2 = 1 - y$. **12.** $(x^2 + y^2 - 1)z = 2x^2 + 2y^2$.

In Exs. 13–18, discuss and sketch the surface. (§ 159.)

13. $x^2 + y^2 = 4x$. **14.** $4y = 1 - x^2$.
15. $2yz = a^2$. **16.** $y^2 - z^2 = a^2$.
17. $x^2z = (1 - x)^2$. **18.** $x^2z - x^2 + z = 0$.

In Exs. 19–24, show that the surface is a cylinder with generators parallel to a coordinate plane. Sketch the surface.

19. $y^2 = 1 - x - z$. **20.** $z^2 = 6 - y - 2x$.
21. $xy + yz = 1$. **22.** $y^2 + xy + zy = 1$.
23. $(x + y)^2 = 1 - z$. **24.** $y^2 + (x + z)^2 = 1$.

160. Cylindrical coordinates. Given a point P (Fig. 177), let us drop a perpendicular from P to a point M in the xy-plane. The position of P is evidently determined if we know the polar coordinates r, θ of M together with the z-coordinate MP. This combination of polar coordinates in the xy-plane with the rectangular z is the so-called *cylindrical* system. The coordinates are written in the order (r, θ, z).

The only formulas required for transformation from cylindrical to rectangular coordinates, or vice versa, are those of § 43.

161. Spherical coordinates. The position of a point P is fixed if we know the length ρ of its radius vector, the angle θ from the x-axis to the horizontal projection OM of the radius

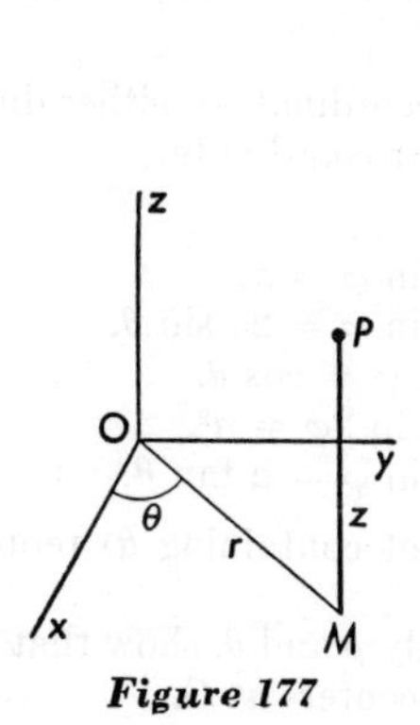

Figure 177

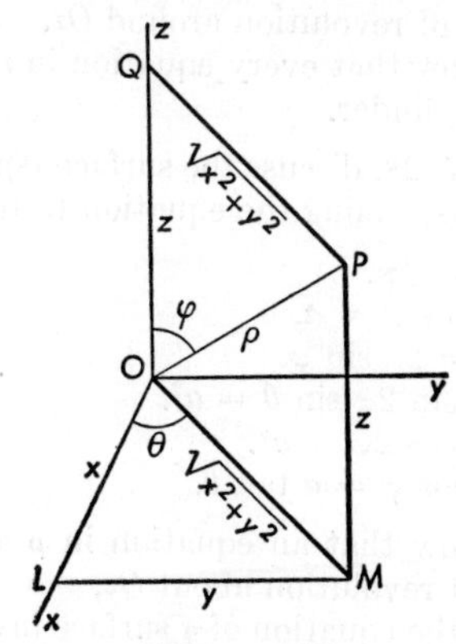

Figure 178

vector, and the angle φ from the z-axis to the radius vector. The coordinates (ρ, θ, φ) are the *spherical coordinates* of P.

In Fig. 178, $OL = OM \cos\theta$, $OM = QP = OP \sin\varphi$, $LM = OM \sin\theta$, $MP = OQ = OP \cos\varphi$. That is,

$$\boldsymbol{x = \rho \cos\theta \sin\varphi, \qquad y = \rho \sin\theta \sin\varphi, \qquad z = \rho \cos\varphi;}$$

$$\boldsymbol{\rho = \sqrt{x^2 + y^2 + z^2}, \qquad \tan\theta = \frac{y}{x}, \qquad \tan\varphi = \frac{\sqrt{x^2 + y^2}}{z}.}$$

A moment's thought shows that the system of coordinates used in geography is merely a slight modification of the system just described. The "longitude" of a point, measured from some standard meridian, is our angle θ; the "latitude," in the northern hemisphere, is the complement of φ.

Exercises

In Exs. 1–14, discuss the surface (cylindrical coordinates), either directly or after transforming the equation to rectangular coordinates.

1. $\theta = \frac{1}{3}\pi$.
2. $r = 3$.
3. $r \sin\theta = 2$.
4. $r \cos\theta = 2$.
5. $r = 2a \cos\theta$.
6. $r = 2a \sin\theta$.
7. $r \cos\theta = z$.
8. $r^2 \sin 2\theta = z$.
9. $r^2 = z$.
10. $r^2 = 4(1 - z^2)$.
11. $r^2 + z^2 = 2az$.
12. $r^2 + z^2 = 2ar$.
13. $z = a \cot\theta$.
14. $z = a(1 - \tan\theta)$.

15. Show that every equation in r and z (i.e., not involving θ) represents a surface of revolution around Oz.

16. Show that every equation in r and θ (not containing z) represents a vertical cylinder.

In Exs. 17–28, discuss the surface (spherical coordinates), either directly or after transforming the equation to rectangular coordinates.

17. $\varphi = \frac{1}{4}\pi$.
18. $\rho = a$.
19. $\rho \cos\varphi = a$.
20. $\rho \sin\varphi = a$.
21. $\rho = 2a \sin\varphi$.
22. $\rho \sin\varphi = 2a \sin\theta$.
23. $\rho^2 \sin 2\varphi \sin\theta = a^2$.
24. $\cot\varphi = \cos\theta$.
25. $\rho^2 \cos 2\varphi = a^2$.
26. $\rho^2 \sin 2\varphi = a^2$.
27. $\rho \cos\varphi = a \tan\theta$.
28. $\rho \sin\varphi = a \tan\theta$.

29. Show that an equation in ρ and φ (not containing θ) represents a surface of revolution about Oz.

30. If the equation of a surface involves only ρ and θ, show that sections by planes through the z-axis are circles with centers at O.

31. Sketch the surface $\rho = 2a \cos\theta$. (Ex. 30.)

162. Curves. Two surfaces intersect in general in a curve. If the equations of the two surfaces be considered as simultaneous, their locus consists of all points lying on both surfaces.

The locus of two simultaneous equations is a curve, the curve of intersection of the surfaces represented by the two equations separately.

Of course the curve of intersection may be a plane curve, but ordinarily its points do not all lie in a plane, in which case it is called a *twisted curve*, or *skew curve.*

An infinite number of surfaces may evidently be passed through a given curve. Thus, while there is only one equation representing a given surface, *a curve may be represented by an infinite number of different pairs of equations*, by the equations of any two surfaces having that curve as their intersection. From the given pair of equations it is often possible to derive a simpler pair from which the form and properties of the curve are more readily seen. Indeed, if $u = 0$, $v = 0$ is any curve C and if k is constant, then the locus of the equation $u + kv = 0$ is a surface containing the curve C.

163. Projecting cylinders. If a perpendicular be dropped from a point P to a plane, the foot P' of the perpendicular is the *projection* of P upon the plane, and the line PP' is the *projecting line.* If all the points of a curve in space be projected upon any plane, their projections form a certain curve in that plane: this curve is called the *projection* of the given curve, and the cylinder whose generators are the projecting lines is the *projecting cylinder* of the curve upon the plane. Figure 179 shows the projection $A'P'B'$ of a curve APB upon the xy-plane, together with that portion of the projecting cylinder that lies between the curve and its projection.

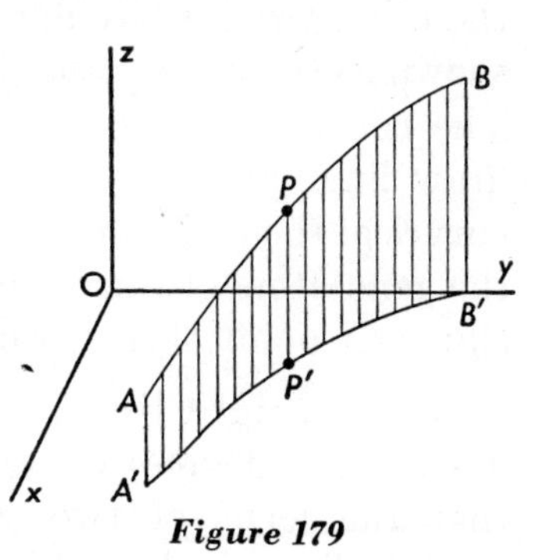

Figure 179

Upon eliminating z between the equations of a curve we obtain a third equation which is satisfied whenever both of the original equations are satisfied, and therefore represents a surface through the given curve. Further, since the new equation does not contain z, it represents a cylinder perpendicular to the xy-plane — i.e., the xy-projecting cylinder of the curve.

To obtain the xy-projecting cylinder of a given curve, eliminate z between the equations of the curve; similarly for the other projecting cylinders.

An especially useful method of drawing a curve in space is to exhibit it as the intersection of two of its projecting cylinders.

Example: Draw the curve

$$4x^2 + y^2 + 3z^2 = 5,$$
$$2x^2 + y^2 + z^2 = 3.$$

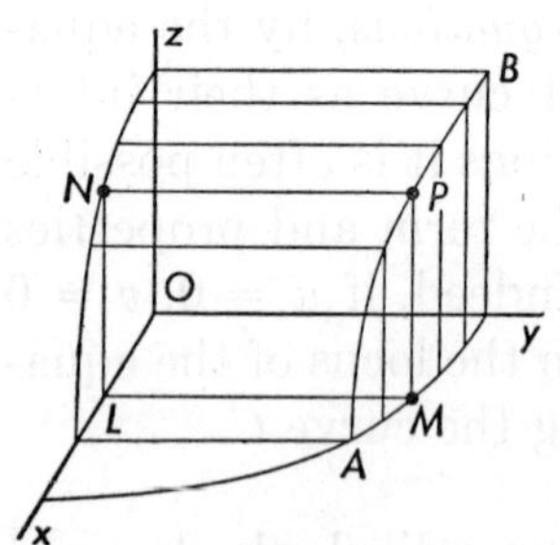

Figure 180

Eliminating x, y, and z in turn, we get the three projecting cylinders

$$(1) \qquad y^2 - z^2 = 1,$$
$$(2) \qquad x^2 + z^2 = 1,$$
$$(3) \qquad x^2 + y^2 = 2.$$

Thus one projection of the curve is an arc of a hyperbola, while the other two are circular arcs. In Fig. 180 the curve APB is shown as the intersection of the circular cylinders (2) and (3). The construction is as follows: through any point L on Ox draw lines parallel to Oy and Oz intersecting the directing curves of the cylinders at M and N. Through M and N draw the generators of the cylinders respectively parallel to Oz and Oy: their point of intersection P is a point of the curve.

164. Straight lines. It will be shown in § 167 that every linear (first degree) equation represents a plane. It follows that the locus of two simultaneous linear equations is a straight line, the line of intersection of the planes represented by the given equations.

Example: Draw the line

$$4x + 4y + 3z = 14,$$
$$x - 2y + 3z = 2.$$

Eliminating z and y in turn, we find the xy- and zx- projecting planes:

$$x + 2y = 4, \qquad 2x + 3z = 6.$$

The xy-traces of these planes are the lines LQ, MP_1, intersecting at P_1; the yz-traces are the lines LP_2, NP_2, intersecting at P_2. Thus the line P_1P_2 is the given line.

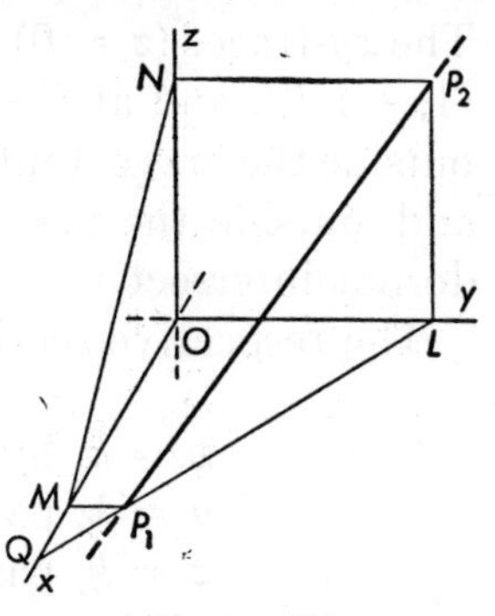

Figure 181

165. Solid with composite boundary. In calculus, it is frequently necessary to picture a portion of space bounded by two or more surfaces, in addition (usually) to the coordinate planes. As a rule the best plan is to cut through the whole figure by a set of planes parallel to a coordinate plane, exactly as in § 156. But now, instead of picking out the sections that are simplest for some one surface, we try to select those that are simplest, on the average, for all the surfaces involved.

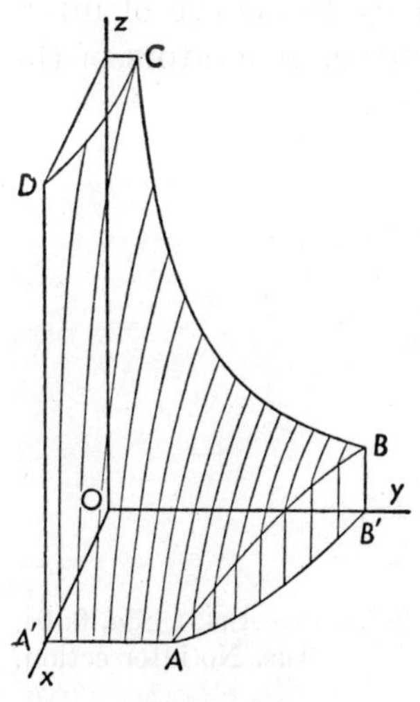

Figure 182

Example: Sketch the solid that is common to those shown in Figs. 172, 176.

The bounding surfaces are the coordinate planes and the surfaces

$$x^2 + yz = 1, \tag{1}$$
$$y^2 = 4 - 3x. \tag{2}$$

The xy-traces ($z = 0$) of these surfaces intersect at A: (1, 1, 0), (1, −1, 0), and at $(-1, \pm\sqrt{7}, 0)$, the last three of which are outside the first octant. The yz-traces intersect at B: $(0, 2, \frac{1}{2})$ and, outside the first octant, at $(0, -2, -\frac{1}{2})$. The xz-traces do not intersect.

The respective sections are:

$x = k$, hyperbolas and straight lines;
$y = k$, parabolas and straight lines;
$z = k$, parabolas and parabolas.

We begin by drawing the relevant portions of the traces of the two surfaces: in the xy-plane, the arcs $A'A$, AB'; in the yz-plane, the arcs $B'B$, BC; in the xz-plane, the arc $A'D$. Choosing sections $y = k$, we reproduce Fig. 172 in the interval $0 < y < 1$ (out to A). In the interval $1 < y < 2$, the sections of the surface (1) are cut off by the curve of intersection AB; from the points of that curve, generators of the cylinder (2) are dropped to the xy-plane.

Exercises

In Exs. 1–14, draw the curve.

1. $x^2 + z^2 = a^2$, $y = x$.
2. $x^2 + y^2 = a^2$, $y^2 + z^2 = 2a^2$.
3. $y^2 + z^2 = a^2$, $x + y = a$.
4. $x^2 + z^2 = 2a^2$, $y = a$.
5. $x^2 + y^2 + 3z^2 = 4$, $2x^2 - y^2 + 9z^2 = 8$.
6. $x^2 + y^2 + 2z^2 = 2a^2$, $3x^2 - y^2 + 2z^2 = 2a^2$.
7. $x^2 + 2y^2 + z^2 = 5$, $x^2 - y^2 - 2z^2 = 2$.
8. $3x^2 + 3y^2 + 2z^2 = 8$, $3x^2 = y^2 + 6z^2$.
9. $x^2 - 3y^2 + z^2 = 9a^2$, $x^2 + 9y^2 + 4z^2 = 9a^2$. *Ans.* $(\pm 3a, 0, 0)$.
10. $x^2 + y^2 + z^2 = 1$, $x^2 + 2y^2 + z^2 = 5$. *Ans.* No intersection.
11. $z = x^2 + y^2$, $z^2 = x^2 + y^2$.
12. $z = xy$, $z^2 = y$.
13. $xyz = 1 - y^2$, $xy = z$.
14. $y = xz$, $xyz = 1$.
15. If $u = 0$ and $v = 0$ are any two equations in x, y, and z, what is the locus of the equation $u^2 + v^2 = 0$?

16. Sketch the surface $x^2 + y^2 + z^2 = 2xy$. (Ex. 15.)

In Exs. 17–20, draw the straight line.

17. $x + 3y + z = 3,\ 2x + 4y + z = 4.$
18. $x + y + 4z = 4,\ 2x + 3y + 3z = 6.$
19. $x + 2y - z = 0,\ 3x - y - 2z = 0.$
20. $6x - y - 2z = 0,\ y = 3x.$

In Exs. 21–34, draw the solid in the first octant bounded by the given surfaces.

21. $x^2 + y^2 = a^2,\ y^2 = az.$
22. $y^2 + z^2 = a^2,\ x^2 + y^2 = a^2.$
23. $x^2 = ay,\ z = y,\ y = a.$
24. $y^2 + z^2 = a^2,\ y = x,\ 2y = x.$
25. $x^2 = y,\ z = x + y,\ y = 1.$
26. $x + y + z = 2,\ z = 1 - x^2.$
27. $z^2 + ay = a^2,\ y^2 + z^2 = ax.$
28. $z = 1 - x^2,\ y = 1 - xz.$
29. $y = x^2 z,\ z^2 = x,\ z = x^2.$
30. $z = x^2 y^2,\ x = 2y,\ x = 2.$
31. $y + z = 1,\ x = y - zy.$
32. $az = xy,\ x^2 + ay = 4a^2.$
33. $xy = az,\ z = x,\ x = a.$
34. $x^2 + z^2 = a^2,\ y = x + z.$

From a sheet of paper 1 ft. square, portions are removed as indicated in Exs. 35–38. Exhibit graphically the remaining area as a function of x and y, drawing only the portion of the surface that has a meaning in the problem.

35. Fig. 183.
36. Fig. 184.
37. Fig. 185.
38. Fig. 186.

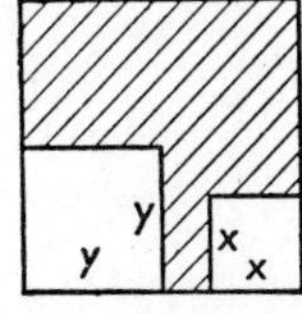

Figure 183

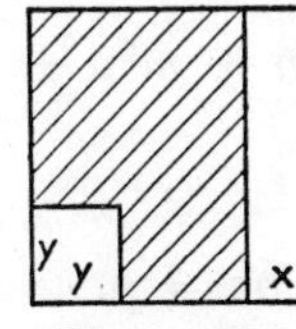

Figure 184

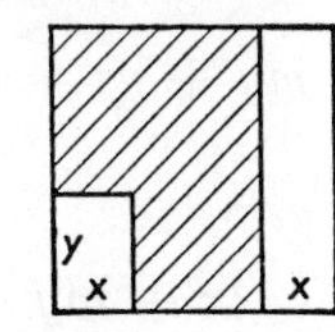

Figure 185

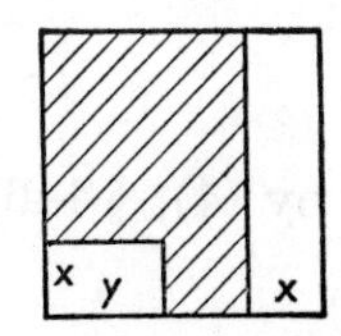

Figure 186

39. A box of volume 1 cu. ft. is to be placed in a cubical container of edge 2 ft. Show graphically the depth of the box as a function of its other two dimensions.

CHAPTER 19 *The Plane*

166. Normal form. Given a plane RST (Fig. 187), let $P:(x, y, z)$ be any point of that plane, and N the foot of the perpendicular from O upon the plane. Denote the length of the normal ON by p, and its direction cosines by l, m, n. We shall consider the segment p as *directed:* positive if the given plane has a positive z-intercept, so that ON slopes upward, negative if ON slopes downward. With this convention, the plane is determined when l, m, n, and p are given.

By (3), § 146, the coordinates of N are (lp, mp, np), so that by § 148 the direction components of NP are $x - lp, y - mp, z - np$. By § 151 (corollary), the lines ON and NP are perpendicular if and only if

$$l(x - lp) + m(y - mp) + n(z - np) = 0:$$

this is therefore the equation of the plane. Simplifying, we get

$$lx + my + nz = l^2p + m^2p + n^2p,$$

or by (4), § 146,

$$lx + my + nz = p. \tag{1}$$

This is the *normal form* of the equation of the plane.

Our convention as to the sign of p fails if the plane is perpendicular to the xy-plane (no z-intercept) or passes through the origin (z-intercept zero). In the former case (Fig. 188), let us say that p shall have the same sign as the y-intercept. The above derivation goes through unchanged, except that now $\gamma = 90°$ and $n = 0$: the result of course is

$$lx + my = p.$$

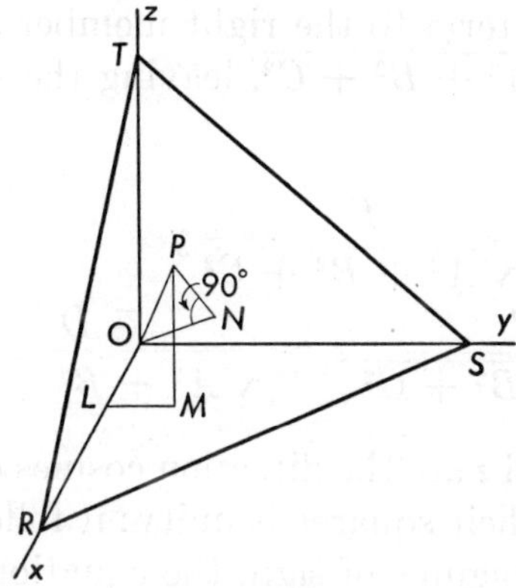

Figure 187

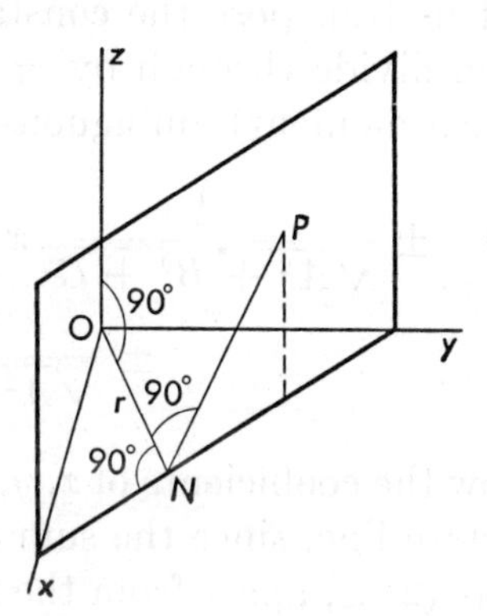

Figure 188

This does not cover the case in which the plane is parallel to the yz-plane (no z- or y-intercept). In this case, let p have the same sign as the x-intercept. Now, $m = n = 0$, $l = 1$, and the equation is

$$x = p.$$

Finally, given a plane through the origin, let l, m, n be the direction cosines of the perpendicular line through O. Since this line is perpendicular to every line OP in the plane, and since the direction components of OP, by § 146, are x, y, z, the equation of the plane is (§ 151, corollary)

$$lx + my + nz = 0.$$

But this is the form to which (1) reduces when, as here,

$$p = 0.$$

Thus, formula (1) applies to any plane.

THEOREM: *A plane always has an equation of the first degree.*

167. General form; reduction to normal form. Every equation of the first degree can be written in the form

$$(1)\qquad Ax + By + Cz + D = 0.$$

Let us transpose the constant term to the right member and then divide through by $\pm\sqrt{A^2 + B^2 + C^2}$, leaving the sign (for a moment) ambiguous:

$$(2)\qquad \pm\frac{A}{\sqrt{A^2 + B^2 + C^2}}x \pm \frac{B}{\sqrt{A^2 + B^2 + C^2}}y \pm \frac{C}{\sqrt{A^2 + B^2 + C^2}}z = \frac{\mp D}{\sqrt{A^2 + B^2 + C^2}}.$$

Now the coefficients of x, y, and z are the direction cosines of a certain line, since the sum of their squares is unity: it follows that (2) is, apart from the ambiguity of sign, the equation of a plane in the normal form. This reduction can always be carried out, because $\sqrt{A^2 + B^2 + C^2}$ cannot be zero.

THEOREM: *Every equation of the first degree represents a plane.*

Further, the conventions laid down in § 166, as to the sign of p, yield a systematic procedure for reduction to normal form.

RULE: *To reduce to the normal form the equation*

$$(a)\qquad Ax + By + Cz + D = 0, \qquad C \neq 0,$$

divide by $\sqrt{A^2 + B^2 + C^2}$ *and choose signs so that the coefficient of* z *is positive;*

$$(b)\qquad Ax + By + D = 0, \qquad B \neq 0,$$

divide by $\sqrt{A^2 + B^2}$ *and choose signs so that the coefficient of* y *is positive;*

$$(c)\qquad Ax + D = 0,$$

divide by A.

168. Perpendicular line and plane. The coefficients A, B, C in the general equation of the plane are, by (2), § 167, proportional to the direction cosines of the normal.

THEOREM: *If a line and plane are perpendicular, the direction components of the line may be taken as coefficients of* x, y, z *in the equation of the plane, and vice versa.*

Thus, if a line has direction components a, b, c, then a plane is perpendicular to the line if and only if the equation of the plane can be written in the form

$$ax + by + cz = k.$$

Example: Write the equation of a plane through (0, 4, 5), and perpendicular to the line joining (2, −3, 5) to (−1, −4, 3).

The numbers 3, 1, 2 are direction components of the line; by the above theorem, these numbers may be taken as coefficients in the equation of the plane. The result is

$$3x + y + 2z = 14,$$

the right member being necessarily the value assumed by the left member when the coordinates (0, 4, 5) are substituted. (Compare with the example of § 38.)

Exercises

In Exs. 1–12, reduce the equation to normal form; determine the direction cosines of the normal and the distance of the plane from the origin.

1. $4x + y + 8z = 36$. Draw the figure.
2. $x + 2y + 2z = 6$. Draw the figure.

3. $2x + y - 2z = -12$.
4. $2x - 3y + 6z = -21$.
5. $3x - 4y - 5z = 30$.
6. $x - y - 2z = 24$.
7. $x + y + z = 0$.
8. $x - 2y + 2z = 0$.
9. $2y + 7 = 0$.
10. $7x - y + 10 = 0$.

11. $4y + 3z = 30$. Draw the figure.
12. $x + z = 6$. Draw the figure.

In Exs. 13–24, find the equation of the plane.

13. At a distance 4 from O, with normal having direction components 1, 2, −2. *Ans.* $x + 2y - 2z = \pm 12$.

14. At a distance 2 from O, with normal having direction components 8, −1, 4. *Ans.* $8x - y + 4z = \pm 18$.

15. Through (1, 2, 3) perpendicular to the radius vector of that point. *Ans.* $x + 2y + 3z = 14$.

16. Through (−2, 1, 4) perpendicular to the radius vector of that point. *Ans.* $2x - y - 4z = -21$.

17. With the point (5, 3, −1) as the foot of the normal from O.

18. With the point (4, −3, −2) as the foot of the normal from O.

19. Through (1, 7, −4) perpendicular to the radius vector of the point (2, 1, −2). *Ans.* $2x + y - 2z = 17$.

20. Through (4, −3, 2) perpendicular to the radius vector of the point (1, 0, 3). *Ans.* $x + 3z = 10$.

21. Through (2, 0, −1) perpendicular to the line joining (3, 4, 4), (−1, 2, 1). *Ans.* $4x + 2y + 3z = 5$.

22. Through (2, −2, 1) perpendicular to the line joining (7, −3, 2), (4, −2, 3). *Ans.* $3x - y - z = 7$.

23. At a distance 4 from O perpendicular to the line joining (−2, 3, 1), (−5, 1, −5). *Ans.* $3x + 2y + 6z = \pm 28$.

24. At a distance 3 from O perpendicular to the line joining (6, 4, 1), (4, 5, 0). *Ans.* $2x - y + z = \pm 3\sqrt{6}$.

25. One side of a right triangle joins the points (2, 4, −3), (1, 0, 5), with the right angle at the latter point. Find the locus of the third vertex. *Ans.* $x + 4y - 8z = -39$.

26. One side of a right triangle joins the points (7, −2, 4), (1, 1, −1), with the right angle at the latter point. Find the locus of the third vertex. *Ans.* $6x - 3y + 5z = -2$.

27. Solve Ex. 25 by another method.

28. Solve Ex. 26 by another method.

29. The base of an isosceles triangle joins the points (4, 3, 7), (−2, 1, 5). Find the locus of the third vertex by § 168. *Ans.* $3x + y + z = 11$.

30. The base of an isosceles triangle joins the points (4, 3, 0), (2, 1, 1). Find the locus of the third vertex by § 168. *Ans.* $4x + 4y - 2z = 19$.

31. Solve Ex. 29 by another method (§ 144).

32. Solve Ex. 30 by another method (§ 144).

33. Derive the normal form from the fact that, in Fig. 187, the projection of the broken line $OLMPN$ upon the normal must equal ON.

34. Derive the normal form from the fact that the midpoint of OP is equidistant from O and N. Does the derivation hold in all cases?

35. Derive the normal form by applying the Theorem of Pythagoras to the triangle ONP. Does the derivation hold in all cases?

36. Prove the corollary, § 159, by use of the normal form.

169. Parallel planes. If the equations of two planes differ only in the constant term, the same must be true after reduction to the normal form. Hence the normals to the two planes have the same direction cosines and are parallel, from which it follows that *the planes are parallel.*

Conversely, *if two planes are parallel, their equations can be made to differ only in the constant term.*

Example: Write the equation of a plane through (3, 1, 4) and parallel to the plane

$$2x - 5y + 6z = 7.$$

The answer is

$$2x - 5y + 6z = 25,$$

the right member being written down at once from the fact that it must be the value assumed by the left member when the coordinates (3, 1, 4) are used for x, y, z, respectively.

170. Plane through a given point. If the plane

$$Ax + By + Cz + D = 0$$

is to pass through the point (x_1, y_1, z_1), we must have

$$Ax_1 + By_1 + Cz_1 + D = 0.$$

By subtraction, we find that:

The equation of any plane through the point (x_1, y_1, z_1) may be *written in the form*

$$A(x - x_1) + B(y - y_1) + C(z - z_1) = 0. \tag{1}$$

171. Plane determined by three points. The general equation of the plane contains three essential constants, since at least one of the quantities A, B, C, D must not be zero, and can therefore be divided out. We conclude that a plane is determined by three points (not in a straight line), or by any set of three independent conditions which, when expressed analytically, give three consistent equations to determine the essential constants.

A direct method of finding the equation of a plane determined by three points is to substitute the coordinates of the points in turn in the equation

$$Ax + By + Cz + D = 0,$$

thus obtaining three equations to solve for three of the constants in terms of the fourth. The solution may, however, be expedited by use of (1), § 170.

Example: Find the equation of the plane through the points (2, 4, 3), (1, 3, 1), (−1, −1, −4).

The equation of any plane through (2, 4, 3) is

$$(1) \qquad A(x - 2) + B(y - 4) + C(z - 3) = 0.$$

Substitute the coordinates of the other points in (1):

$$-A - B - 2C = 0,$$
$$-3A - 5B - 7C = 0.$$

Solving for A and C in terms of B, we find

$$A = 3B, \qquad C = -2B.$$

Using these expressions for A and C in (1), we find the equation of the desired plane to be

$$3B(x - 2) + B(y - 4) - 2B(z - 3) = 0,$$

or

$$3x + y - 2z = 4.$$

172. Perpendicular planes. Two planes

$$A_1x + B_1y + C_1z + D_1 = 0,$$
$$A_2x + B_2y + C_2z + D_2 = 0$$

are perpendicular if and only if their normals are perpendicular to each other. The direction cosines of the normals are proportional to A_1, B_1, C_1 and A_2, B_2, C_2 respectively. We can now apply the condition of perpendicularity (§ 151) to these lines.

THEOREM: *Two planes*

$$A_1x + B_1y + C_1z + D_1 = 0,$$
$$A_2x + B_2y + C_2z + D_2 = 0$$

are perpendicular if and only if

$$A_1A_2 + B_1B_2 + C_1C_2 = 0. \tag{1}$$

Example: Find the equation of a plane through the points (1, 1, 2), (2, 4, 3), and perpendicular to the plane

$$x - 3y + 7z + 5 = 0. \tag{2}$$

The equation of any plane through (1, 1, 2) is (§ 170)

$$A(x - 1) + B(y - 1) + C(z - 2) = 0. \tag{3}$$

Substituting the coordinates (2, 4, 3) in (3), we get

$$A + 3B + C = 0. \tag{4}$$

The condition of perpendicularity (1) applied to the planes (2) and (3) gives

$$A - 3B + 7C = 0. \tag{5}$$

From (4) and (5) we find

$$A = -4C, \qquad B = C,$$

whence the equation of the plane is

$$-4C(x - 1) + C(y - 1) + C(z - 2) = 0,$$

or

$$4x - y - z = 1.$$

Exercises

In Exs. 1–12, find the equation of the plane.

1. Through (2, 1, −3) parallel to the plane $3x + 4y + z = 4$.
Ans. $3x + 4y + z = 7$.

2. Through (1, 0, 4) parallel to the plane $2x - 5y + 3z = 6$.

3. Through (−1, −2, 3) parallel to the plane $x + y + 4z = 5$.

4. Through (4, 3, 1) parallel to the plane $x + 3z = 8$.

5. Parallel to the plane $x + 4y - 8z = 18$ and (*a*) half as far from the origin; (*b*) at a distance 4 from the origin; (*c*) at the same distance from the origin. *Ans.* (*a*) $x + 4y - 8z = \pm 9$.

6. Parallel to the plane $6x - 3y + 2z = 21$ and (*a*) twice as far from the origin; (*b*) one unit nearer the origin; (*c*) at a distance 4 from the origin. *Ans.* (*b*) $6x - 3y + 2z = \pm 14$.

7. Parallel to the plane $2x + y + 2z = 15$ and (*a*) 2 units nearer the origin; (*b*) 2 units farther from the origin; (*c*) at a distance 4 from the given plane. *Ans.* (*c*) $2x + y + 2z = 15 \pm 12$.

8. Parallel to the plane $x + y - 2z = 12$ and (*a*) 3 units nearer the origin; (*b*) 3 units farther from the origin; (*c*) at a distance 2 from the given plane. *Ans.* (*c*) $x + y - 2z = 12 \pm 2\sqrt{6}$.

9. Parallel to the plane $4x + y + 8z = 11$ and passing at a distance 2 from $(1, 5, -3)$. *Ans.* $4x + y + 8z = -15 \pm 18$.

10. Parallel to the plane $x - 2y + 2z = 7$ and passing at a distance 4 from $(4, 1, 3)$. *Ans.* $x - 2y + 2z = 8 \pm 12$.

11. Parallel to the plane $4x - 3y + 5z = 8$ and (*a*) twice as far from $(1, 2, 1)$; (*b*) half as far from $(1, 2, 1)$; (*c*) at a distance 6 from $(1, 2, 1)$.

12. Parallel to the plane $4x + 4y - 2z = 11$ and (*a*) twice as far from $(1, -1, -3)$; (*b*) half as far from $(1, -1, -3)$; at a distance unity from $(1, -1, -3)$.

In Exs. 13–16, find the distance between the planes.

13. $x + y - 2z = 5$, $x + y - 2z = 17$.

14. $8x - y - 4z = -4$, $8x - y - 4z = 23$.

15. $2x + 2y + z + 11 = 0$, $4x + 4y + 2z = 11$.

16. $3x + y + z + 4 = 0$, $6x + 2y + 2z = 1$.

17. Show that the planes $x + y - z + 1 = 0$, $x + y - z - 1 = 0$, $3x - 2y + z + 3 = 0$, $3x - 2y + z + 2 = 0$, $x + 4y + 5z + 20 = 0$, $x + 4y + 5z = 1$ form a box, and find its volume. *Ans.* $V = 1$.

18. Show that: *The directed distance from the plane*

$$Ax + By + Cz + D = 0$$

to the point (x_1, y_1, z_1) is

$$d = \frac{Ax_1 + By_1 + Cz_1 + D}{\pm\sqrt{A^2 + B^2 + C^2}},$$

where the ambiguous sign is chosen like the sign of C if $C \neq 0$, *or like the sign of B if* $C = 0$, $B \neq 0$, *or like the sign of A if* $B = C = 0$.

In Exs. 19–22, find the distance from the plane to the point. (Ex. 18.)

19. $x + y + 2z = 4$, $(2, 0, -5)$. *Ans.* $-2\sqrt{6}$.

20. $3x + y - z = -23$, $(2, 5, 1)$. *Ans.* $-3\sqrt{11}$.

21. $6x - 3y - 2z = -4$, $(4, 1, 2)$. *Ans.* -3.

22. $8x + 4y + z = -9$, $(4, -3, 7)$. *Ans.* 4.

In Exs. 23–24, find the equations of the planes bisecting the angles between the given planes. [Cf. Example (*c*), § 46.]

23. $x + 2y + 2z = 0$, $6x - 2y + 3z = 14$.
Ans. $11x - 20y - 5z = 42$, $25x + 8y + 23z = 42$.

24. $x + y + z = 2$, $5x - y + z + 4 = 0$.
Ans. $x - 2y - z + 5 = 0$, $4x + y + 2z = 1$.

In Exs. 25–30, find the equation of the plane through the given points.

25. $(3, 2, 1)$, $(1, -3, -2)$, $(3, 11, 4)$. *Ans.* $2x + y - 3z = 5$.

26. $(6, 6, 1)$, $(-6, 3, -1)$, $(1, 3, 0)$. *Ans.* $3x + 2y - 21z = 9$.

27. $(4, 3, -1)$, $(0, 4, 5)$, $(6, -3, 7)$. *Ans.* $2x + 2y + z = 13$.

28. $(2, 0, 3)$, $(5, 6, -1)$, $(-1, 4, -3)$. *Ans.* $2x - 3y - 3z = -5$.

29. $(3, 2, -2)$, $(4, 1, 3)$, $(-1, 6, 1)$. *Ans.* $x + y = 5$.

30. $(a, 0, 0)$, $(0, b, 0)$, $(0, 0, c)$. *Ans.* $\frac{x}{a} + \frac{y}{b} + \frac{z}{c} = 1$.

In Exs. 31–38, find the equation of the plane.

31. Through the points $(1, 3, 1)$, $(4, 6, -2)$, perpendicular to the plane $x + y - z = 3$. *Ans.* $2x + y + 3z = 8$.

32. Through the points $(-2, 1, 2)$, $(-2, -3, -1)$, perpendicular to the plane $x - 3y - 2z = 4$. *Ans.* $x + 3y - 4z = -7$.

33. Through $(8, 3, 1)$, $(1, \frac{1}{2}, -1)$, perpendicular to the yz-plane.

34. Through $(1, 4, 1)$, $(-8, 3, 7)$, perpendicular to the xz-plane.

35. Through $(3, 0, -1)$, perpendicular to the planes $2x - y - 4z - 9 = 0$, $x + y + 2z = -1$. *Ans.* $2x - 8y + 3z = 3$.

36. Through $(2, 2, 3)$, perpendicular to the planes $2x - 2y - 4z + 1 = 0$, $3x + y + 6z = 14$. *Ans.* $x + 3y - z = 5$.

37. Perpendicular to the planes $y = 3x + z$, $x + 5y + 3z = 0$, and passing at a distance $\sqrt{6}$ from the origin. *Ans.* $x + y - 2z = \pm 6$.

38. Perpendicular to the planes $z = 4y - x$, $3x + 4y + z = 2$, and passing at a distance 1 from the origin. *Ans.* $4x - y - 8z = \pm 9$.

39. Prove that the equation

$$\begin{vmatrix} x & y & z & 1 \\ x_1 & y_1 & z_1 & 1 \\ x_2 & y_2 & z_2 & 1 \\ x_3 & y_3 & z_3 & 1 \end{vmatrix} = 0$$

represents the plane determined by the points (x_1, y_1, z_1), (x_2, y_2, z_2), (x_3, y_3, z_3).

40. In Ex. 39, what happens if the minors of all the elements in the first row are zero? Explain geometrically.

41. Solve Ex. 25, using Ex. 39.

42. Solve Ex. 26, using Ex. 39.

CHAPTER 20 *The Straight Line*

173. Planes through a given line. The following result, stated in § 164, is a consequence of the fact that every first degree equation represents a plane and two intersecting planes determine a line.

THEOREM: *The locus of two different consistent simultaneous equations of the first degree is a straight line.*

If the two equations are inconsistent, the planes represented are parallel and do not intersect; the two equations, taken together, then represent no locus.

Given the equations

$$(1) \qquad \begin{cases} A_1x + B_1y + C_1z + D_1 = 0, \\ A_2x + B_2y + C_2z + D_2 = 0 \end{cases}$$

of a straight line, consider the equation formed by multiplying one of the given equations by an *arbitrary constant* c and adding:

$$(2) \quad A_1x + B_1y + C_1z + D_1 + c(A_2x + B_2y + C_2z + D_2) = 0.$$

Since this equation is of first degree, it represents, as c ranges through all real values,* a *family of planes*. All these planes have one property in common: they pass through the line (1). To see this, let $P\colon(x_1, y_1, z_1)$ be any point of that line. Since the coordinates of P satisfy both of the equations (1), we

* Compare with §§ 125, 129.

have the relations

$$(3) \qquad \begin{cases} A_1x_1 + B_1y_1 + C_1z_1 + D_1 = 0, \\ A_2x_1 + B_2y_1 + C_2z_1 + D_2 = 0. \end{cases}$$

Now, substitute the coordinates of P in (2): the result is

$$A_1x_1 + B_1y_1 + C_1z_1 + D_1 + c(A_2x_1 + B_2y_1 + C_2z_1 + D_2) = 0,$$

which is true, for all values of c, by virtue of (3). Thus all the planes (2) pass through P, and by the same argument, through all points of the line (1).

THEOREM: *If*

$$(4) \quad A_1x + B_1y + C_1z + D_1 = 0, \quad A_2x + B_2y + C_2z + D_2 = 0$$

are the equations of any line, the equation

$$(5) \quad A_1x + B_1y + C_1z + D_1 + c(A_2x + B_2y + C_2z + D_2) = 0$$

represents a family of planes through the given line.

Although there is no value of c that will produce the plane

$$(6) \qquad A_2x + B_2y + C_2z + D_2 = 0,$$

we could equally well have written, instead of (5), the equation

$$(7) \quad k(A_1x + B_1y + C_1z + D_1) + A_2x + B_2y + C_2z + D_2 = 0,$$

which includes (6) as the case $k = 0$. In this sense we shall say that we include (6) in the family (5) by special agreement.

Given any plane through the line (4), excepting (6), let Q be any point of that plane, not in the line (4). The coordinates of Q, substituted in (5), unfailingly determine a value of c, which means that our plane belongs to the family (5). Since (6) is already included, we have proved that the family (5) *includes all planes through the line* (4).

Any two equations of the family (5) may of course be used to represent the line (4), instead of the original pair. For instance, in § 164 we found it convenient to use, instead of the given pair, the equations of two projecting planes.

Example: Find the equation of the plane through the line

$$2x - 2y + 3z = 1, \qquad x - 5y - z = 2$$

and perpendicular to the plane

$$(8) \qquad 2x + y + 4z = 5.$$

The required equation must be of the form

$$2x - 2y + 3z - 1 + c(x - 5y - z - 2) = 0,$$

or

$$(9) \quad (2 + c)x - (2 + 5c)y + (3 - c)z - 1 - 2c = 0.$$

By § 172, planes (8) and (9) will be perpendicular if

$$2(2 + c) - 1(2 + 5c) + 4(3 - c) = 0.$$

This gives $c = 2$; substituting in (9), we have the answer

$$4x - 12y + z - 5 = 0.$$

Exercises

In Exs. 1–14, find the equation of the plane.

1. Through the line $x + 2z = 1$, $y = 5$ and the point (1, 2, 3).

2. Through the line $3x + y = 2$, $z = 1$ and the point (2, −1, 4).

3. Through the line $x + 3y - 2z = 3$, $2x + y + 5z = -3$ and the point (1, 0, −2).

4. Through the line $4x = y + 2z - 5$, $x + y = 7z + 2$ and the point. (−2, 1, 0).

5. Through the line $x + y + z = 0$, $2x - y + 3z = 4$ and perpendicular to the plane $3x + y - 2z = 6$. *Ans.* $5x - y + 7z = 8$.

6. Through the line $y + 2z = 8$, $2x - y = 4$ and perpendicular to the plane $x + 3y + 4z = 6$. *Ans.* $11x - 5y + z = 26$.

7. Through the line of Ex. 5, perpendicular to the yz-plane.

8. Through the line of Ex. 6, perpendicular to the xz-plane.

9. Through the line $x - y + z = 0$, $2x + y + z = 2$ and perpendicular to the plane $x + 2y + 2z = 4$.

10. Through the line $3x + y - z = 4$, $x + 2y + z = 1$ and perpendicular to the plane $2x - y - z = 5$.

11. Through the line $x - 5z = 6$, $y = z$ and passing at a distance $\sqrt{2}$ from the origin. *Ans.* $x - 4y - z = 6$, $x - y - 4z = 6$.

12. Through the line $y - 5z + 1 = 0$, $x + y = 2$ and passing at a distance $\frac{1}{2}\sqrt{2}$ from O. *Ans.* $3x + 4y - 5z = 5$, $4x + y + 15z = 11$.

13. Through the line $x + 2z = -10$, $y - 2z = 8$ and passing at a distance 2 from the origin. *Ans.* $x + 2y - 2z = 6$; $11x + 10y + 2z = -30$.

14. Through the line $y - 2z = 7$, $x + y = 3$ and passing at a distance 2 from (2, 1, 0). *Ans.* $x + 2y - 2z = 10$; $2x + y + 2z = -1$.

15. Show that the line $3x + y - z = 4$, $2x + y + 3z = 7$ lies in the plane $5x + y - 9z = -2$.

16. Show that the line $x + 7y - 3z = 2$, $2x + 4y + z = 5$ lies in the plane $8x + 6y + 11z = 21$.

17. Solve Ex. 15 by another method.

18. Solve Ex. 16 by another method.

In Exs. 19–22, write the equations of the line.

19. Through (2, −3, −2) parallel to the line $3x - 2y + 3z + 3 = 0$, $x + y - 2z = 4$. *Ans.* $3x - 2y + 3z = 6$, $x + y - 2z = 3$.

20. Through (−3, 1, −2) parallel to the line $x + 2y - 3z - 4 = 0$, $x - y - 4z = 0$. *Ans.* $x + 2y - 3z = 5$, $x - y - 4z = 4$.

21. Through (0, −5, −1) parallel to the line $12x - y + 3z + 2 = 0$, $3x + y + z = 4$.

22. Through (2, 7, 3) parallel to the line $15x - 2y + 4z - 7 = 0$, $x + y - z = 0$.

23. What is represented by (5), § 173, when the given planes are parallel?

174. Parametric equations of a line.

Consider the parametric equations

$$(1) \qquad \begin{aligned} x &= x_1 + at, \\ y &= y_1 + bt, \\ z &= z_1 + ct, \end{aligned}$$

with t as a parameter, and a, b, c, x_1, y_1, z_1 as specified constants. Equations (1) represent a straight line because the elimination of t from them yields two linear equations in x, y, and z. Further, the point (x_1, y_1, z_1) lies on the line since for $t = 0$, $x = x_1$, $y = y_1$, $z = z_1$. The numbers a, b, c form a set of direction components for the line (1) since a, b, c are proportional to $x - x_1$, $y - y_1$, $z - z_1$.

Example (*a*): For the line

$$x = 2 + t,\quad y = 5 - 3t,\quad z = -1 + 4t, \tag{2}$$

find the direction cosines, and locate three points on the line.

A set of direction components for (2) is 1, −3, 4, so that direction cosines for (2) are $\frac{1}{\sqrt{26}}, \frac{-3}{\sqrt{26}}, \frac{4}{\sqrt{26}}$. Using $t = 0$, 1, $-\frac{1}{2}$, we immediately compute coordinates of three points on the line, (2, 5, −1), (3, 2, 3), and $(\frac{3}{2}, \frac{13}{2}, -3)$.

Example (*b*): Write parametric equations for the line through the points (4, 3, 5), (2, 3, 8).

A set of direction components for the line is 2, 0, −3. The equations

$$x = 4 + 2t,\quad y = 3,\quad z = 5 - 3t \tag{3}$$

are, therefore, parametric equations of the line. We can equally well use the point (2, 3, 8) in the parametric form and write, instead of (3), the equations

$$x = 2 + 2w,\quad y = 3,\quad z = 8 - 3w, \tag{4}$$

with parameter w.

175. Symmetric equations of a line. Elimination of the parameter t from equations (1), § 174, yields the equations

$$\frac{x - x_1}{a} = \frac{y - y_1}{b} = \frac{z - z_1}{c}, \tag{1}$$

called the symmetric form of the equations of a line.* Equa-

* It may be objected that in (1) there are three equations:

$$\frac{x - x_1}{a} = \frac{y - y_1}{b},\qquad \frac{x - x_1}{a} = \frac{z - z_1}{c},\qquad \frac{y - y_1}{b} = \frac{z - z_1}{c}.$$

These equations, however, are not independent, since any one follows from the other two. These equations of course represent three planes through the line; they are in fact the projecting planes.

tions (1) may be obtained at once, without recourse to the parametric form, as follows. Given a point (x_1, y_1, z_1) on the line, and direction components a, b, c for the line, let (x, y, z) be any point on the line. Then $x - x_1$, $y - y_1$, $z - z_1$, are also direction components of the line, and must be proportional to a, b, c, thus yielding (1).

Example: Write the equations of the line joining the points $(2, 3, 1)$, $(1, -3, -1)$.

By § 148, the direction components are 1, 6, 2; hence the equations of the line are

$$\frac{x-2}{1} = \frac{y-3}{6} = \frac{z-1}{2}. \tag{2}$$

Given the equations of a line in the symmetric form, it is frequently necessary to obtain two equations in the ordinary or general form. As an example, we will do this for the line (2). Disregarding the third member and clearing of fractions, we get

$$6x - y = 9. \tag{3}$$

Disregarding the second member, we find

$$2x - z = 3. \tag{4}$$

Equations (3) and (4) represent the line; if more equations are needed, combinations of these may be made by use of formula (5), § 173.

176. Line parallel to a coordinate plane. If a line is parallel to the yz-plane, the parallel line through the origin intersects the x-axis at right angles, so that $\alpha = 90°$, and $l = 0$; thus any set of direction components must be of the form $0, b, c$. Difficulty therefore arises in using the symmetric form, since one of the denominators is zero. The same

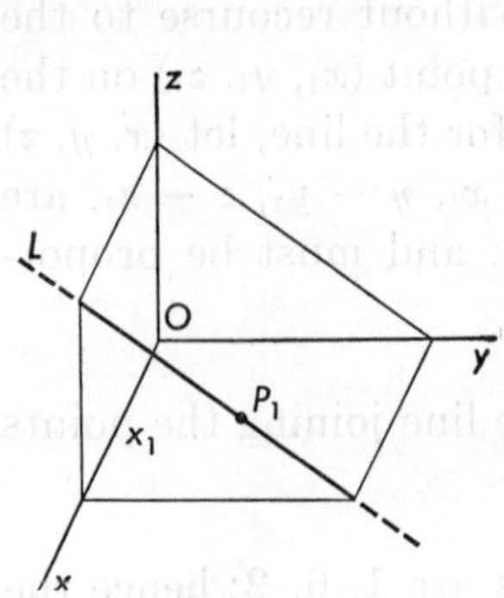

Figure 189

trouble appears, of course, if the line is parallel to the *zx*- or *xy*-plane. No such difficulty arises if the parametric equations of § 174 are employed.

Consider the line L (Fig. 189) through P_1: (x_1, y_1, z_1), with direction components 0, b, c. The *xy*-projecting plane is

$$x = x_1;$$

the *yz*-projecting plane,

$$\frac{y - y_1}{b} = \frac{z - z_1}{c}.$$

Thus the equations of the line are

$$(1) \qquad x - x_1 = 0, \qquad \frac{y - y_1}{b} = \frac{z - z_1}{c}.$$

However, the symmetric form is so useful that we should like to have it always available. Formal substitution in (1), § 175, produces the result

$$(2) \qquad \frac{x - x_1}{0} = \frac{y - y_1}{b} = \frac{z - z_1}{c}.$$

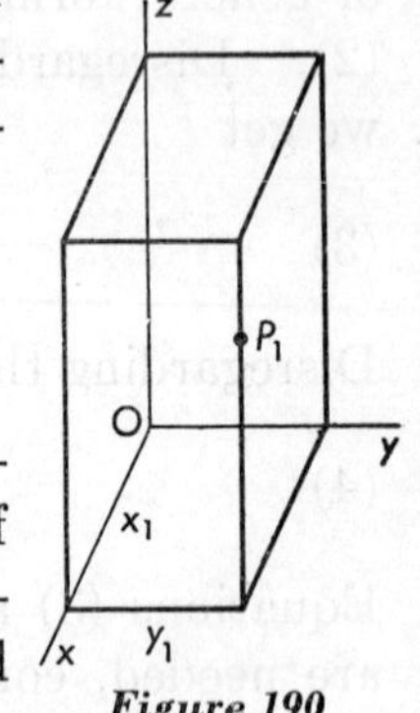

Figure 190

If we agree to consider this as merely a conventional way of writing the equations* of the line — that is, if we consider (2) as *equivalent to* (1) — the symmetric form is preserved when the line is parallel to a coordinate plane.

If the line is parallel to Oz (Fig. 190), its equations are

$$(3) \qquad x - x_1 = 0, \qquad y - y_1 = 0.$$

* Once more we emphasize that division by zero is never permitted. An "equation" containing a zero denominator is not a true equation at all, and can only be interpreted and used in accordance with some special convention, as in the present instance.

By a further convention, we will agree to write

$$\frac{x - x_1}{0} = \frac{y - y_1}{0} = \frac{z - z_1}{1}$$

as *equivalent to* (3). Thus the symmetric form becomes available in all cases.

177. Reduction to the symmetric form. Being given the equations of a line in the general form, it is often necessary to reduce them to the symmetric form, or at least to find a set of direction components.

Example (a): Reduce the equations of the line

$$\begin{aligned} 4x + 4y + 3z &= 14, \\ x - 2y + 3z &= 2 \end{aligned}$$

to the symmetric form (cf. the example of § 164).

Find any two of the projecting planes: for instance,

$$x + 2y = 4, \qquad 2x + 3z = 6.$$

These equations contain one variable in common; *solve for that variable and equate values:*

$$x = -2y + 4 = \frac{-3z + 6}{2}.$$

Divide through by such a number as to reduce the coefficient of each variable to unity, in this case by -6:

$$\frac{x}{-6} = \frac{y - 2}{3} = \frac{z - 2}{4}.$$

Comparing these equations with (1), § 175, we see that the line passes through (0, 2, 2) and has direction components -6, 3, 4.

If the line is parallel to a coordinate plane, the above process fails, since the equations of the projecting planes will not contain one variable in common.

Example (*b*): Reduce the equations of the line

$$2x + 2y + 3z = 8,$$
$$x + 4y + 6z = 10$$

to the symmetric form.

The projecting planes, found by the usual method, are

(1) $$x = 2,$$
(2) $$2y + 3z = 4.$$

Equation (1) shows that the line is parallel to the yz-plane. Equation (2) may be written in the form

$$2y - 4 = -3z,$$

or

(3) $$\frac{y-2}{3} = \frac{z}{-2}.$$

By the convention of § 176, (1) and (3) may be put in the form

$$\frac{x-2}{0} = \frac{y-2}{3} = \frac{z}{-2}.$$

This is the line L of Fig. 189.

Exercises

In Exs. 1–8, write parametric equations of the line through the given points.

1. (3, 4, −1), (1, 1, 4). **2.** (6, 3, 2), (4, 0, 1).
3. (2, 1, 3), (0, −3, 1). **4.** (−7, 1, 4), (3, −4, −1).
5. (3, 1, 2), (3, −1, 5). **6.** (2, 4, −1), (−1, 4, −2).
7. (6, −2, 3), (4, −2, 3). **8.** (5, 4, 3), (5, 1, 3).

In Exs. 9–16, write symmetric equations of the line through the given points.

9. Ex. 1. **10.** Ex. 2. **11.** Ex. 5.
12. Ex. 6. **13.** Ex. 7. **14.** Ex. 8.
15. (2, 2, 1), (1, 3, −2). **16.** (4, 1, 4), (3, −4, 1).

In Exs. 17–22, reduce the equations to symmetric form.

17. $5x + 2y - 2z = 9$, $4x + 4y - z = 3$.

18. $x + 4y + 8z = 0,\ 2x - 2y + z = -5.$
19. $2x + 4y + z = 14,\ x + 2y + 3z = 2.$
20. $6x - y + 2z = 4,\ 3x + 2y + z = 7.$
21. $3x - 4z + 6 = 0,\ 4x + 3z = 17.$ *Ans.* $\dfrac{x-2}{0} = \dfrac{y}{1} = \dfrac{z-3}{0}.$
22. $2y + 3z = 9,\ 3y - 5z = 4.$ *Ans.* $\dfrac{x}{1} = \dfrac{y-3}{0} = \dfrac{z-1}{0}.$

In Exs. 23–28, find the angle between the given lines.

23. $\dfrac{x+1}{1} = \dfrac{y}{2} = \dfrac{z-7}{2};\ \dfrac{x-6}{-3} = \dfrac{y-2}{4} = \dfrac{z}{5}.$ *Ans.* 45°.
24. $\dfrac{x}{2} = \dfrac{y}{3} = \dfrac{z+6}{4};\ \dfrac{x-3}{1} = \dfrac{y+4}{-2} = \dfrac{z}{1}.$ *Ans.* 90°.
25. $\dfrac{x-4}{1} = \dfrac{y+2}{2} = \dfrac{z-4}{3};\ \dfrac{x}{3} = \dfrac{y+8}{2} = \dfrac{z}{-1}.$ *Ans.* $\cos\varphi = \frac{2}{7}.$
26. $\dfrac{x+5}{3} = \dfrac{y-1}{2} = \dfrac{z-4}{-2};\ \dfrac{x}{2} = \dfrac{y}{1} = \dfrac{z-5}{4}.$
27. $x + y - z = 2,\ 2x + 2y - z = 7;\ 2x + y - 2z = 2,\ 3x + y - z = -2.$ *Ans.* $\cos\varphi = \frac{5}{6}.$
28. $x + 3y - z = 5,\ z = x;\ x + y + 3z = 1,\ 2x + 3y + 6z = 3.$
29. Write symmetric equations of the line through (4, 1, 2) parallel to the line

$$\frac{x-1}{3} = \frac{y}{5} = \frac{z-7}{-1}.$$

30. Write symmetric equations of the line through (2, 0, −4) parallel to the line

$$\frac{x}{4} = \frac{y+6}{-7} = \frac{z-2}{3}.$$

31. Show that the lines $3x - 5y - z = 1,\ 4x - 3y + 6z = 5,$ and $x - y + z = 0,\ 4x - 7y - 2z = 6$ are parallel.
32. Show that the lines $2x + y - z - 1 = 0,\ x - y + 3z - 10 = 0$ and $5x - 2y + 8z = 44,\ x + 5y - 11z = -38$ are parallel.
33. Show that for the line

$$a_1x + b_1y + c_1z = d_1$$
$$a_2x + b_2y + c_2z = d_2,$$

a set of direction components is

$$\begin{vmatrix} b_1 & c_1 \\ b_2 & c_2 \end{vmatrix}, \quad -\begin{vmatrix} a_1 & c_1 \\ a_2 & c_2 \end{vmatrix}, \quad \begin{vmatrix} a_1 & b_1 \\ a_2 & b_2 \end{vmatrix}.$$

In Exs. 34–36, use the result proved in Ex. 33.

34. Ex. 27. **35.** Ex. 31. **36.** Ex. 32.

37. Find the equation of the plane determined by the parallel lines of Ex. 31. *Ans.* $3x - 4y + z = 2$.

38. Find the equation of the plane determined by the parallel lines of Ex. 32. *Ans.* $7x + 5y - 7z = -6$.

39. Find the point of intersection of the intersecting lines $2x + y + 2z = 2$, $y = 2x$ and $2x - y + 2z + 2 = 0$, $3x + y + z = 4$.

40. Find the point of intersection of the intersecting lines $x + y - z = 0$, $2x - y - 3z = 1$ and $x + 2y + z = 3$, $3x + 3y - 2z = 2$.

41. Find the equation of the plane determined by the lines of Ex. 39. *Ans.* $18x + y + 10z = 10$.

42. Find the equation of the plane through the lines of Ex. 40. *Ans.* $12x + 9y - 13z = 1$.

43. Find the distance from the point (4, 0, 1) to the line $2x - y - z = 4$, $x - y = 1$. (Pass a plane through the point and the line, then a plane through the line perpendicular to the plane just determined.) *Ans.* $\sqrt{6}$.

44. Find the distance from the point (2, 4, 1) to the line $2x = 3y + 2z - 7$, $y + z = 2$. (Note the suggestion in Ex. 43.) *Ans.* 3.

178. Perpendicular line and plane. From the theorem of § 168 we obtain at once a condition that a line and a plane be perpendicular.

THEOREM: *The plane*

$$Ax + By + Cz + D = 0$$

and the line

$$\frac{x - x_1}{a} = \frac{y - y_1}{b} = \frac{z - z_1}{c}$$

are perpendicular if and only if the quantities A, B, C and a, b, c are proportional.

Example: The equation of the plane through the point (3, 2, −1) perpendicular to the line

$$\frac{x - 2}{2} = \frac{y - 1}{1} = \frac{z + 3}{4}$$

is

$$2x + y + 4z = 4.$$

179. Parallel line and plane. If a line is parallel to a plane, it is perpendicular to a normal of the plane. The normal has as direction components the coefficients A, B, C in the equation of the plane. Thus the condition that two lines be perpendicular yields a condition that a line and plane be parallel.

THEOREM: *The plane*

$$Ax + By + Cz + D = 0$$

and the line

$$\frac{x - x_1}{a} = \frac{y - y_1}{b} = \frac{z - z_1}{c}$$

are parallel if and only if

$$aA + bB + cC = 0.$$

Exercises

In Exs. 1–6, write the equations of the line.

1. Through $(2, 4, -3)$ perpendicular to the plane $5x - y + 7z = 8$.
2. Through $(-1, 0, 5)$ perpendicular to the plane $2x + 3y - z = 4$.
3. Through $(7, 1, 0)$ perpendicular to the plane $4x + 3z = 8$.
4. Through $(4, 0, -3)$ perpendicular to the plane $2y + 5z = -1$.
5. Through $(2, 1, 4)$ perpendicular to the xy-plane.
6. Through $(2, 1, 4)$ perpendicular to the yz-plane.

7. Find the foot of the perpendicular from the origin upon the plane $x + 2y + z = 12$. Draw the figure. *Ans.* $(2, 4, 2)$.

8. Find the foot of the perpendicular from the point $(2, 4, 3)$ upon the plane $x + 2y + 3z = 5$. *Ans.* $(1, 2, 0)$.

In Exs. 9–18, find the equation of the plane.

9. Through $(2, -1, 4)$ perpendicular to the line

$$\frac{x - 1}{5} = \frac{y}{-3} = \frac{z + 2}{1}.$$

10. Through $(0, 4, 3)$ perpendicular to the line

$$\frac{x + 3}{1} = \frac{y - 2}{4} = \frac{z - 5}{-2}.$$

11. Through $(3, 1, -2)$ perpendicular to the line $2x = y = 3z + 3$.

12. Through (4, 0, 1) perpendicular to the line $x - 2 = y = 2z - 3$.

13. Perpendicular to the line $x + 1 = y = 2z$, and passing at a distance 2 from the origin. *Ans.* $2x + 2y + z = \pm 6$.

14. Perpendicular to the line $x = y - 3 = 2(z - 1)$, and passing at a distance 1 from (1, 2, −4). *Ans.* $2x + 2y + z = 2 \pm 3$.

15. Through (2, 1, 1), (3, 4, −2), and parallel to the line $7x + y - z = 15$, $x - y - z = 6$. *Ans.* $3x - 7y - 6z = -7$.

16. Through (1, 2, 0), (2, 0, −4), and parallel to the line $x + 2y = 8$, $y + z = 2$. *Ans.* $2x + 3y - z = 8$.

17. Through the line $x + 2y + z = 0$, $2x - 3y + 3z = 2$, and parallel to the line $2y - z = -1$, $2x + 3z = 2$. *Ans.* $5x + 3y + 6z = 2$.

18. Through the line $y = x + 3z$, $2y = 2x + z - 1$, and parallel to the line $2x = y = 2z$. *Ans.* $5x - 5y + 5z = 2$.

19. Find the distance of the point (4, 2, 3) from the line $y = x + 1$, $y + 2z = 5$. Solve in two ways. *Ans.* 3.

20. Find the distance of the point (1, 3, 0) from the line $y = 2x + z$, $4x - y + 3z = 4$. Solve in two ways. *Ans.* $\frac{1}{3}\sqrt{6}$.

21. Find symmetric equations of the line through (3, 1, 5) intersecting the line

$$\frac{x + 2}{4} = \frac{y + 3}{2} = \frac{z + 2}{-1}$$

at right angles. *Ans.* $\frac{x - 3}{1} = \frac{y - 1}{2} = \frac{z - 5}{8}$.

22. Find symmetric equations of the line through (3, 1, −2) intersecting the line $3x - y - 3z = -6$, $3x + 2y - 12z = 3$ at right angles.

Ans. $\frac{x - 3}{2} = \frac{y - 1}{-1} = \frac{z + 2}{-1}$.

23. Solve Ex. 21 by another method.

24. Solve Ex. 22 by another method.

25. Show that the line $4x - y = 2$, $y - 4z = 18$ and the plane $4x + y - 8z = 2$ are parallel, and find the distance between them. *Ans.* 4.

26. Show that the line $x = 4y + 1 = 19 - 4z$ and the plane $x - 2y + 2z = -8$ are parallel, and find the distance between them. *Ans.* 6.

27. Find the distance between the parallel lines $x = y - 1 = -z + 4$ and $x = y = -z$. *Ans.* $\sqrt{14}$.

28. Find the distance between the parallel lines $2x + z = -1$, $y - 3x = 2$ and $3x + y + 3z = 28$, $9x - y + 3z = 44$. *Ans.* $3\sqrt{5}$.

29. Find the locus of points equidistant from the parallel lines of Ex. 27. *Ans.* $x + 2y + 3z = 7$.

30. Find the locus of points equidistant from the parallel lines of Ex. 28. *Ans.* $8x + 4y + 10z = 43$.

31. Find the shortest distance between the skew lines

$$\frac{x}{1} = \frac{y+3}{1} = \frac{z-2}{1}; \qquad \frac{x}{1} = \frac{y+1}{2} = \frac{z}{-1}.$$

(Pass a plane through each line parallel to the other line.) *Ans.* $\frac{1}{7}\sqrt{14}$.

32. Find the shortest distance between the skew lines $x + y = 2$, $y + 2z = 3$ and $x + y - 3z = 3$, $x + z = -8$. (Note the suggestion in Ex. 31.)

Ans. 2.

CHAPTER 21 *Quadric Surfaces*

180. Quadric surfaces. A surface whose equation is of the second degree is called a *quadric surface.* In this chapter we shall discuss a number of surfaces of particular importance, confining our attention chiefly to quadrics with equations in the simplest standard form.

The quadrics proper are of nine species: the ellipsoid (of which the sphere is a special case), the hyperboloids (two species), the paraboloids (two species), the quadric cylinders (three species), and the quadric cone. In addition there are the degenerate forms: two parallel, coincident, or intersecting planes, a single straight line, or a point.

181. The sphere. A *sphere* is the locus of a moving point that remains at a constant distance from a fixed point. The constant distance is the *radius* and the fixed point the *center.*

The equation of a sphere of given center and radius can be written down at once by the distance formula. The equation of a sphere of radius a is, if the center is the origin,

$$x^2 + y^2 + z^2 = a^2; \tag{1}$$

if the center is the point (h, k, l),

$$(x - h)^2 + (y - k)^2 + (z - l)^2 = a^2. \tag{2}$$

Every equation of the form

$$Ax^2 + Ay^2 + Az^2 + Gx + Hy + Iz + K = 0 \quad (A \neq 0) \tag{3}$$

can be reduced to the form (2) by completing the squares in x, y, and z; hence *every equation of the form* (3) *represents a sphere* (exceptionally, a point, or no locus).

Exercises

In Exs. 1–6, find the center and radius of the sphere.

1. $x^2 + y^2 + z^2 - 4x - 6y + 2z - 11 = 0$.
2. $x^2 + y^2 + z^2 + 8x - 2y + 1 = 0$.
3. $x^2 + y^2 + z^2 = 6y - 8z$.
4. $x^2 + y^2 + z^2 = 12x$.
5. $2x^2 + 2y^2 + 2z^2 + 14x - 6y = 1$.
6. $3x^2 + 3y^2 + 3z^2 - x + 2y + 4z - 2 = 0$.

In Exs. 7–13, find the equation of the sphere.

7. Of radius 4 with center at $(1, -2, 3)$.
8. Of radius 3 with center at $(0, 4, -1)$.
9. With center at O tangent to the plane $x + 2y - 3z = 28$.
10. With center at O tangent to the plane $x + 2y - 2z = 15$.
11. With center at $(-2, 3, 5)$ tangent to the plane $4x + 8y = z - 7$.
12. With center at $(1, 2, 3)$ tangent to the plane $3x + 4z = 5y - 15$.
13. With center on the line $x = y = z$ and passing through $(5, 3, 0)$, $(-1, 4, 1)$. *Ans.* $(x - 2)^2 + (y - 2)^2 + (z - 2)^2 = 14$.
14. Find the tangent plane to the sphere $x^2 + y^2 + z^2 = 9$ at $(1, 2, -2)$. *Ans.* $x + 2y - 2z = 9$.
15. Find the tangent plane to the sphere of Ex. 13 at $(5, 3, 0)$. *Ans.* $3x + y - 2z = 18$.
16. Find the tangent plane to the sphere $x^2 + y^2 + z^2 - 2x + 4y = 76$ at $(0, 2, 8)$. *Ans.* $x - 4y - 8z = -72$.
17. Find the tangent plane to the sphere $x^2 + y^2 + z^2 = a^2$ at any point (x_1, y_1, z_1) on it. *Ans.* $x_1x + y_1y + z_1z = a^2$.
18. Prove that a sphere is determined by four points not in a plane.
19. Find the equation of a sphere through $(0, 0, 0)$, $(1, 0, 0)$, $(0, 4, 0)$, $(-1, 2, 1)$. *Ans.* $x^2 + y^2 + z^2 - x - 4y + z = 0$.
20. A point moves so that the sum of the squares of its distances from two fixed points is constant. Show that its locus is a sphere.
21. A point moves so that the square of its distance from a fixed point is proportional to its distance from a fixed plane. Show that its locus is a sphere.
22. Find the equation of a sphere of radius 9 touching the plane $x + 8y - 4z = 9$ at $(1, 1, 0)$.
23. Find the equation of a sphere touching the plane $x + y + 2z = 1$ at $(0, -1, 1)$ and passing through $(-2, 0, 0)$. *Ans.* $(x + 1)^2 + (y + 2)^2 + (z + 1)^2 = 6$.

182. The ellipsoid. The locus of the equation

$$\frac{x^2}{a^2} + \frac{y^2}{b^2} + \frac{z^2}{c^2} = 1$$

is an *ellipsoid*. This surface is symmetric with respect to all three coordinate planes and lies within the box

$$-a \leqq x \leqq a, \qquad -b \leqq y \leqq b, \qquad -c \leqq z \leqq c.$$

The segments of length $2a$, $2b$, $2c$ cut off on the coordinate axes are the *axes* of the ellipsoid, the point O is the *center*.

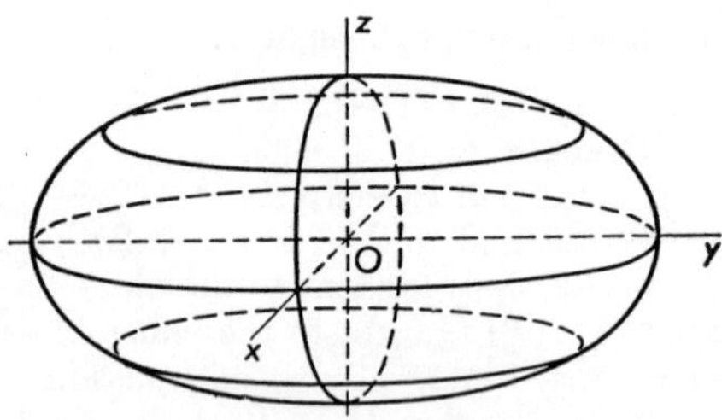

Figure 191

When two of the semiaxes a, b, c are equal, so that the sections by planes perpendicular to the third axis are circles, the surface becomes an *ellipsoid of revolution:* if the equal axes are shorter than the third, a *prolate spheroid;* if longer, an *oblate spheroid.* When $a = b = c$, the surface is a sphere.

Example: An equation such as

$$x^2 + 4y^2 + 9z^2 - 2x + 16y - 19 = 0$$

can, by completing squares, be rewritten as

$$(x - 1)^2 + 4(y + 2)^2 + 9z^2 = 36,$$

in which form it is readily recognized as the equation of an ellipsoid with center at $(1, -2, 0)$, and with semiaxes of lengths 6, 3, 2.

183. The hyperboloid of one sheet. The equation

$$\frac{x^2}{a^2} + \frac{y^2}{b^2} - \frac{z^2}{c^2} = 1 \tag{1}$$

represents a *hyperboloid of one sheet* (Fig. 192). The sections parallel to the yz- and zx-planes are hyperbolas; parallel to the xy-plane, ellipses. The intercepts on Ox are $\pm a$; on Oy, $\pm b$; on Oz, imaginary. The surface is a connected, open surface extending to infinity in both directions along the z-axis.

If $a = b$, the elliptic sections become circular and the surface is the *hyperboloid of revolution of one sheet.* If $a = c$ or $b = c$, but $a \neq b$, the surface is not one of revolution: the result merely means that the sections $y = k$ or $x = k$ are equilateral hyperbolas.

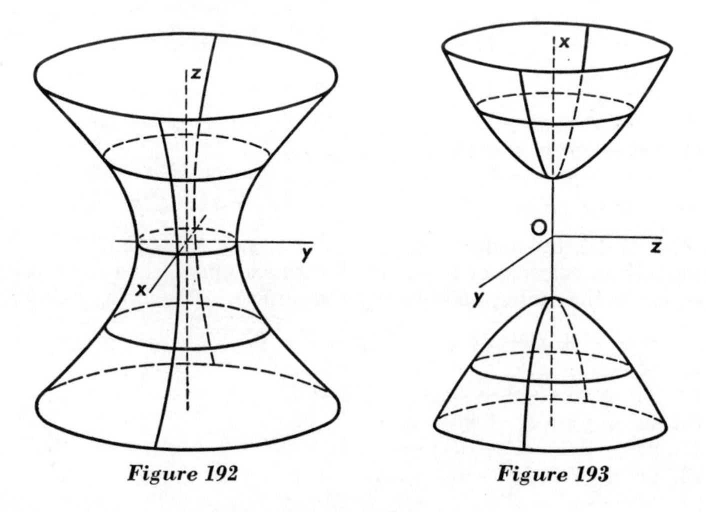

Figure 192 *Figure 193*

184. The hyperboloid of two sheets. The equation

$$\frac{x^2}{a^2} - \frac{y^2}{b^2} - \frac{z^2}{c^2} = 1 \tag{1}$$

represents a *hyperboloid of two sheets* (Fig. 193). A study of the sections shows that the surface consists of two disconnected sheets, one in the region $x \geqq a$, the other in the region $x \leqq -a$, each opening out larger along Ox as x increases numerically. Note the unconventional arrangement of the axes in Fig. 193.

When $b = c$, the elliptic sections become circles, and the surface becomes the *hyperboloid of revolution of two sheets.*

Exercises

In Exs. 1–16, classify the surface, state whether it is a surface of revolution, and make a sketch.

1. $\frac{x^2}{4} + \frac{y^2}{1} + \frac{z^2}{9} = 1.$

2. $\frac{x^2}{1} + \frac{y^2}{16} + \frac{z^2}{9} = 1.$

3. $\frac{x^2}{9} - \frac{y^2}{4} + \frac{z^2}{4} = 1.$

4. $\frac{z^2}{2} + \frac{y^2}{2} - \frac{x^2}{16} = 1.$

5. $\frac{x^2}{9} - \frac{y^2}{4} - \frac{z^2}{4} = 1.$

6. $\frac{y^2}{4} - \frac{x^2}{1} - \frac{z^2}{1} = 1.$

7. $x^2 - y^2 + z^2 = 1.$

8. $z^2 = 1 - 4x^2 + 4y^2.$

9. $x^2 - y^2 + z^2 + 1 = 0.$

10. $z^2 = 1 + 4x^2 + 4y^2.$

11. $2x^2 + 2y^2 = 4 - z^2.$

12. $3z^2 + 3x^2 = y^2 + 27.$

13. $z^2 = 9 + x^2 + 4y^2.$

14. $y^2 = 4(4 - x^2 - z^2).$

15. $x^2 + 5y^2 + 8z^2 = 0.$

16. $4x^2 + y^2 + z^2 = -4.$

In Exs. 17–24, by completing squares in x, y, z, show that the equation represents an ellipsoid or hyperboloid with axes parallel to the coordinate axes, locate the center, and identify the surface. (Example, p. 284.)

17. $x^2 - y^2 + z^2 + 2x + 4y - 4z - 3 = 0.$

18. $x^2 - y^2 - z^2 - 6x + 2z - 8 = 0.$

19. $2x^2 + 4y^2 + z^2 + 4y - 4z + 1 = 0.$

20. $4x^2 + y^2 + z^2 - 4y + 2z - 4 = 0.$

21. $9x^2 - 2y^2 - z^2 - 18x + y - 4z + 15 = 0.$

22. $4x^2 + y^2 - 4z^2 + 12x - 8y + 4z + 8 = 0.$

23. $16x^2 + y^2 - z^2 - 8x - 2y - 30 = 0.$

24. $x^2 + 3y^2 + z^2 + x + 3y + 4z - 4 = 0.$

25. A point moves so that the sum of its distances from two fixed points is constant. Show that its locus is a prolate spheroid. Take the fixed points as $(\pm c, 0, 0)$. Compare with § 67.

26. A point moves so that the difference of its distances from two fixed points is constant. Show that its locus is a hyperboloid of revolution of two sheets. Compare with § 73.

185. The elliptic paraboloid. The locus of the equation

$$\frac{x^2}{a^2} + \frac{y^2}{b^2} = \frac{z}{c} \tag{1}$$

for either positive or negative values of c is an *elliptic paraboloid.*

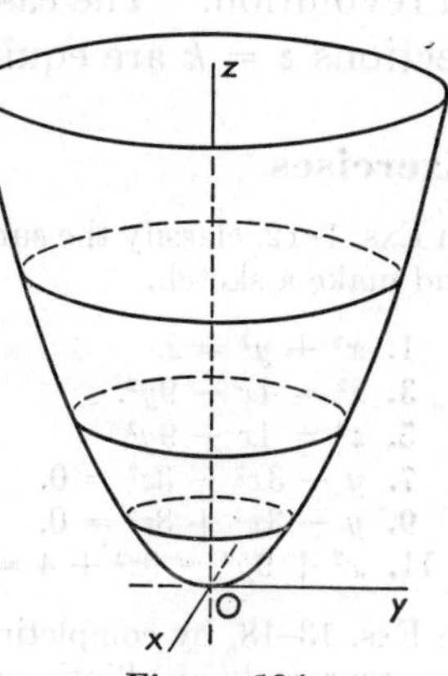

Figure 194

From equation (1) we find that the elliptic paraboloid has two planes of symmetry; it also has one line of symmetry, called the *axis* of the surface. The axis intersects the surface in a single point, called the *vertex.* The surface lies entirely on one side of the xy-plane, and extends to infinity along Oz.

When $a = b$, the surface is the *paraboloid of revolution.*

186. The hyperbolic paraboloid. The surface

$$\frac{x^2}{a^2} - \frac{y^2}{b^2} = \frac{z}{c} \tag{1}$$

is a *hyperbolic paraboloid.* The sections $y = k$ are parabolas opening upward or downward, according to the sign of c; sections $x = k$ are parabolas opening in the opposite direction. Thus the surface is "saddle-shaped," as shown in Fig. 195.

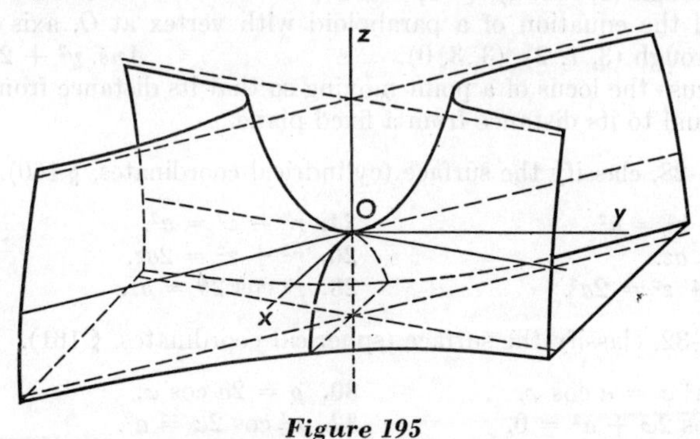

Figure 195

The hyperbolic paraboloid cannot in any case be a surface of revolution. The case $a = b$ is merely the case in which the sections $z = k$ are equilateral hyperbolas.

Exercises

In Exs. 1–12, classify the surface, state whether it is a surface of revolution, and make a sketch.

1. $x^2 + y^2 = z.$
2. $x^2 + z^2 - 9y = 0.$
3. $z^2 = 4x - 9y^2.$
4. $y^2 = 9x - 16z^2.$
5. $z^2 = 4x + 9y^2.$
6. $y^2 = 9x + 16z^2.$
7. $y - 3x^2 - 3z^2 = 0.$
8. $4z + x^2 = 9y^2.$
9. $y - 3x^2 + 3z^2 = 0.$
10. $x^2 + 4y + 4z^2 = 0.$
11. $x^2 + 9y^2 - 9z^2 + 4 = 0.$
12. $4x^2 + y^2 + z^2 = 16.$

In Exs. 13–18, by completing squares, determine whether the given equation represents an elliptic or hyperbolic paraboloid, and locate the vertex or saddle point, whichever is pertinent.

13. $x^2 + 4y^2 - 4x - 8y - 2z + 14 = 0.$ *Ans.* $V\!:(2, 1, 3).$
14. $4y^2 - z^2 + x - 4y + 6z - 9 = 0.$ *Ans.* $SP\!:(1, \frac{1}{2}, 3).$
15. $x^2 - 3z^2 - x + 4y - 3z - 2 = 0.$
16. $y^2 + 4z^2 - x + 4y - 8z + 10 = 0.$
17. $x^2 + z^2 + y - 6z = 0.$
18. $x^2 - 2z^2 + 4x - y - 4z + 2 = 0.$
19. In Figs. 194–195, is the constant c positive or negative?
20. Find the equation of a paraboloid with vertex at O, axis Oy, and passing through $(1, -2, 1)$, $(-3, -3, 2)$. *Ans.* $x^2 - 3z^2 = y.$
21. Find the equation of a paraboloid with vertex at O, axis Ox, and passing through $(3, 1, 2)$, $(3, 3, 0)$. *Ans.* $y^2 + 2z^2 = 3x.$
22. Discuss the locus of a point moving so that its distance from a fixed point is equal to its distance from a fixed plane.

In Exs. 23–28, classify the surface (cylindrical coordinates, § 160).

23. $z^2 - r^2 = a^2.$
24. $r^2 - z^2 = a^2.$
25. $r^2 = az.$
26. $r^2 + z^2 = 2az.$
27. $2r^2 + z^2 = 2a^2.$
28. $r^2 \cos 2\theta = az.$

In Exs. 29–32, classify the surface (spherical coordinates, § 161).

29. $\rho \sin^2 \varphi = a \cos \varphi.$
30. $\rho = 2a \cos \varphi.$
31. $\rho^2 \cos 2\varphi + a^2 = 0.$
32. $\rho^2 \cos 2\varphi = a^2.$

187. Quadric cylinders. If a cylinder has its generators perpendicular to one of the coordinate planes, its equation has only two variables in it; for example,

$$f(x, z) = 0.$$

Then by examining sections $y = k$, we see that the cylinder will have an equation of second degree if and only if its right section is a conic. This situation is a special case of the following theorem, which can be proved with the aid of rotation of axes in three dimensions.

THEOREM: *A cylinder is a quadric surface if and only if its right section is a conic.*

A quadric cylinder is called *elliptic*, *parabolic*, or *hyperbolic*, according to the nature of its right section.

188. Cones. A *cone* is the surface generated by a moving line that always passes through a fixed point, called the *vertex*, and intersects a fixed curve, called the *directing curve*. Thus the surface is completely covered by straight lines, or *generators*, all passing through a fixed point. Like the cylinder, a cone may or may not be a quadric surface.

189. The elliptic cone. The locus of the equation

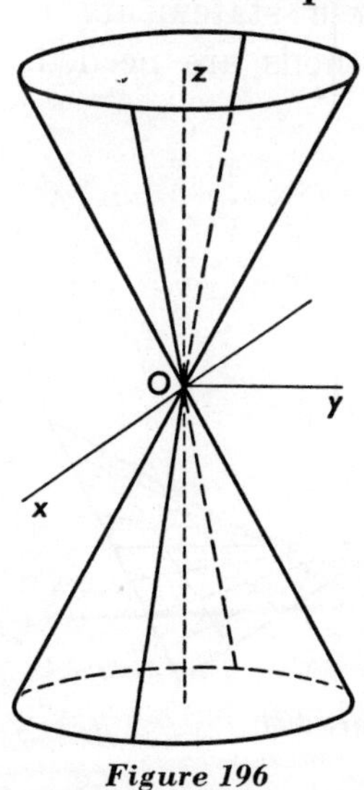

Figure 196

$$\frac{x^2}{a^2} + \frac{y^2}{b^2} - \frac{z^2}{c^2} = 0 \tag{1}$$

is a cone with vertex at the origin. To see this, let $P:(x_1, y_1, z_1)$ be any point of the surface (except O), so that

$$\frac{x_1^2}{a^2} + \frac{y_1^2}{b^2} - \frac{z_1^2}{c^2} = 0. \tag{2}$$

Let Q be any other point of the straight line OP. Then the coordinates of Q must be proportional to those of P: i.e., they must have the form (kx_1, ky_1, kz_1). Substituting these coordinates in (1), we get

$$k^2\left(\frac{x_1^2}{a^2} + \frac{y_1^2}{b^2} - \frac{z_1^2}{c^2}\right) = 0,$$

which is true by (2). Thus the entire line OP lies in the surface.

A study of the sections shows that the surface consists of two open sheets extending to infinity along the z-axis.

By suitable coordinate transformations, the equation of every quadric cone can be reduced to the form (1). Hence there is only a single species of quadric cone. Since the surface is most clearly visualized by means of its elliptic sections, it is usually called the *elliptic cone*. The line through the centers of the elliptic sections is the *axis* of the cone, and sections by planes perpendicular to the axis are *right sections*.

When $a = b$, the right sections are circles, and the surface is the *circular cone*, or *cone of revolution*.

190. Ruled surfaces. A surface which can be generated by a moving straight line is called a *ruled surface*. The straight lines are called *rulings* or *generators*.

A cylinder is a ruled surface all of whose rulings are parallel, and a cone is a ruled surface all of whose rulings are concurrent.

The hyperboloid of one sheet and the hyperbolic paraboloid are doubly-ruled; that is, each is covered by two distinct families of straight lines. Proof of these statements is omitted to save space; no advanced methods are needed.

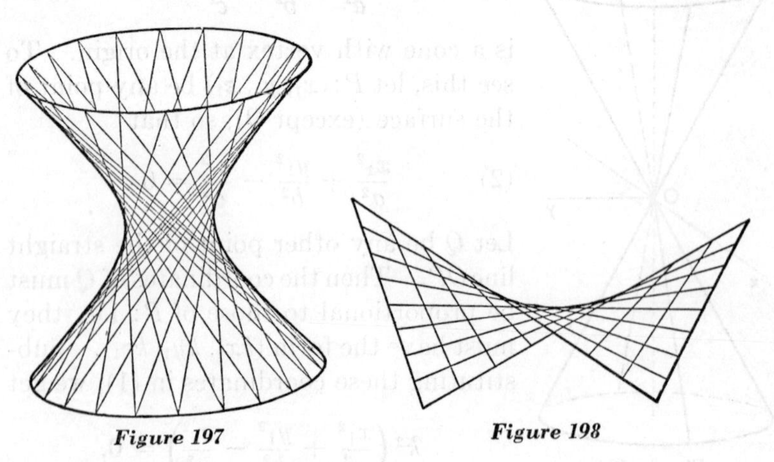

Figure 197 *Figure 198*

Figs. 197–198 exhibit rulings on these surfaces. More striking verification can be obtained by placing a straight edge in appropriate positions on the plaster models of such surfaces, models which are usually in the possession of the mathematics department.

Except for the four surfaces mentioned above, the quadric surfaces are not ruled.

191. Transformation of coordinates. We have seen (Ex. 19, p. 242) that a pure translation of axes in space may be effected by formulas strictly analogous to those of § 61. Rotation of axes in three dimensions is not as simple as in two dimensions, and is omitted in this book.

Rotation of axes in one of the three coordinate planes, holding the third axis fixed, is precisely a two-dimensional rotation accomplished by the method of § 77.

Example: Show that the surface (Fig. 173, p. 247)

$$x^2 = yz \tag{1}$$

is an elliptic cone.

We know by § 80 that rotation of the axes through 45° replaces the product term by the difference of two squares. Hence, taking $\varphi = 45°$ in § 77, put

$$x = x_1, \qquad y = \tfrac{1}{2}\sqrt{2}(y_1 - z_1), \qquad z = \tfrac{1}{2}\sqrt{2}(y_1 + z_1).$$

This reduces (1) to the form

$$2x_1{}^2 + z_1{}^2 - y_1{}^2 = 0.$$

This shows that the surface (1) is an elliptic cone, the axis being the line whose equations (in the original system) are

$$x = 0, \qquad y = z.$$

Thus the (elliptic) sections by planes $y + z = k$ would give a clear picture of the surface.

Exercises

In Exs. 1–12, classify and draw the surface.

1. $x^2 + z^2 = a^2$. **2.** $z^2 = 4ay$.

3. $\dfrac{x^2}{a^2} - \dfrac{y^2}{b^2} = 1$. **4.** $\dfrac{x^2}{a^2} + \dfrac{z^2}{b^2} = 1$.

5. $2zy = a^2$. **6.** $x^2 - z^2 = 0$.

7. $x^2 - 4y^2 + z^2 = 0$. **8.** $9x^2 - 4y^2 - 36z^2 = 0$.

9. $4x^2 - y^2 - 4z^2 = 0$. **10.** $3y^2 - 2x^2 + 3z^2 = 0$.

11. $4z^2 = 9x^2 - 9y^2$. **12.** $25z^2 = 4x^2 + 4y^2$.

13. A moving point is equidistant from the point $(a, 0, 0)$ and the line $x + a = 0$, $y = 0$. Find the equation of its locus. *Ans.* $z^2 = 4ax$.

14. A fixed line and fixed plane are perpendicular to each other; a point moves so that its distance from the line bears a constant ratio to its distance from the plane. Find the equation of its locus. (Take the line as x-axis, the plane as yz-plane.)

15. Show that the surface $Ax^2 + By^2 + Cz^2 + Dyz + Ezx + Fxy = 0$ is a cone with vertex at the origin. (Compare § 189.) Discuss exceptions.

In Exs. 16–19, classify the surface (cylindrical coordinates, § 160).

16. $r = z$. **17.** $r^2 \cos 2\theta = z^2$.

18. $r^2 \sin 2\theta = z^2$. **19.** $z = a \tan \theta$.

20. Show that an equation in φ and θ (spherical coordinates, § 161) represents a cone with vertex at the origin. Discuss exceptions.

In Exs. 21–23, show that the surface (spherical coordinates) is a circular cone.

21. $\varphi = \frac{1}{3}\pi$. **22.** $\cot^2 \varphi = \cos 2\theta$. **23.** $\cot^2 \varphi = \sin 2\theta$.

In Exs. 24–26, prove the statement by rotating one of the coordinate planes through 45°. (Adapt the formulas of § 77.)

24. The surface $z^2 = 2xy$ is a circular cone.

25. The surface $az = 2xy$ is a hyperbolic paraboloid.

26. The surface $x^2 + yz = 1$ is a hyperboloid of one sheet. (Fig. 172.)

27. The cost of a freight haul is roughly proportional to weight and distance: $C = kwd$. Draw the graph of C. (Ex. 25.)

28. The time required to pave a highway is (roughly) proportional to the length and inversely proportional to the number of men employed: $t = \dfrac{kl}{n}$. Show that the graph of t is a hyperbolic paraboloid. (Ex. 25.)

192. Generation of surfaces of revolution. The equation of a surface of revolution is easily derived when the axis and the generating curve are given, if the axis is parallel to a coordinate axis and the generating curve lies in a plane parallel to a coordinate plane.

Example: Derive the equation of the surface generated by revolving the parabola

$$(1) \qquad y^2 = -8(x - 2), \qquad z = 0$$

about the x-axis. See Fig. 199.

In the rotation, every point of the generating curve describes a circle parallel to the yz-plane, with its center on the x-axis. Let Q be any point on the generating curve, and P any point of the circle described by Q, so that P is a general point on the surface. If the coordinates of P are denoted by (x, y, z), those of Q are $(x, k, 0)$, where k is the y-coordinate of Q, the distance LQ in Fig. 199. We must obtain an equation containing x, y, z, but not involving k.

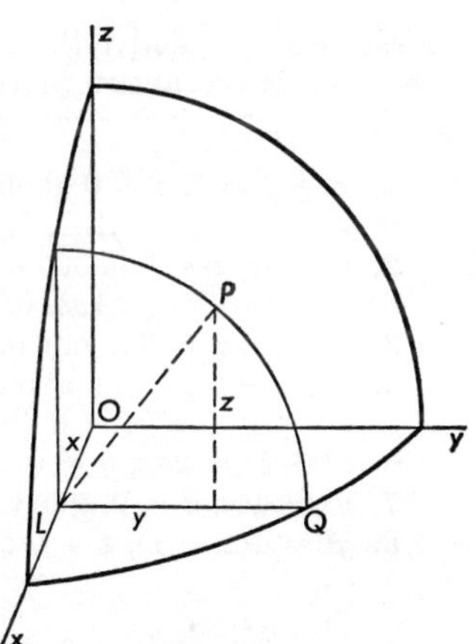

Figure 199

Since Q lies on the generating curve, the coordinates of Q must satisfy the equations of that curve. Thus we obtain the relation

$$(2) \qquad k^2 = -8(x - 2),$$

and note as a check that the z-coordinate of Q must be zero.

We need another equation involving k, so that we can eliminate k with the aid of the relation (2). Since P and Q lie on a circle with center at L, the radii LP and LQ must be equal. Since L has the same x coordinate as P and since L is on the x-axis, the coordinates of L are $(x, 0, 0)$. Then $LQ = k$ and $LP = \sqrt{y^2 + z^2}$, from which it follows that

$$(3) \qquad k^2 = y^2 + z^2.$$

The elimination of k from equations (2) and (3) yields the desired equation of the surface of revolution

$$(4) \qquad y^2 + z^2 = -8(x - 2).$$

As a check, note that sections x = constant yield circles, and that the xy-trace of the surface (4) is the generating parabola (1).

Exercises

In each exercise, derive the equation of the surface generated by revolving the given curve about the designated axes or other lines, drawing the pertinent figure similar to Fig. 199. Check your results.

1. $x + y = 1$, $z = 0$ about (*a*) Ox; (*b*) Oy.
Ans. (*a*) $y^2 + z^2 = (1 - x)^2$; (*b*) $x^2 + z^2 = (1 - y)^2$.

2. $y = 3x$, $z = 0$ about (*a*) Ox; (*b*) Oy.
Ans. (*a*) $9x^2 - y^2 - z^2 = 0$; (*b*) $9x^2 - y^2 + 9z^2 = 0$.

3. $z = 2y$, $x = 0$ about (*a*) Oy; (*b*) Oz.

4. $z = x + 4$, $y = 0$ about (*a*) Ox; (*b*) Oz.

5. $x^2 + y^2 = a^2$, $z = 0$ about Oy.

6. $x^2 + z^2 = 2ax$, $y = 0$ about Ox.

7. $y^2 = 4az$, $x = 0$ about (*a*) its axis; (*b*) the tangent at the vertex.

8. $y^2 = 4(x - 1)$, $z = 0$ about (*a*) its axis; (*b*) its directrix.
Ans. (*b*) $y^4 - 16x^2 + 8y^2 - 16z^2 + 16 = 0$.

9. $\dfrac{x^2}{a^2} + \dfrac{z^2}{b^2} = 1$, $y = 0$ about each of its axes in turn.

10. $z^2 - y^2 = a^2$, $x = 0$ about each of its axes in turn.

11. $x = 4$, $z = 0$ about Oy.

12. $z = 6$, $y = 0$ about Ox.

13. $z = y^3$, $x = 0$ about Oz.

14. $xy^2 = 4$, $z = 0$ about Oy.

15. $z = y^3$, $x = 0$ about Oy.

16. $xy^2 = 4$, $z = 0$ about Ox.

17. $y = z = 4$ about Ox.

18. $x = 3$, $y = 4$ about Oz.

19. $z = x + 4$, $y = 4$ about the line $y = 4$, $z = 0$.
Ans. $(y - 4)^2 + z^2 = (x + 4)^2$.

20. $(x - 4)^2 + z^2 = 4$, $y = 2$ about the line $x = 4$, $y = 2$.
Ans. $(x - 4)^2 + (y - 2)^2 + z^2 = 4$.

21. $z^2 = y - 1$, $x = 2$ about its axis. *Ans.* $(x - 2)^2 + z^2 = y - 1$.

22. $x = 3$, $z = 1$ about the line $x = 4$, $z = 2$.

23. $x = 4$, $y = 3$ about the line $x = 2$, $y = 4$.

24. $x = 3y$, $y = z$ about (*a*) Ox; (*b*) Oy; (*c*) Oz.

Ans. (*a*) $9y^2 + 9z^2 = 2x^2$.

25. The line through (0, 0, 0) and (1, 2, 3) about each of the axes in turn.

Ans. About Oy, $2x^2 + 2z^2 = 5y^2$.

26. $x + y = a$, $z = a$ about Oy.

Ans. $x^2 + z^2 - y^2 = 2a(a - y)$.

Index

Plane Analytic Geometry

(THE REFERENCES ARE TO PAGES.)

Solid Analytic Geometry

(THE REFERENCES ARE TO PAGES.)